Miscible Displacement

Fred I. Stalkup Jr.
Consulting Research Engineer
ARCO Oil & Gas Co.

Second Printing
Henry L. Doherty Memorial Fund of AIME
Society of Petroleum Engineers of AIME
New York   1984   Dallas

Dedication

This work is dedicated to my wife, Jane, for her encouragement, understanding, and devotion.

ISBN 0-89520-319-7

SPE Monograph Series

The Monograph Series of the Society of Petroleum Engineers of AIME was established in 1965 by action of the SPE Board of Directors. The Series is intended to provide authoritative, up-to-date treatment of the fundamental principles and state of the art in selected fields of technology. The Series is directed by the Society's Monograph Committee, one of more than 50 Society-wide committees, through a committee member designated as Monograph Coordinator. Technical evaluation is provided by the Review Committee. Below is a listing of those who have been most closely involved with the preparation of this monograph.

Monograph Coordinators

Herbert L. Stone, Exxon Production Research Co.
J.W. Watts III, Exxon Production Research Co.
Curtis Granberry, Exxon Production Research Co.

Monograph Review Committee

Elmond L. Claridge, chairman, U. of Houston
L.W. Holm, Union Oil Co. of California
Steven W. Poston, Texas A&M U.
T.M. Geffen, Amoco Production Co.
Robert J. Blackwell, Exxon Production Research Co.
L.L. Cargile, Mobil Producing Nigeria

Acknowledgments

Numerous colleagues have supported, encouraged, and advised me in the course of writing this monograph. Their efforts have made the monograph much better than it otherwise would have been. I especially want to acknowledge the contributions of E.L. Claridge, Review Committee chairman, for sharing freely his years of experience and expertise in miscible flooding, for his invaluable consultation and direction, and for his encouragement; R.J. Blackwell, T.M. Geffen, L.W. Holm, L.L. Cargile, and S.W. Poston, Review Committee members, for the hours spent reviewing drafts and for their frank opinions, stimulating debate, and constructive suggestions; and W.F. Yellig, F.M. Orr Jr., J.W. Watts, G.K. Youngren, and L.G. Chorn for reviewing selected chapters and for the many and valuable suggestions for improving those chapters. A very special thank you goes to Lynne K. Smith, who typed and proofread the many versions with dedication, cheerfulness, and humor.

Contents

Preface

This is the third SPE monograph to treat comprehensively a given mechanism of oil production. The other two are Vol. 3, *The Reservoir Engineering Aspects of Waterflooding*, by Forrest F. Craig Jr., and Vol. 7, *Thermal Recovery*, by Michael Prats.

As is true of all works in a field where technology is still undergoing development, this monograph and the previous ones represent the state of the art at the time of publication. Most of what is stated will be valid indefinitely, but some statements inevitably will be shown by later information to be incomplete, or in a few cases, to be simply erroneous. The author and the review committee have tried to minimize such errors. Furthermore, with regard to many matters implicit in a work of this nature, experimental evidence is incomplete and differing opinions exist. These are described generally by the author. He has, in fact, pointed out in Chap. 9 significant areas of research in this field, which will help to resolve these differences of opinion.

The author, Fred I. Stalkup, is a recognized expert in the field. He has published more than 15 papers, including the recent survey articles "Carbon Dioxide Miscible Flooding: Past, Present, and Outlook for the Future," (*J. Pet. Tech.*, Aug. 1978, Pages 1102–12) and "Status of Miscible Displacement," *J. Pet. Tech.*, April 1983, Pages 815–826. After receiving a PhD degree in chemical engineering from Rice U. in 1961, his professional career has been entirely with ARCO Oil & Gas Co. in their research center near Dallas. During this period he has served as director of process development research, reservoir engineering research, reservoir mathematics research, and enhanced oil recovery research. He currently is consulting research engineer.

Many of the articles referred to in this monograph are available either in the 1965 edition of SPE Reprint Series No. 8, *Miscible Processes*, or in the revised 1984 edition. In choosing articles for the latter reprint book, an effort was made to select those considered most useful as direct source materials in association with the monograph. It is recommended that engineers intending to apply the principles described in this monograph obtain and use the two reprint books as supplementary guides.

It is expected that this monograph will be used both for training purposes and as a professional reference book by petroleum engineers. The review committee recommends the monograph for these purposes.

Houston
January 1983

MISCIBLE DISPLACEMENT
REVIEW COMMITTEE

Chapter 1
Introduction

1.1 Scope and Objectives of the Monograph

This monograph is limited to miscible flooding with hydrocarbon solvents, flue gas, nitrogen, and carbon dioxide. Alcohol flooding is not included because the cost of alcohol and the slug sizes required for effective oil recovery appear to make this method mostly of academic interest. Although there are some aspects of miscible displacement in the processes based on surfactant-containing injection fluids and although miscible displacement plays a role in polymer and alkaline flooding, it is only one of the mechanisms important for oil recovery with these processes. A separate monograph is more appropriate to do justice to the overall subject of chemical flooding. Even so, the principles of miscibility and miscible displacement discussed here apply equally well to the miscible displacement aspects of the chemical flooding processes.

There were four objectives in writing this monograph. One objective was to organize and to document in one publication the information currently available from laboratory research, engineering studies, and field testing to provide a more comprehensive future source document on the subject. There is an extensive body of literature on miscible flooding. Although a volume of reprints[1] was published that contained a collection of the more important single publications at the time and although a few works have given broad-brush summaries,[2,3] organized and comprehensive information is available only from specialists in the field or perhaps from proprietary company manuals available to employees.

Another objective was to review and to teach those principles and basic phenomena that are general to miscible displacement regardless of the specific process. An understanding of these principles is necessary to achieve an awareness of the important factors that should be taken into account to evaluate and design projects intelligently.

A third objective was to give up-to-date engineering assistance in design and performance projections for miscible floods. The reader should be aware, however, that our ability to calculate performance is limited by the predictive models that are currently available, by knowledge gaps that still remain of some process mechanisms, and by our ability to describe the reservoir. In some instances, adequate predictive methods will not even be available for the job at hand and only "ballpark" methods can be applied. No attempt has been made to make this monograph a design manual with step-by-step or "cookbook" procedures to follow. Instead, the intent is to educate the reader to enable him to select tools and methods most appropriate for his needs.

A final objective was to give an up-to-date assessment of how the various miscible processes have performed in field trials. Included in this assessment are a discussion of the range of applicability of each process and guidelines for identifying potential projects.

The monograph is intended to be a document for research, staff, and field engineers. Emphasis, however, is given more to the reservoir engineering aspects of miscible flooding. Material that may be of more interest to the specialist is referenced.

Miscible flooding is a complex technology. Although it has been a subject of research, development, and field trials for more than 30 years, there is still some disagreement in the interpretation of laboratory and field-test data and in the selection of predictive methods. This monograph attempts to give a balanced treatment of the subject and points out conflicting views where it is appropriate to do so. However, the monograph is necessarily written from the author's perspective, and this should be kept in mind by the reader.

1.2 Organization of the Monograph

The subject matter begins with a review of miscible-displacement phase behavior and flow behavior fundamentals before proceeding to a discussion of design and projection methods and a review of field test experience. Chap. 2 is devoted to principles of phase behavior and miscibility. Current concepts of the phase behavior required for achieving first-contact miscibility with liquid petroleum gas (LPG) and dynamic miscibility with hydrocarbon gases, flue gas, nitrogen, and carbon dioxide are reviewed and illustrated with pseudoternary diagrams. The significance of pressure/composition diagrams is discussed. Methods for experimentally measuring miscibility conditions and for calculating phase behavior are described, and a preferred experimental method is recommended.

Chap. 3 discusses phenomena that affect displacement efficiency. Topics include mixing of fluids by dispersion, viscous fingering and its effect both on areal sweep efficiency and solvent slug integrity, gravity segregation and vertical displacement efficiency, factors affecting unit displacement efficiency, and the significance of reservoir heterogeneities.

Chap. 4 deals with reservoir engineering methods for designing miscible floods and projecting reservoir flooding performance. Advantages and limitations of current mathematical models are discussed.

The next four chapters describe field experience with first-contact miscible, condensing-gas drive, vaporizing-gas drive, and carbon dioxide miscible floods. Short summaries are given of selected projects, and an assessment of the field behavior results is given for each process. These chapters also contain guidelines for screening reservoirs and identifying potential projects. Some discussion is devoted to potential solvent sources.

The concluding chapter summarizes the current state of technology and identifies important areas where further study is needed. An assessment of the potential for miscible flooding is included.

1.3 Overview of Conventional Oil-Recovery Methods

After discovery, most oil reservoirs typically undergo a period of production called "primary recovery" in which natural energy associated with a reservoir is used to recover a portion of the oil. Mechanisms such as liquid expansion and rock compaction initially help drive reservoir fluid into the wellbores as pressure declines. When pressure falls below the oil's bubble point, additional recovery results from gas liberation and expansion. Some reservoirs have a gas cap, and gas-cap expansion, perhaps coupled with gravity drainage, also may help drive oil to producing wells. Other reservoirs may be connected to aquifers that provide active or partially active water drives. Water encroachment from an aquifer both displaces oil from reservoir pore space and helps moderate the pressure decline caused by fluid withdrawal.

From the early days of oil production up until about the early 1930's, most reservoirs were produced by primary recovery mechanisms until an uneconomical oil rate was reached. At this point they were abandoned. When this occurred, reservoir pressure generally had been depleted to a low value, or the WOR had become excessive in those reservoirs with strong natural water drives. Primary recovery efficiency varied greatly from reservoir to reservoir, as it does today, depending on the

mechanism or combination of mechanisms responsible for production, the type of reservoir, rock properties, and oil properties. A typical range for recovery efficiency is 5 to 20% OOIP.

Fluid injection into one or more wells gradually became accepted as a method for increasing oil recovery and productivity above primary production levels. Water and/or natural gas, at pressures where the gas is immiscible with oil, have been the injection fluids used almost exclusively for this purpose in the past. Injecting gas into the gas cap, water into the aquifer near the water/oil contact, or either fluid into the oil column are common fluid-injection techniques. At first, the improved recovery was thought to result from moderating or preventing a decline in reservoir pressure so the producing rate could be maintained at a higher level for a longer time than would have been possible from primary producing mechanisms only. The technique was called "pressure maintenance." Later it was recognized that in addition to supporting reservoir pressure, the injection fluids also displaced some oil from the rock pore space and drove it to producing wells. Today fluid injection following some short or prolonged period of primary production usually is called "secondary recovery." Ultimate oil recovery resulting from both primary recovery and secondary recovery by water or immiscible gas injection generally is in the range of 20 to 40% OOIP, although there are instances where ultimate recovery is significantly higher or lower.

The ultimate recovery achievable by immiscible gas injection or waterflooding is limited primarily by three factors: (1) volumetric sweepout of the injection fluid, (2) capture of the displaced oil at producing wells, and (3) displacement efficiency of the injection fluid in rock that is swept. Volumetric sweepout of the reservoir volume is always less than 100% because of permeability stratification, viscous fingering, gravity segregation, and incomplete areal sweepout. In addition, all the oil displaced from the swept reservoir volume is not captured by producers because some goes to resaturate unswept rock that has been partially depleted of fluids by primary recovery. Even in the swept volume, oil displacement is incomplete, and a residual oil saturation is left behind the advancing injection fluid. In some instances, rock pore structure may contribute to incomplete displacement by the injection fluid (e.g., dead-end pores filled with oil), but immiscibility between oil and water and between oil and natural gas is the primary cause of high residual oil saturations in most rocks. The advantage of miscible injection fluids is that less residual oil, usually significantly less, is left in the swept reservoir rock.

1.4 Immiscible Displacement

Oil and water do not mix. If these two fluids are poured into a vessel and allowed to settle, two distinct liquid phases are apparent, separated by a sharp interface. The oil and water are said to be immiscible.

Similarly, oil and natural gas also are immiscible, although natural gas has some solubility in oil. This solubility depends on pressure level, and as soon as the solubility limit is exceeded, two phases form—a gas phase and a liquid phase, separated by an interface.

When two immiscible phases flow simultaneously in a

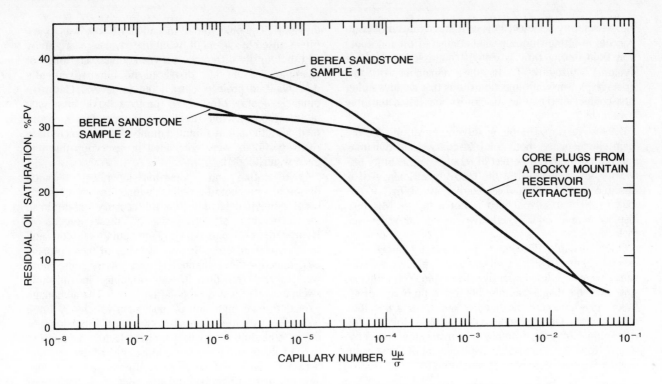

Fig. 1.1—Dependence of residual oil saturation on capillary number.

porous medium, as for example in the displacement of oil by water, the flow behavior in a given unit volume of rock is determined by the relative permeability characteristics of the rock. Oil relative permeability decreases with decreasing oil saturation, but oil permeability reduces to zero and oil stops flowing while the oil saturation is finite. This limiting saturation is called the residual oil saturation or, perhaps more precisely, the relative permeability endpoint residual oil saturation to distinguish it from the material balance or average residual oil saturation left in a reservoir after waterflooding. Similarly, water permeability decreases to zero at a finite value of water saturation called the irreducible water saturation. Oil and water both flow at saturations between the two endpoints, and the shape of the relative permeability curves between the endpoints determines how much fluid must be injected to reduce the oil saturation in the swept rock to a given level.

In addition to being affected by rock pore structure, residual oil saturation and relative permeability are affected by rock wettability and by interfacial tension (IFT) at the oil/water interface. Craig[4] defined wettability as the tendency of one fluid to spread on or to adhere to a solid surface in the presence of other immiscible fluids. It can have a pronounced effect on immiscible displacement. For example, relative permeabilities for oil-wet rocks are more unfavorable for waterflooding than are relative permeabilities for water-wet rocks. Other factors being equal, this causes earlier water breakthrough in the oil-wet rock and requires that more water be injected to achieve a given reduction in oil saturation than in water-wet rocks.

The effect of oil/water IFT on residual oil saturation is illustrated in Fig. 1.1. Here the residual oil saturation is plotted vs. capillary number, the product of Darcy velocity and oil viscosity divided by IFT. Capillary

number is an approximate measure of the ratio of viscous to capillary forces. Over ranges of velocity, oil viscosity, and oil/water IFT found in conventional waterflooding, residual oil saturation is insensitive to capillary number. Fig. 1.1 shows that a drastic reduction in oil/water IFT, by several orders of magnitude or more, is required to achieve a significant reduction in waterflood residual oil saturation.

Although the discussion of this section centered on waterflooding, similar behavior is found when a gas immiscibly displaces oil.

1.5 Miscible Displacement

Two fluids are miscible when they can be mixed together in all proportions and all mixtures remain single phase. Gasoline and kerosene are examples of two liquids that are miscible. Because only one phase results from mixtures of miscible fluids, there are no interfaces and consequently no IFT between the fluids.

It is apparent from Fig. 1.1 that if IFT between oil and displacing fluid is eliminated completely (i.e., the capillary number becomes infinite), residual oil saturation can be reduced to its lowest possible value. This is the objective of miscible displacement.

Some injection fluids for miscible displacement mix directly with reservoir oils in all proportions and their mixtures remain single phase. They are said to be miscible on first contact or "first-contact miscible." Other injection fluids used for miscible flooding form two phases when mixed directly with reservoir fluids—i.e., they are not first-contact miscible. However, with these fluids, in-situ mass transfer of components between reservoir oil and injection fluid forms a displacing phase with a transition zone of fluid compositions that ranges from oil to injection fluid composition, and all compositions within the transition zone of this phase are contiguously miscible. Miscibility achieved by in-situ mass transfer of components resulting from repeated contact of oil and injection fluid during flow is called "multiple contact" or dynamic miscibility.* In the remainder of this monograph, miscible injection fluids that achieve either first-contact or dynamic miscibility are called miscible "solvents."

Because mixtures in the oil/solvent transition zone remain single phase, both for first-contact and dynamic miscibility, there is no effect of relative permeability between solvent and oil on the recovery efficiency of a miscible displacement. The relative wettability of the rock to oil and solvent also is not a factor. Miscible displacement is equally effective in oil- or water-wet rock.**

Often miscible flooding is practiced by injecting a limited volume or slug of solvent and displacing the solvent slug with a less expensive drive fluid. The solvent slug may be displaced miscibly by appropriate drive fluids, in which case the flood is said to be a miscible slug process. Other practice is to displace the solvent slug immiscibly; for example, a propane slug driven by water. However, immiscible displacement of the solvent slug leaves a residual solvent saturation in the reservoir.

1.6 History of Miscible Flooding

The search for an effective and economical solvent along with development and field testing of miscible-flood processes has continued since the early 1950's. Early focus was on hydrocarbon solvents, and three types of hydrocarbon-miscible processes were developed: the first-contact miscible process, the vaporizing-gas drive process, and the condensing-gas drive process. Dynamic miscibility is achieved in the latter two processes.

Propane or LPG mixtures typically were the solvents used in first-contact hydrocarbon miscible flooding, whereas natural gas at high pressure and natural gas with appreciable concentrations of intermediate-molecular-weight hydrocarbons were injection fluids in vaporizing-gas drive and condensing-gas drive floods. Flue gas also was found to achieve dynamic miscibility with some oils at high pressures by the vaporizing gas-drive mechanism. The high cost of propane, LPG, or enriched hydrocarbon gas dictated that these solvents be injected as slugs, which usually were driven with natural gas.

Hydrocarbon miscible processes received extensive field testing in the 1950's and 1960's in the U.S. and Canada.† More than 150 projects were initiated in this time period. The majority were small-scale pilot tests involving one or at most a few injection wells; however, a number of large projects were undertaken involving several thousand acres or more. A few projects tested flue-gas injection.

Most floods were in essentially horizontal reservoirs, although a number of projects were in reservoirs of high relief where solvent was injected updip to displace oil downdip. The objective of these latter floods was to improve miscible sweepout by taking advantage of gravity and the density difference between solvent and oil. For example, solvent has displaced oil almost vertically downward in projects carried out in several Canadian pinnacle reefs. Most field projects have employed hydrocarbon miscible flooding as a secondary recovery method, although in a limited number of projects tertiary recovery floods were conducted in reservoirs that had been waterflooded.

Field testing and supporting laboratory research disclosed a number of problems with hydrocarbon miscible flooding that acted to limit oil recovery and diminish the economic attractiveness of these processes. Hydrocarbon solvents typically are both less viscous and less dense than reservoir oils. Because of these properties, injection fluid channeling and gravity segregation were more severe than in waterflooding, and miscible sweepout often was disappointingly low. The advantage gained from a high unit-displacement efficiency often was offset by a low volumetric sweepout. Breakdown of the solvent slugs also was a limiting factor. Mixing between solvent and oil at the leading edge of the slug and between solvent and gas at the trailing edge of the slug, aggravated by fingering, quickly diluted small slugs of less than a few percent PV to concentrations that no longer were miscible with the reservoir oil. Despite these disadvantages, there have been some notable successes, both technical and economic, in hydrocarbon miscible/flue-gas projects. Oil recovery of more than 60% OOIP apparently will be achieved in several floods.

Results of laboratory research on miscible flooding with alcohols were published in the late 1950's and early 1960's.[1] Low-molecular-weight alcohols, such as isopropyl and tertiary butyl alcohols, are first-contact miscible with both reservoir oil and brine. A process scheme was envisioned in which a small slug of alcohol would be driven through the reservoir with brine, miscibly displacing oil and water ahead and achieving a volumetric sweepout comparable to that achieved in a waterflood. In practice, however, alcohol fingered into the oil, and laboratory research found that uneconomically large slugs were required to prevent dilution of the alcohol with oil and brine to immiscible concentrations. There was at least one field trial of this process.[1]

Research to develop a process with both good unit displacement efficiency and sweepout resulted in publications in the late 1950's and early 1960's describing soluble oils and micellar solutions as injection fluids.[5] These fluids contained oil, water, surfactants, inorganic salts, and cosolvents such as alcohols and low-molecular-weight ethers. As in alcohol flooding, the process scheme was to drive a relatively small slug of the micellar solution through the reservoir with brine. To achieve a favorable viscosity ratio between the micellar slug and drive brine, water-soluble polymers were added to the brine. It was necessary also to control brine salinity at values optimal for this process. Slug viscosity was controlled by composition and concentration of constituents in the slug.

Since the early publications, there has been extensive research on surfactant-containing solutions of various formulations for oil displacement. These fluids can be

---

*Actually, one or more subordinate immiscible phases may coexist with the major phase in displacements that achieve dynamic miscibility as discussed in Secs. 2.5 through 2.7. The transition zone referred to in the previous sentence occurs in the major displacing phase.

**However, relative permeability and wettability between aqueous and nonaqueous phases can affect miscible displacement efficiency as, for example, in the water-alternate-gas process or in a capillary transition zone.

†References for hydrocarbon miscible flood field tests are given in Chaps. 5, 6, and 7.

formulated to be miscible with oil, at least initially, or miscible with the drive-polymer solution.[5] Generally, the surfactant-containing slug is not miscible with both oil and drive polymer and, depending on the specific formulation, may not be miscible with either. Designs seek to achieve ultralow IFT's with the nonmiscible phase or phases. Floods with surfactant-containing injection fluids are more appropriately classified as low-IFT rather than miscible displacements, even when the injection fluid is formulated to have initial miscibility with the oil. Reservoir mixing most likely rapidly dilutes initial compositions to the point where two or more phases of low IFT result.

There has been limited field testing of surfactant-containing injection fluids to date, and most tests have been only pilot-scale. Testing has disclosed a problem in maintaining the integrity and effectiveness of small surfactant solution/micellar solution slugs.

Recent activity in miscible flooding has focused on the CO_2-miscible process.* CO_2 has a low viscosity, similar to that of hydrocarbon miscible solvents. As in hydrocarbon miscible flooding, volumetric sweepout in CO_2 flooding is affected by an unfavorable viscosity ratio. CO_2 density is similar to that of oil in many reservoirs, which minimizes CO_2/oil segregation, but there is enough density contrast with brine for gravity segregation to occur when there is mobile reservoir brine.

Despite its low viscosity, CO_2 can be an attractive injection fluid. Dynamic miscibility can be achieved in many reservoirs because of a relatively low operating pressure requirement. In addition, both supply and cost of CO_2 for miscible flooding may be more favorable in the future than for hydrocarbon-miscible solvents. This is because large quantities of CO_2 are available from natural deposits and from manufacturing and power-generating plants as by-products. CO_2 from some of these sources, particularly natural deposits, potentially may be developed and transported to favorably located oil fields at acceptable costs.

There has been limited field testing of the CO_2-miscible process.

1.7 Applicability of Miscible Flooding

Oil recovery by miscible flooding has not been as widely applicable as waterflooding. Waterflooding can be practiced successfully from both a technical and economic viewpoint in a broad spectrum of reservoirs. This has not been the case for miscible flooding with hydrocarbons, flue gas, nitrogen, or CO_2. One limiting factor has been the pressure/composition requirement for miscibility. This requirement can limit the number of prospective reservoirs severely, for example when natural gas, flue gas, and nitrogen are used, or it may be only moderately restrictive, as is the case when CO_2 is used.

High cost, however, has been a more important limitation. When a reservoir is evaluated for miscible flooding, the projected economics must be compared with the economics projected for other operating alternatives. Continued primary depletion, pressure maintenance, waterflooding, or even abandonment are all potential alternatives. Miscible floods can be substantially more costly than these alternatives, primarily because of solvent cost; but equipment and operating costs can be higher also. Moreover, much of the added cost is concentrated in the early project life or front end, while revenue from the additional oil recovery is concentrated more in the middle of the project. Although oil recovery in miscible flooding may be greater than for an alternative operating plan, this positive factor on project economics may be offset by the higher costs and front-end investment. This has been the experience of many past field projects.

Solvent availability is another important factor affecting the application of miscible flooding. Large ultimate volumes and supply rates of solvent can be required. For example, a CO_2-flood project could require a CO_2 reserve ranging from several tens of billion cubic feet to more than a trillion cubic feet delivered at rates from several tens to several hundred million standard cubic feet per day.

For miscible flooding to be a competitive process in a given reservoir, several conditions must be satisfied: (1) an adequate volume of solvent must be available at a rate and cost that will allow favorable economics, (2) the reservoir pressure required for miscibility between the solvent and oil in question must be attainable, and (3) incremental oil recovery must be sufficiently large and timely for project economics to withstand the added costs.

Incremental oil recovery is determined largely by reservoir properties, which must be evaluated carefully to determine whether potential exists for a miscible flood. More detailed and comprehensive reservoir engineering, cost determination and control, and project monitoring are necessary for the typical miscible-flood project than for waterflooding. There is a correspondingly greater requirement for trained manpower.

In the future, hydrocarbon solvents in the U.S. could be too expensive and in too short supply for miscible flooding except in unusual situations. However, supply and cost in some of the other major oil-producing countries as well as conservation policies could be favorable for projects with hydrocarbon solvents. Flue gas and nitrogen will have some limited application in deeper, higher pressure reservoirs where the miscibility pressure is attainable. CO_2 has the most potential for future miscible-flood projects in the U.S., but this potential depends heavily on the availability and cost of future CO_2 supplies.

References

1. *Miscible Processes*, Reprint Series, SPE, Dallas (1972) No. 8.
2. Smith, C.R.: *Mechanics of Secondary Oil Recovery*, Reinhold Publishing Co., New York City (1966).
3. *Secondary and Tertiary Oil Recovery Processes*, Interstate Oil Compact Commission, Oklahoma City (Sept. 1974).
4. Craig, F.F.: *The Reservoir Engineering Aspects of Waterflooding*, Monograph Series, SPE, Dallas (1971) **3**.
5. *Improved Oil Recovery by Surfactant and Polymer Flooding*, D.O. Shah and R.S. Schechter, (eds.), Academic Press Inc., New York City (1977).

*References for the CO_2-flood process are given in Chap. 8.

Chapter 2

Principles of Phase Behavior and Miscibility

2.1 Representation of Three-Component Phase Behavior With a Triangular Diagram

Reservoir fluids are complex multicomponent mixtures of hundreds of different hydrocarbons and some nonhydrocarbons. According to the Gibbs phase rule,[1] $F=n-P+2$ degrees of freedom must be specified for a mixture containing n components and consisting of P phases before the phase behavior can be defined completely. However, the exact composition of a reservoir fluid is never known; and even if it were, such a rigorous specification of the phase behavior would be impossibly cumbersome.

An approximate method for representing the phase behavior of multicomponent mixtures utilizes the triangular diagram. Such a diagram is shown in Fig. 2.1. The phase behavior of three-component mixtures can be represented exactly on a triangular diagram, whereas its use for multicomponent mixtures requires that these mixtures be approximated by three pseudocomponents. Representation of phase behavior by the triangular diagram method will be illustrated in this section for a simple three-component hydrocarbon system. Extension to multicomponent mixtures is described in Sec. 2.2.

Each corner of a triangular diagram represents 100% of a given component. The opposite side of the triangle represents 0% of that component. For example, the uppermost corner of the triangle in Fig. 2.1 represents 100% methane (C_1), while the opposite, or bottom, side of the triangle represents 0% methane. Any concentration of methane between zero and 100% is represented at a proportionate distance between the bottom of the triangle and the upper corner. Similarly, the lower right corner represents 100% n-butane (nC_4), and the lower left corner represents 100% decane (C_{10}). With this manner of specifying component concentrations, mixtures can be plotted on the diagram. Thus, Mixture S in Fig. 2.1 contains 68% methane, 21% n-butane, and 11% decane.

For the phase relations shown in Fig. 2.1, the mixture with overall composition represented by Point S is a two-phase mixture. This means that if the three components were mixed together in a pressure vessel at 2,500 psi and 160°F in the relative proportions specified by Point S and allowed to equilibrate, two phases would result: an equilibrium gas phase with composition y and an equilibrium liquid phase with composition x. The dashed line connecting the equilibrium gas and liquid compositions is called a tie line. Since the gas and liquid are in equilibrium with each other, they are fully saturated—i.e., the gas is saturated with condensible components and therefore is at its dewpoint, while the liquid is saturated with vaporizable components and is at its bubble point.

For the phase relations of Fig. 2.1, the dewpoint curve through all the dewpoint compositions joins the bubble-point curve through all the bubble-point compositions at the plait point. At the plait point, also called the critical point, the composition and properties of equilibrium gas and liquid become identical. The phase boundary curve so defined separates the single-phase and two-phase regions of the diagram. At the pressure and temperature of the diagram, any system of the three components whose composition is inside the phase-boundary curve will form two phases. Any system with a composition lying outside of this curve will be in a single phase. The single-phase gas region lies above the dewpoint curve, while the single-phase liquid region lies below the bubble-point curve.

Pressure and temperature influence the size of the two-phase region. When pressure is reduced, the size of the two-phase region increases, as shown in Fig. 2.2. Eventually, the plait point disappears with continued reduction of pressure, as the two-phase region intersects the right side of the diagram, indicating that at this limiting pressure the methane and n-butane no longer form single-phase mixtures for all mixture compositions. An

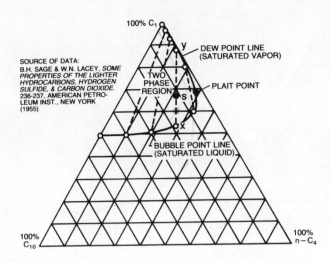

Fig. 2.1—Phase relations for methane/n-butane system at 160°F and 2,500 psia (mol%) (after Ref. 2).

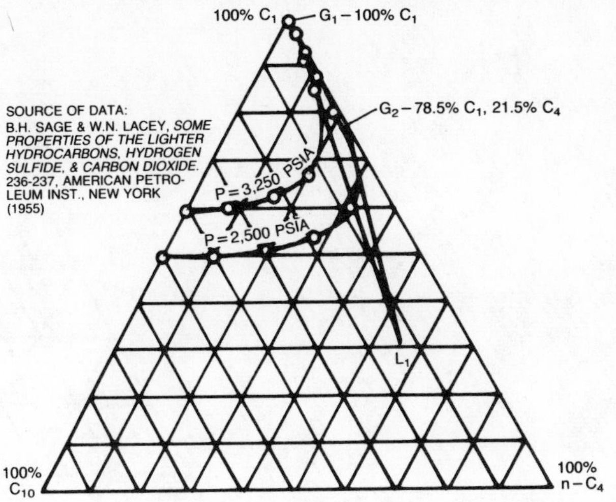

Fig. 2.2—Required butane injected gas for miscibility with reservoir fluid (operation above the cricondenbar), $T = 160°F$ (mol%) (after Ref. 2).

increase in temperature increases the size of the two-phase region.

One additional principle is important to recognize when representing phase behavior relations with a triangular diagram: The compositions of all the mixtures of two fluids are represented by the straight line connecting the compositions of the two fluids. For example, all the mixtures of nC_4 and Bubble-Point Fluid X in Fig. 2.1 lie along the straight line (not shown) connecting Composition X with the lower right corner of the diagram, which represents 100% nC_4. All of these mixtures fall entirely within the single-phase region, indicating that nC_4 is completely miscible with Bubble-Point Liquid X. Methane, on the other hand, is immiscible with Liquid X, since some mixtures of these two fluids lie within the two-phase region.

2.2 Representation of Multicomponent Phase Behavior With a Pseudoternary Diagram

The phase behavior relations of a reservoir fluid can be represented approximately on a triangular diagram by grouping the components of the reservoir fluid into three pseudocomponents. Such a diagram is called a "pseudoternary diagram." One possible grouping that has been used frequently is: a volatile pseudocomponent composed of nitrogen and methane, an intermediate pseudocomponent composed of hydrocarbons of intermediate volatility such as ethane through hexane, and a relatively nonvolatile pseudocomponent composed of $C_7 +$; i.e., hydrocarbons with molecular weight greater than that of hexane. Other groupings have been reported, most notably where the intermediate pseudocomponent contained the ethane through pentane hydrocarbons and the low-volatility pseudocomponent contained the hexane and heavier hydrocarbons.

The representation of mixture compositions and phase behavior in this manner is approximate since the individual components within a pseudocomponent group have different volatilities and will not be distributed within that group in the same way in the gas and liquid

phases.[2] For this reason, the composition and properties of the pseudocomponent do not remain constant for all mixtures. Also, the position of the phase boundary curve on the triangular coordinates and the slope of the tie lines depend on the overall mixture composition, which cannot be defined adequately by the simple pseudocomponent grouping.

These deviations from the behavior of true three-component systems are demonstrated in Fig. 2.3, which shows the phase behavior for a reservoir fluid.[2] The equilibrium gas and liquid compositions connected by the dashed tie lines were obtained from analysis of phases after static equilibrium had been reached in a pressure vessel. Different overall mixture compositions in the plait-point region of the diagram were achieved by a multiple-batch-contact procedure in which the equilibrium gas from a previous contact was saved and mixed with fresh reservoir fluid, L_1. The detailed composition analysis shown in this figure for Equilibrium Gas Y and Equilibrium Liquid X illustrates how the composition of the intermediate pseudocomponent varies between the gas and liquid phases. As expected, the more volatile components, such as ethane, are concentrated in the gas phase, while the less volatile components are concentrated in the liquid phase. The changing character of the $C_7 +$ fraction is reflected in the molecular weight of this fraction in the equilibrium gas and liquid phases.

The one solid tie line shown in Fig. 2.3 was obtained from a flow experiment, where Gas G_1 displaced Reservoir Fluid L_1 from a long, sandpacked tube. An increment of gas produced from a separator at the end of the tube was recombined with an increment of separator liquid produced concurrently with the gas sample. This mixture was allowed to reach equilibrium in a static cell at the temperature and pressure of Fig. 2.3, and the resulting equilibrium vapor and liquid phases were analyzed to determine the solid tie line. The continuous, rather than batch, contacting of liquid by gas in this latter experiment produced equilibrium gas and liquid samples that appear to fall on the phase boundary curve obtained

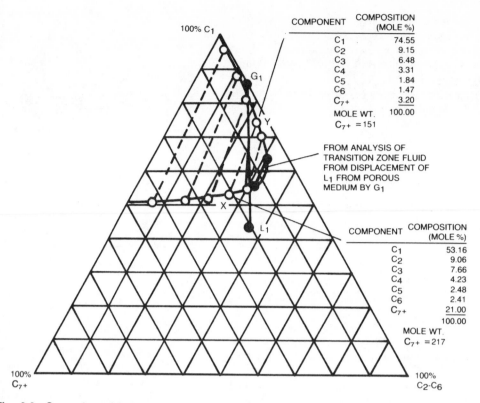

100% C₁

COMPONENT	COMPOSITION (MOLE %)
C_1	74.55
C_2	9.15
C_3	6.48
C_4	3.31
C_5	1.84
C_6	1.47
C_{7+}	3.20
MOLE WT.	100.00
C_{7+} =151	

G_1

Y

FROM ANALYSIS OF
TRANSITION ZONE FLUID
FROM DISPLACEMENT OF
L_1 FROM POROUS
MEDIUM BY G_1

X

L_1

COMPONENT	COMPOSITION (MOLE %)
C_1	53.16
C_2	9.06
C_3	7.66
C_4	4.23
C_5	2.48
C_6	2.41
C_{7+}	21.00
	100.00
MOLE WT.	
C_{7+} =217	

100% C_{7+}

100% C_2-C_6

Fig. 2.3—Comparison of tie line slopes—batch contact vs. data from a displacement, exemplified by University Block 31 data, p = 4,000 psia, T = 140°F (after Ref. 2).

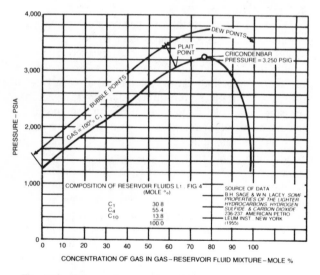

Fig. 2.4—Pressure/composition diagram for mixtures of C_1 with a C_1/nC_4/C_{10} liquid (after Ref. 2)

in the batch-contacting experiments. However, Fig. 2.3 indicates that (1) the compositions of the C_2 through C_6 and the C_{7+} groups were internally different in the flow experiments compared with the batch experiments and (2) that the slope of tie lines near the plait point may be different, depending on whether compositions near the plait point are the result of multiple-batch contacting of fluids or the result of continuous contacting in a porous medium. Since the slope of the tie line that passes through the plait point is critical for achieving dynamic miscibility* in some of the miscible displacement proc-

*See Sec. 1.5 for definition of "dynamic miscibility."

esses, this conclusion needs to be verified with the more accurate techniques reported recently[3-5] for sampling during flow experiments and for analysis of the samples.

The addition of the intermediate-molecular-weight hydrocarbons or CO_2 to some reservoir fluids may cause multiple phases to form, rather than resulting in the simple gas/liquid equilibrium illustrated on the pseudoternary diagram. Phase behavior has been reported where solid or semisolid asphaltenic material precipitates,[6-8] where two liquid phases coexist,[7,9-11] where two liquids and a gas phase are in equilibrium,[7,9,11,12] and even where two liquids, a gas, and a solid are present.[7,9,12] This complex phase behavior is discussed more fully in later sections.

Although the representation of solvent/reservoir fluid phase behavior by the pseudocomponent, triangular diagram method has shortcomings, it nevertheless is a highly useful tool for the conceptual understanding of miscible processes.

2.3 Other Useful Methods for Displaying Phase Behavior

In addition to triangular diagrams, pressure/composition (P-X) and pressure/temperature (P-T) diagrams are useful methods for displaying phase behavior data. Fig. 2.4 is a pressure/composition diagram for the simple, three-component system, methane-n-butane-decane. This figure shows the phase behavior for mixtures of methane with fluid of Composition L_1 in Fig. 2.2. Saturation pressures at constant temperature are plotted on the ordinate for mixtures spanning the composition range from pure Fluid L_1 to pure methane. Mixture composition, expressed as a percent of the injection gas added to Fluid L_1, is plotted on the abscissa.

Initially, as methane is added, mixtures have bubble points at the saturation pressure. At sufficiently high concentrations of methane, saturated mixtures exhibit dewpoints. The curve connecting bubble-point mixtures joins the curve connecting dewpoint mixtures at the critical mixture or plait point. For pressures above the bubble- and dewpoint curves, mixtures are single phase, and the highest pressure at which two phases can exist is called the cricondenbar. Above the cricondenbar, all mixtures are single phase. Below the bubble- and dewpoint curves, two phases, gas and liquid, coexist. For example, at 2,500 psi, mixtures of methane and reservoir fluid that contain more than 42% but less than 95% methane are in the two-phase region. On the ternary diagram of Fig. 2.2, these are the mixtures on the line connecting methane and Fluid L_1 that fall within the two-phase region at 2,500 psi.

Pressure/composition diagrams for reservoir oils are determined in high pressure visual cells by adding injection fluid to the reservoir oil and determining the appropriate saturation pressure—i.e., bubble point or dewpoint. The method is less cumbersome and more rapid for obtaining an overall perspective of solvent/oil phase behavior than determining pseudoternary diagrams, since saturation pressures and visual observations can be determined relatively easily and routinely, and equilibrium samples and compositions are not obtained. However, at a given pressure, the P-X method examines only those overall compositions that result from batch mixing the injection gas under study with reservoir fluid. Thus, the P-X diagram does not contain information about all the mixtures that might be of interest in a miscible displacement. Also, as discussed in the preceding section, the P-X diagram mixtures obtained by batch contacting may not correspond exactly to the mixtures that will result from the complicated phenomena causing mixing in a flow setting.

Pressure/composition diagrams for mixtures of miscible displacement injection fluids and reservoir oils may exhibit considerably more complicated behavior than that shown in Fig. 2.4 for the simple hydrocarbon system. Some examples of complicated phase behavior that was studied by the P-X method are shown in Figs. 2.5, 2.6, 2.7, and 2.8 for mixtures of CO_2 and reservoir oils. In addition to a region of vapor/liquid equilibrium, several of these figures show a region of liquid/liquid equilibrium and a region of liquid/liquid/vapor equilibrium. A small amount of a fourth, asphaltic or asphaltene, solid phase also has been reported over a limited region of some P-X diagrams[7,11-16] (not shown in Figs. 2.5, 2.6, 2.7, and 2.8).

Pressure/temperature diagrams are also useful for studying the miscibility relationships between solvent and drive gas. In addition, they are valuable in the design of handling and transmission systems for solvents. Such a diagram is shown in Fig. 2.9 for ethane and heptane mixtures. In this figure, Curves AC and BC_7 represent the vapor pressure curves of pure ethane and heptane. Curves $A_1C_1B_1$, $A_2C_2B_2$, and $A_3C_3B_3$ trace the bubble- and dewpoint curves for various mixtures of ethane/heptane. For example, Curve $A_3C_3B_3$ defines bubble- and dewpoint pressures at various temperatures for a mixture containing 9.78 wt% ethane. The dashed line in this figure connects the plait points for all mix-

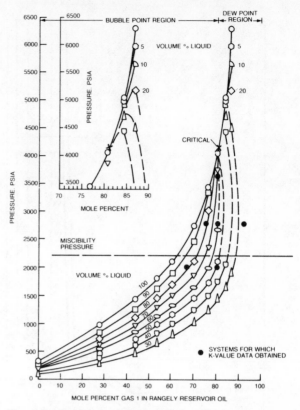

Fig. 2.5—Pressure/composition diagram for Type I phase behavior (after Ref. 15).

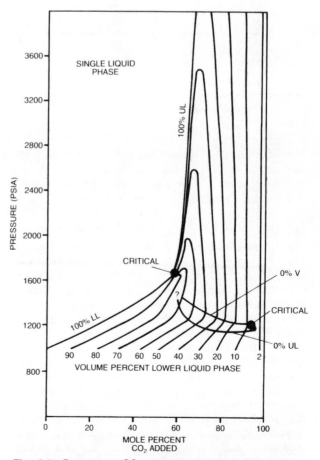

Fig. 2.6—Pressure vs. CO_2 concentration phase diagram for Wasson crude at 105°F. Not indicated in the diagram are regions where a small volume fraction precipitate is observed (after Ref. 11).

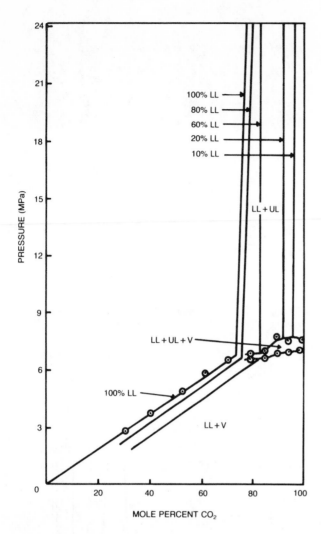

Fig. 2.7—Phase behavior of binary mixtures of CO_2 and Maljamar separator oil (after Ref. 13).

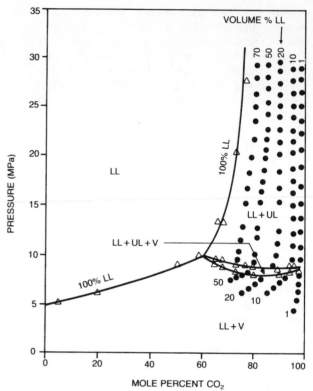

Fig. 2.8—CO_2/Oil A phase equilibria 314.2°K (after Ref. 16; composite of Figs. 5 and 14).

tures of ethane and heptane. Above the dashed curve all mixtures of these two components are single phase, whereas beneath this curve two phases, gas and liquid, may coexist, depending on mixture composition.

2.4 The First-Contact Miscible Process

The simplest and most direct method for achieving miscible displacement is to inject a solvent that mixes with the oil completely, in all proportions, such that all mixtures are single phase. Intermediate-molecular-weight hydrocarbons, such as propane, butane, or mixtures of LPG, are solvents that in the past have been used most often for first-contact miscible flooding.

Fig. 2.10 illustrates the phase behavior requirement for first-contact miscibility. LPG solvent on this pseudoternary diagram is represented by the C_{2-6} pseudocomponent. All mixtures of LPG and oil in this diagram lie entirely within the single-phase region. In fact, for the phase behavior of Fig. 2.10, LPG could be diluted with methane to Composition A, and the resulting mixtures would remain first-contact miscible with reservoir oil. Composition A is the intersection of the right side of the triangle (representing all methane/LPG compositions) and the tangent to the phase

boundary curve that passes through the oil composition.

For first-contact miscibility to be achieved between solvent and oil, the displacement pressure must be above the P-X diagram cricondenbar, since all solvent/oil mixtures above this pressure are single phase. When propane and butane are liquids at reservoir temperature and pressure, the saturation pressures of mixtures of oil with these solvents will vary monotonically between the oil bubble-point pressure and the solvent bubble point. The cricondenbar in this case is the higher of these two bubble points.

When the propane/butane solvent is a gas at reservoir temperature, P-X diagram phase behavior can be more complicated, and the cricondenbar may occur at mixtures intermediate between pure solvent and pure oil. This is illustrated in Fig. 2.11, which shows two P-X diagrams for a recombined south Texas oil—one with propane as solvent, the other with a propane-methane mixture as solvent. Reservoir temperature in this example is above the critical temperature of propane. (Actually, the reservoir fluid composition is slightly different for the two P-X diagrams. Each reservoir fluid has the same stock-tank oil base, but the bubble-point pressures are slightly different.)

When pure propane was added to the reservoir fluid with the lower bubble point, bubble-point pressures decreased monotonically with increasing propane mole percent until there was about 94 mol% propane in the mixture. Further additions of propane gave dewpoint mixtures. Although dewpoint pressures then increased to a maximum at about 97 mol% propane in the overall mixture, the highest saturation pressure for the P-X diagram occurred at the reservoir fluid bubble point.

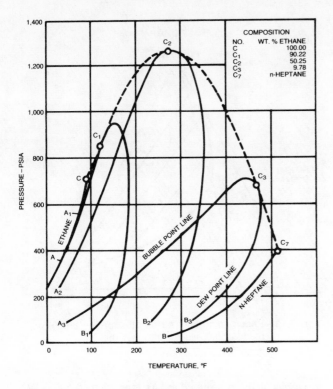

Fig. 2.9—Pressure/temperature diagram for the ethane/n-heptane system (after Ref. 12).

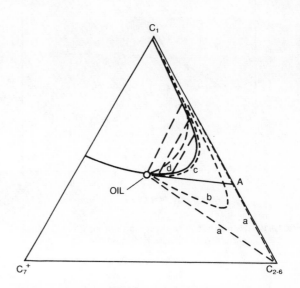

Fig. 2.10—First-contact miscibility and dilution of solvent slug.

When the mixture of propane with 12 mol% methane was added to the reservoir fluid with the higher bubble point, saturation pressure maxima occurred in both the bubble- and dewpoint segments of the P-X diagram. The cricondenbar in this case occurred at the maximum of the bubble-point curve.

As the concentration of methane in the injection fluid increases, the cricondenbar increases and ultimately becomes impractically high for first-contact miscibility. When this happens, dynamic miscibility can be achieved by the condensing-gas drive or vaporizing-gas drive mechanisms unless miscibility pressure is also impractically high for these methods (see Secs. 2.5 and 2.6). Fig. 2.12 shows a partial P-X diagram for natural gas and a west Texas oil. The cricondenbar in this instance is higher than 10,000 psi.

The cricondenbar for CO_2, another injection fluid useful for miscible flooding, is too high for first-contact miscibility. This is illustrated by the P-X diagrams shown in Figs. 2.5, 2.6, 2.7, and 2.8 for CO_2 and various reservoir oils. However, as explained in Sec. 2.7, dynamic miscibility can be achieved with CO_2 at pressures far lower than the cricondenbar.

Intermediate-molecular-weight hydrocarbon solvents for first-contact miscible flooding will precipitate some of the asphalt from asphaltic crudes. The tendency toward precipitation decreases as molecular weight of the hydrocarbon solvent increases. Bossler and Crawford discuss the phenomenon in detail.[8] Fig. 2.13 shows a pseudoternary diagram for a California crude oil/propane/asphalt system. Precipitation occurs only over a certain range of propane concentrations. For example, with an initial asphalt concentration P_1 in the oil, precipitation begins at about 50% propane and continues

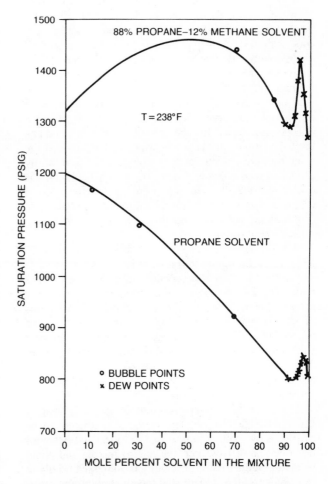

Fig. 2.11—Pressure/composition diagrams for two recombined south Texas oils.

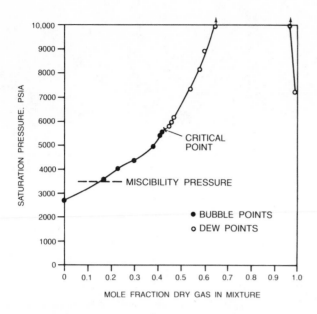

Fig. 2.12—Pressure/composition diagram for Block 31 Devonian reservoir fluid with natural gas.

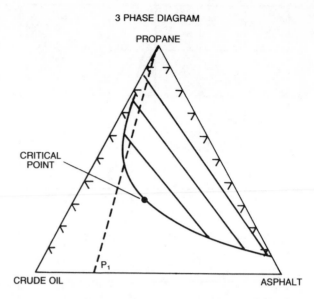

Fig. 2.13—Phase diagram for a California crude/propane/asphalt system (after Ref. 8).

to high propane concentrations. Strictly speaking, a first-contact miscible displacement does not occur, since some compositions fall within a two-phase region. However, in the solvent/oil mixed zone, propane is directly miscible with the deasphalted oil, which in turn is miscible with the original crude; miscible displacement is achieved, although precipitation of asphalt occurs within part of the mixed zone. Severe asphalt precipitation may reduce permeability and affect well injectivities and productivities. It may also cause plugging in producing wells.

In past practice, LPG solvents that are first-contact miscible with reservoir fluids have been too expensive to inject continuously. Instead, the solvent was injected in a limited volume, or slug, that was small relative to the reservoir pore volume, and the slug in turn was miscibly displaced with a less expensive fluid such as natural gas or flue gas. Ideally, with such a process scheme, solvent miscibly displaces reservoir oil while drive gas miscibly displaces the solvent, propelling the small solvent slug through the reservoir.

Miscibility between solvent and driving gas normally determines the minimum pressure required for miscible displacement in the first-contact miscible slug process with LPG solvents. Pressure at the trailing edge of the solvent slug must be above the cricondenbar for mixtures of solvent and drive gas. Fig. 5.1 in Chap. 5 shows cricondenbars for methane and several pure component LPG slug materials. Fig. 2.10 illustrates pseudoternary diagram phase behavior where first-contact miscibility is achieved between solvent and reservoir oil and between solvent and drive gas (methane).

A number of factors discussed in the next chapter cause miscible displacement in the slug process to be considerably more complex than the idealized behavior described here. However, the role of mixing and phase behavior on the loss of miscibility is addressed briefly at this point. As the solvent slug travels through the reservoir, it mixes with oil at the leading edge and with the drive gas at the trailing edge. As long as the solvent at the midpoint of the slug remains undiluted, the profile of compositions from oil through the slug to the drive gas is similar to the dotted line labeled "a" in Fig. 2.10. Eventually, the midpoint of the slug is diluted below its original concentration, and a composition profile similar to dotted Curve "b" results. With continued passage through the reservoir, small slugs may be diluted to composition profile, Curve "c," which just intersects the two-phase region. At this point the miscible displacement is lost, since subsequent mixing dilutes the slug into the two-phase region as illustrated by Curve "d." The size and shape of the two-phase region, as dictated by temperature, pressure, and fluid compositions, determine the degree to which the solvent slug can be diluted by mixing before miscibility is lost.

2.5 The Condensing-Gas Drive Process

Injection gases with compositions between A and B in Fig. 2.14 still can miscibly displace the reservoir oil even though they are not first-contact miscible with it. In this situation, dynamic miscibility results from the in-situ transfer of intermediate-molecular-weight hydrocarbons, predominantly ethane through butane, from the injected gas into the reservoir oil. For example, assume that gas of Composition B is injected to displace the oil in this figure. Gas Composition B is defined by extending the limiting tie line through the plait point, P, until it intersects the right side of the triangle. (The right side of the triangle represents all mixtures of methane and intermediate-molecular-weight hydrocarbons.) Oil and Gas B are not miscible initially, because most of their mixtures fall within the two-phase region. Suppose Mixture M_1 within the two-phase region results after the first contact of reservoir oil by Gas B. According to the tie line passing through M_1, Liquid L_1 and Gas G_1 are in equilibrium at this point in the reservoir. Subsequent injection of additional Gas B pushes the mobile

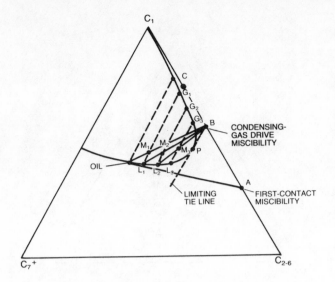

Fig. 2.14—Condensing-gas drive miscibility.

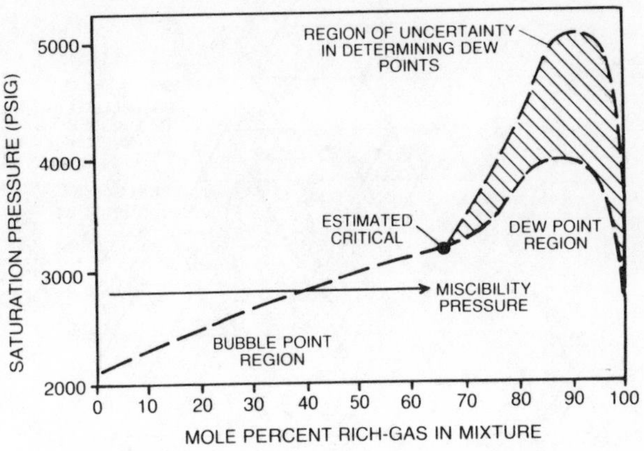

Fig. 2.15—Pressure/composition diagram for Reservoir Fluid No. 2 and Rich Gas No. 2 at 240°F (after Ref. 66).

equilibrium Gas G_1 ahead into the reservoir, leaving Equilibrium Liquid L_1 for Gas G_1 to contact. Gas G_1 and Liquid L_1 mix to give a new overall Mixture M_2 at this location. However, Equilibrium Gas G_2 and Equilibrium Liquid L_2 result from Mixture M_2, and Liquid L_2 lies closer to the plait point than the Liquid L_1 left after the first contact. By continued injection of Gas B, the composition of liquid at the wellbore is altered progressively in a similar manner along the bubble-point curve until it reaches the plait-point composition. The plait-point fluid is directly miscible with injection gas. By this multiple-contacting mechanism, reservoir oil is enriched with intermediate-molecular-weight hydrocarbons until it becomes miscible with the injected gas.

This mechanism for the in-situ generation of miscibility is called variously the condensing-gas drive process or the enriched-gas drive process. The multiple-contacting mechanism creates a transition zone of contiguously miscible liquid compositions from reservoir oil composition through Compositions L_1, L_2, L_3, P on the bubble-point curve to injected gas composition. Sufficient gas/oil contacts must occur before the miscible transition zone is developed. At the same time, the multiple-contacting mechanism establishes a transition zone of gas compositions from G_1 to P along the dewpoint curve, and, although reservoir oil is displaced miscibly, two-phase gas/liquid flow can occur in the transition zone.

If a gas is injected that contains less intermediate hydrocarbon than is dictated by the intersection of the limiting tie line with the right side of the diagram, the oil cannot be enriched to the point of miscibility. For example, the tie line connecting Equilibrium Gas G_1 and Equilibrium Liquid L_1, when extended, passes through Gas Composition C. If Gas C were injected to displace reservoir oil, enrichment of the oil to Composition L_1 would occur; but then enrichment would cease, since further contacts would always result in G_1 and L_1 as the equilibrium gas and liquid. If an extended tie line passes through the injected gas composition, an immiscible displacement will result. *For dynamic miscibility to be*

achieved by the condensing-gas drive method with an oil whose composition lies to the left of the limiting tie line on a pseudoternary diagram, the enriched-gas composition must lie to the right of the limiting tie line.

For clarity, two major simplifications were made to describe the development of condensing-gas drive dynamic miscibility: batch contacting of oil by rich gas and the idea that only Phase G moves during the development of miscibility. In a porous medium, of course, the contacting of fluids is continuous and results from both fluid convection through a tortuous pore network and from diffusion. The simplification of a flowing gas phase contacting a nonflowing liquid phase is not strictly accurate either, but one-dimensional simulations[17] and Helfferich's general theory[18] show that the qualitative description captures the essence of the compositional part of the process.

Two variables can be adjusted in a condensing-gas drive design to achieve miscibility—reservoir pressure and gas composition. For a given injection gas composition there is a minimum pressure, called the minimum miscibility pressure (MMP), above which dynamic miscibility can be achieved. Because increasing reservoir pressure reduces the size of the two-phase region, a lower concentration of intermediate-molecular-weight hydrocarbons in the injection gas will achieve miscibility as reservoir pressure increases.

For the simple phase behavior shown in Fig. 2.14, oil that is contacted repeatedly by a gas with composition to the right of the limiting tie line can be enriched to the point of complete miscibility; and the condensing-gas drive process, like the first-contact miscible process, has the potential for displacing all the oil in rock actually contacted by the enriched gas. In practice, some small oil saturation may be left to condensing-gas drive miscible flooding because of complex phase behavior or because some oil may remain uncontacted in dead-end or occluded pore volume (see Ref. 19 and Sec. 3.7).

Condensing-gas drive miscibility pressure is below both the cricondenbar and plait-point pressure on a P-X diagram. This is illustrated in Fig. 2.15, which shows

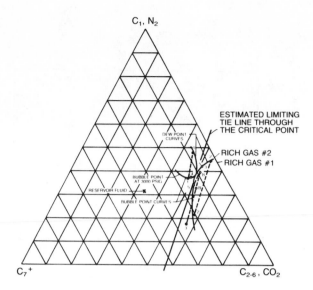

Fig. 2.16—Experimental equilibrium gas and liquid compositions at 240°F, 3,000 psig (after Ref. 66).

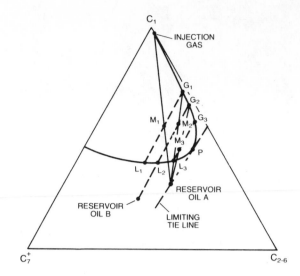

Fig. 2.17—Vaporizing-gas drive miscibility.

the P-X diagram for a north Louisiana oil and an enriched hydrocarbon gas containing 52 mol% of ethane through pentanes. As enriched gas was added to the reservoir oil, mixture bubble-point pressures increased steadily until the plait point was reached at about 3,200 psig. Saturation pressures then rose steeply in the dewpoint segment of the diagram, causing the cricondenbar to be as high as 4,000 to 5,000 psig. Miscibility pressure was determined to be about 2,800 psi from displacement experiments in a sandpacked tube.

Actual dynamic or multicontact phase behavior may be more complicated than shown in the simple pseudoternary diagram of Fig. 2.14. For example, Fig. 2.16 shows the results of multicontact equilibrium cell experiments with the reservoir oil and enriched gas used to determine the P-X diagram of Fig. 2.15. These equilibrium cell experiments were conducted at a pressure of 3,000 psig, about 200 psig above the dynamic miscibility pressure. Enriched gas was mixed with reservoir oil until two phases resulted. The equilibrium liquid from this first mixture was saved and contacted with fresh enriched gas. This procedure was repeated to simulate development of the dynamic miscible transition zone. What appeared to be two two-phase regions joined together at a plait point were defined by this procedure. A point was reached where further contact of equilibrium liquid with enriched gas gave mixtures that were mostly gas in equilibrium with a small amount of dewpoint liquid. In other words, successive multiple contacts of equilibrium liquid by enriched gas altered the liquid to a plait-point composition as expected for a simple condensing-gas drive, but subsequent contacts by enriched gas caused overall compositions to fall in what appeared to be a dewpoint region. The transition zone of contiguously miscible compositions passed from bubble-point fluids through the plait point to dewpoint fluids.

Shelton and Yarborough[7] give another example of complex phase behavior for enriched gas-reservoir oil systems. In a pressure/composition diagram study, they found regions of liquid/liquid, liquid/liquid/solid, liquid/liquid/vapor/solid, and liquid/vapor/solid equilibrium as well as a region of vapor/liquid equilibrium.

In some field projects the enriched gas has been injected continuously, but in most projects an enriched-gas slug was driven with natural gas. Since the composition of the enriched-gas slug initially is closer to the two-phase region than the composition of first-contact miscible slugs, less dilution of the enriched-gas slug can occur before immiscible displacement results.

2.6 The Vaporizing-Gas Drive Process

Another mechanism for achieving dynamic miscible displacement relies on the in-situ vaporization of intermediate-molecular-weight hydrocarbons from the reservoir oil into the injected gas to create a miscible transition zone. This method for attaining miscibility has been called both the high-pressure gas process and the vaporizing-gas process. Miscibility can be achieved by this method with natural gas, flue gas, or nitrogen as injection gases, provided that the miscibility pressure is physically attainable in the reservoir. CO_2 also attains dynamic miscibility by a multiple-contacting mechanism that vaporizes or extracts hydrocarbons from reservoir oil. CO_2, however, extracts much higher-molecular-weight hydrocarbons than do natural gas, flue gas, and nitrogen, which predominantly extract the C_2 to C_5 hydrocarbons. The CO_2 miscible process is discussed separately in Sec. 2.7. This section is concerned only with methane, natural gas, flue gas, and nitrogen as miscible injection fluids.

Consider first methane/natural gas as injection fluids.[2,20] Fig. 2.17 illustrates the mechanism by which vaporizing-gas drive miscibility is achieved with these gases. In this example, Reservoir Oil A contains a high percentage of intermediate-molecular-weight hydrocarbons, and its composition lies on the extension of the limiting tie line through the plait point. Injection gas and

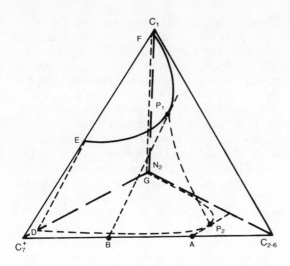 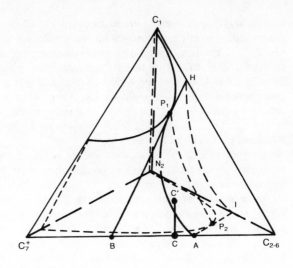

Fig. 2.18—Conceptual phase behavior for the vaporizing-gas drive process with nitrogen as injection gas.

reservoir oil are not initially miscible. Consequently, the injection gas initially displaces oil immiscibly away from the wellbore but leaves some oil undisplaced behind the gas front. Suppose the relative proportions of injection gas and undisplaced oil after this first contact are such as to give the overall Composition M_1. According to the tie line passing through M_1, Liquid L_1 and Gas G_1 are in equilibrium at this point in the reservoir. Subsequent injection of gas into the reservoir pushes the Equilibrium Gas G_1, left after the first contact, further into the reservoir, where it contacts fresh reservoir oil. Liquid L_1 is left behind as a residual saturation. As a result of this second contact, a new overall composition, M_2, is reached with corresponding equilibrium gas and liquid, G_2 and L_2. Further injection causes Gas G_2 to flow ahead and contact fresh reservoir oil, and the process is repeated. In this manner, the composition of gas at the displacing front is altered progressively along the dew-point curve until it reaches the plait-point composition. The plait-point fluid is directly miscible with the reservoir oil. Although this description for the miscibility mechanism of the vaporizing-gas drive process was given for multiple-batch contacts of gas and oil, the process of gas enrichment is, of course, continuous; a transition zone of contiguously miscible compositions is established from reservoir oil composition to injected gas composition.

As long as the reservoir oil composition lies on or to the right of the limiting tie line, miscibility can be attained by the vaporizing-gas drive mechanism with natural gas that has a composition lying to the left of the limiting tie line. If the oil composition should lie to the left of the limiting tie line, gas enrichment will occur only to the composition of equilibrium gas lying on the tie line that can be extended to pass through the oil composition. For example, if Reservoir Oil B in Fig. 2.17 were being displaced, the injection gas would be enriched to the composition of Equilibrium Gas G_2 but could not be enriched past this composition, since any further contacting of reservoir oil by Gas G_2 would result only in mixtures that lie on the tie line passing through G_2. The requirement that the oil composition must lie to the right of the limiting tie line also implies that only oils that are

undersaturated with respect to methane can be miscibly displaced by methane or natural gas. Thus, oil of composition L_2 on the bubble-point curve of Fig. 2.17 could not develop vaporizing-gas drive miscibility with methane/natural gas.

Inspection of Fig. 2.17 shows that as the concentration of intermediate-molecular-weight hydrocarbons in the reservoir oil decreases, the oil composition moves toward the left side of the pseudoternary diagram, and higher pressures are required to develop miscibility. Increasing pressure both decreases the size of the two-phase region and changes the slopes of tie lines by increasing the vaporization of intermediate-molecular-weight hydrocarbons into the vapor phase. Both changes have the effect of moving the extension of the limiting tie line toward the left side of the diagram until, finally, with increasing pressure, oil composition again lies to the right of the limiting tie line. Unfortunately, for a great many oils the miscibility pressure with methane/natural gas is unattainably high for reservoir flooding.

Fig. 2.3 shows experimental pseudoternary diagram phase behavior for University Block 31 reservoir fluid and natural gas. This figure illustrates many of these concepts for a real system. Natural Gas G_1 can attain vaporizing-gas drive miscibility with Reservoir Fluid L_1 at the pressure of Fig. 2.3 because Reservoir Fluid L_1 lies to the right of the limiting tie line.

Miscibility by the vaporizing-gas drive mechanism can also be developed with nitrogen and with flue gas (about 88% nitrogen and 12% CO_2).[21-24] Flue gas is generated by burning hydrocarbons with air. An example is power-plant stack gas. Since CO_2 is soluble in reservoir brine, it tends to be stripped or partially stripped from the flue gas front. Koch and Hutchinson[21] speculate that the gas front may become essentially CO_2-free and develop miscibility in much the same manner and at much the same MMP as if nitrogen had been the injection gas. There is a lack of published data to judge the validity of this view.

Fig. 2.18 illustrates *conceptual* phase behavior for a nitrogen/reservoir-fluid system. A tetrahedron or pseudoquaternary diagram was chosen for this illustration rather than a pseudoternary diagram to illustrate more clearly the effect of nitrogen on phase behavior and miscibility. The reader should be aware that there are no published data to define the pseudoquaternary phase boundaries and tie lines for such a system and that the concepts that follow are deduced from general phase behavior considerations and results of displacement experiments. The concepts that follow are similar to those advanced by Koch and Hutchinson.[21]

The front face of the tetrahedron shown in Fig. 2.18a represents pseudoternary diagram phase behavior of the hydrocarbon system in the absence of nitrogen. This is the by now familiar representation using C_1, C_{2-6}, and C_7^+ as pseudocomponents. The far corner of the tetrahedron represents 100% nitrogen, and the bottom face represents pseudoternary phase behavior of nitrogen with the C_{2-6} and C_7^+ fractions of the oil. The tetrahedron contains a bubble-point surface (EDP_2P_1), a dew-point surface (FGP_2P_1), and a locus of plait-point compositions (P_1P_2). For a given temperature and pressure, vapor/liquid equilibrium ratios, or K-values, are greater for nitrogen than for methane.[25,26] For this reason, the two-phase region of the triangular diagram with N_2, C_{2-6}, and C_7^+ as the pseudocomponents is larger than the two-phase region on the triangular diagram with C_1, C_{2-6}, and C_7^+ as pseudocomponents (Fig. 2.18a), and phase behavior with nitrogen is more unfavorable than phase behavior with methane for achieving vaporizing-gas drive miscibility with a methane-depleted oil. For example, methane can achieve vaporizing-gas drive miscibility with methane-depleted Oil B, whereas nitrogen cannot because Oil Composition B lies to the left of the limiting tie line of the $N_2/C_{2-6}/C_7^+$ pseudoternary diagram. A higher pressure is necessary for dynamic miscibility between nitrogen and Oil B. At the pressure and temperature shown in Fig. 2.18a, the concentration of intermediate-molecular-weight hydrocarbons in the methane-free oil must be greater than that of Composition A for vaporizing-gas drive miscibility with nitrogen.

According to the pseudoternary diagram concept of vaporizing-gas drive miscibility with methane depicted in Fig. 2.17, oil composition must lie to the right of the limiting tie line through the plait point. For the conceptual pseudoquaternary diagram phase behavior with nitrogen depicted in Fig. 2.18b, a surface, P_1AIH, is generated by the limiting tie lines passing through the critical locus, P_1P_2. If reservoir oil composition lies to the right of this surface (or to the right of Curve P_1A for nitrogen-free oils), no tie line or extension of a tie line will pass through the oil composition, and vaporizing-gas drive miscibility should be attained with nitrogen. There are insufficient data to determine whether this criterion is too restrictive—i.e., whether, with an oil composition near but to the left of this surface, nitrogen could be enriched to the critical locus before the tie line through the oil composition is encountered.

Koch and Hutchinson[21] found that higher methane concentrations in the reservoir oil improved attainment of vaporizing-gas drive miscibility with nitrogen—at least for one stock-tank oil that was recombined with different amounts of separator gas to give simulated reservoir fluids with different bubble points. In a series of displacements in long sandpacks at constant pressure,

they observed immiscible displacement of the lower bubble-point oils but miscible displacement of oils with the higher bubble points. Fig. 2.18b offers a possible explanation of this behavior. Nitrogen cannot displace methane-depleted Oil C miscibly because its composition lies to the left of Curve P_1A. As methane is added to this oil, compositions at some point begin to lie to the right of P_1A. For example, Composition C' in the $C_1/C_{2-6}/C_7^+$ plane results from the addition of methane to Oil C and lies to the right of Curve P_1A.

Koch and Hutchinson speculated that methane in a sense was acting like an intermediate-molecular-weight hydrocarbon when nitrogen was the injection gas. They reasoned that the difference between the MMP's for nitrogen and methane should become smaller as the methane content of the reservoir fluid increased and that with some oils of high saturation pressure only a slight increase above the methane miscibility pressure would be required with nitrogen/flue gas. For the oils they tested, nitrogen miscibility pressure was only about 300 psi higher than the methane miscibility pressure for a 2,760-psig bubble-point oil but was more than 900 psi higher for a 905-psig bubble-point oil.* More data are needed, however, to confirm or disprove this hypothesis.

With either methane/natural gas or nitrogen/flue gas there is some immiscible displacement before the gas front is enriched to the point of miscibility with oil because some oil must be left behind the invading gas front to supply the necessary intermediate-molecular-weight hydrocarbons. However, the path length travelled before a miscible gas front composition is reached is negligible compared with interwell distances[2,5,10,19] and may be negligible in many laboratory displacements. The data of Fig. 2.19, for example, show that dynamic miscibility with natural gas must have been achieved over a small fraction of a 6-ft sandstone core because ultimate recovery in these displacements, where viscous fingering and gravity tonguing were not factors, is almost 98% of the initial oil in place. There may be exceptions, however.[5] McNeese[22] published results of experiments that he believed showed that tens of feet were required to attain dynamic miscibility between flue gas and the reservoir fluid under test.

As the transition zone travels through a porous medium, mixing constantly acts to bring the dewpoint fluids into the two-phase region. As a result, some fallout of dewpoint liquid** from the transition zone occurs throughout the displacement, leading to a small amount of oil left behind the gas front all along the displacement path.[2,21,22] (See Sec. 3.7 for further discussion of this phenomenon; also, Refs. 11, 62, and 19 for CO_2 flooding.)

Pressure/composition diagram phase relations for the University Block 31 reservoir fluid with natural gas and flue gas are shown in Figs. 2.12 and 2.20. These figures show that the two-phase region at a given pressure occurs over a wider range of oil/injection gas mixtures with flue gas.

Miscibility pressure for the University Block 31 reser-

---

*There was a 36°F difference in temperature between the two sets of displacements, however.

**For the simple phase behavior shown in Fig. 2.17, the liquid compositions are defined by the bubble-point curve—e.g., L_3, L_2, L_1, etc.

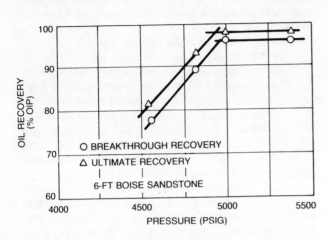

Fig. 2.19—Reservoir oil displaced by a lean hydrocarbon gas in gravity-stable displacements.

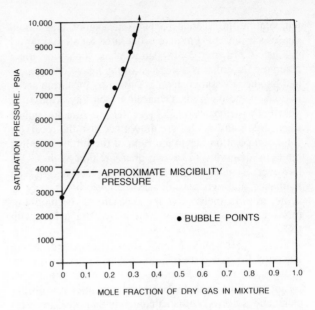

Fig. 2.20—Pressure/composition diagram for Block 31 Devonian reservoir fluid with flue gas.

voir fluid was determined from displacement tests to be about 3,500 psi with natural gas and no more than a few hundred psi higher with flue gas. According to the Koch and Hutchinson hypothesis,[21] this small difference in miscibility pressures may be caused by the relatively high methane content (43 mol%) and saturation pressure (2,760 psig) of the reservoir fluid. In the vaporizing-gas drive process, dynamic miscibility occurs below both the cricondenbar and the plait-point pressure on a P-X diagram. This is illustrated in Figs. 2.12 and 2.20 where the cricondenbars with both gases are greater than 10,000 psi and the plait-point pressure is about 5,600 psi for natural gas and greater than 10,000 psi with flue gas.

2.7 The CO_2 Miscible Process

CO_2 is not first-contact miscible with reservoir oils at realistically attainable reservoir pressures. However, displacement tests in long cores and in sandpacked slim tubes indicate that dynamic miscible displacement is possible above an MMP.[7,10,27-30] Usually the MMP is substantially lower than the miscibility pressure for dry hydrocarbon gas, flue gas, or nitrogen, although there may be exceptions at high temperature.[31] This is a major advantage of the CO_2 miscible process because dynamic miscibility can be achieved at attainable pressures in a broad spectrum of reservoirs.

Fig. 2.21 illustrates why dynamic miscibility with CO_2 can be achieved at a lower pressure than with dry hydrocarbon gas. This example is for a three-component system and is overly simplified for reservoir fluids, but it illustrates an important concept.

Fig. 2.21 shows conceptual ternary diagram phase behavior for CO_2 and methane with mixtures of a heavy hydrocarbon and various low-to-moderate-molecular-weight hydrocarbons. The heavy hydrocarbon has a sufficiently high molecular weight that it does not become first-contact miscible with CO_2 at attainable reservoir pressures. The ternary diagrams are drawn for a

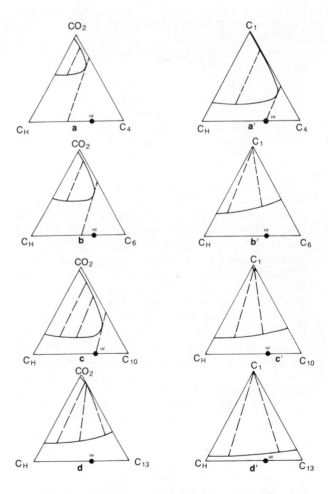

Fig. 2.21—Conceptual phase behavior for CO_2 and methane with simple hydrocarbons at constant pressure.

temperature high enough that only vapor and liquid phases are present. Pressure is the same for all diagrams.

Figs. 2.21a and 2.21a' show phase behavior for a reservoir fluid that is a mixture of the heavy component and butane. Its composition is shown by the circle at the bottom of the diagram. Dynamic miscibility is possible with both methane and CO_2 at the pressure for which Figs. 2.21a and 2.21a' are drawn because the reservoir fluid composition lies to the right of the limiting tie line on both diagrams. The two-phase region with CO_2, however, is much smaller than the two-phase region with methane, and whereas dynamic miscibility barely is achieved with methane at the pressure of the diagram, this pressure is obviously well above the MMP with CO_2.

Figs. 2.21b and 2.21b' show conceptual phase behavior for a reservoir fluid that is a mixture of the heavy component and hexane. At this pressure methane is not miscible with hexane, and dynamic miscibility cannot be achieved with methane; a higher pressure is required. CO_2 and hexane are miscible, however, although the two-phase region with hexane is larger than the two-phase region was with butane in Fig. 2.21a. The pressure for which the diagram is drawn is still above the minimum miscibility pressure for CO_2 but not by as wide a margin as when butane was the second component of the reservoir fluid.

Figs. 2.21c and 2.21d show phase behavior when decane and tricosane are the second components of the reservoir fluid. Dynamic miscibility is just barely achievable at the diagram pressure for the $CO_2/C_{10}/C_H$ system, since the reservoir fluid composition lies only slightly to the right of the limiting tie line. Dynamic miscibility is not attainable with CO_2 and the C_{13}/C_H reservoir fluid of Fig. 2.21d; a higher pressure is required.

Fig. 2.21 illustrates this important concept: *CO_2 achieves dynamic miscibility at lower pressures than is possible with methane by extracting from the oil hydrocarbons of higher molecular weight than the predominantly C_2-through-C_5 hydrocarbons that methane vaporizes to achieve vaporizing-gas drive miscibility.* For miscible displacement to be possible with methane at attainable pressures, a reservoir fluid must be rich in intermediate-molecular-weight hydrocarbons such as ethane through pentane. This is a serious restriction that severely limits application of the process. CO_2 achieves dynamic miscibility by extracting hydrocarbons from the gasoline and gas/oil fractions of the crude[13,27,29] as well as the intermediate-molecular-weight hydrocarbons. Several investigators have reported finding C_5-through-C_{30} hydrocarbons in the CO_2-rich phase of displacement and static equilibrium cell experiments with reservoir oils.[11,13,27,29] This was true at low temperatures where the CO_2-rich phase was a liquid, as well as at higher temperatures where the CO_2-rich phase was a vapor. In fact, development of dynamic miscibility with CO_2 does not require the presence of intermediate-molecular-weight hydrocarbons in the reservoir fluid.[27,29,30] This extraction of a broad range of hydrocarbons from the reservoir oil often causes dynamic miscibility to occur at moderate pressures.

The phase behavior of CO_2 and reservoir oils and the path of transition-zone compositions over which CO_2 achieves dynamic miscibility are only partly understood at this time. The remainder of this section presents additional concepts of CO_2 phase behavior and miscibility with reservoir fluids. These concepts are based on the behavior of CO_2 and hydrocarbons in well-characterized binary and ternary systems and on limited studies of CO_2/reservoir oil phase behavior.[7,10-15,27-29,32-46] They are consistent with the bulk of experimental work reported to date, but it should be emphasized that what follows is not firmly established fact and that additional investigation is needed to establish or modify these concepts. Another word of caution: The descriptions that follow utilize pseudoternary diagrams to illustrate concepts. Such diagrams at best are only approximate representations of actual phase behavior.

Metcalfe and Yarborough[28] classify CO_2/reservoir fluid phase behavior into two broad types according to the characteristics of the pressure/composition diagram. Type I behavior shows only vapor and liquid phases coexisting in the multiphase region of the P-X diagram. According to Metcalfe and Yarborough[28] and Orr *et al.*,[13] this condition usually exists at temperatures higher than about 120°F. Type II phase behavior occurs relatively close to the critical temperature of CO_2—generally at temperatures below 115 to 120°F. Type II systems show some mixtures separating into equilibrium vapor and liquid phases, while others separate into two coexisting liquid phases; and in a small region of the P-X diagram, three phases coexist, two liquids and a gas. With some oils a small amount of asphaltenes also may precipitate over a portion of the P-X diagram.[7,12,47]

The temperature for transition from Type I to Type II phase behavior probably depends on the average molecular weight of the oil, although the exact form of the dependence is not known.[13] As average molecular weight of the oil increases, so does the maximum temperature for Type II behavior.

Type I Phase Behavior

Fig. 2.22 shows hypothetical P-X and pseudoternary diagrams for Type I phase behavior. The ternary diagram pseudocomponents are CO_2, the oil fraction containing those hydrocarbons that are readily extracted by CO_2, and the oil fraction containing high-molecular-weight hydrocarbons that are extracted very little by CO_2. The pressure/composition diagram of Fig. 2.22a is qualitatively similar to P-X diagrams for dry hydrocarbon gas and reservoir oils. A saturation curve of bubble-point fluids and a saturation curve of dewpoint fluids join at a critical point. Lines of constant volume percent liquid show retrograde behavior in the dewpoint region. This type phase behavior has been reported by several authors.[14,15,32,33] Fig. 2.5 shows Type I pressure/composition behavior for an actual reservoir fluid.

When Type I phase behavior prevails, dynamic miscibility can be achieved by the vaporizing-gas drive mechanism. At sufficiently high pressure, hydrocarbons extracted from the oil enrich the CO_2 gas phase to such a degree that the composition of gas at the displacing front becomes miscible with reservoir oil. This is illustrated by Figs. 2.22b, 2.22c, and 2.22d, which show

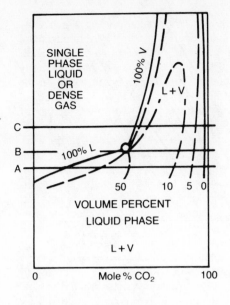

a) PRESSURE-COMPOSITION DIAGRAM

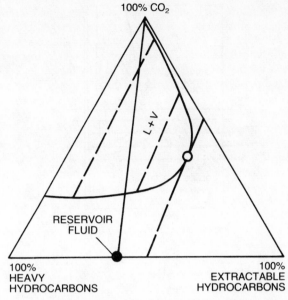

b) PSEUDO TERNARY DIAGRAM AT PRESSURE A

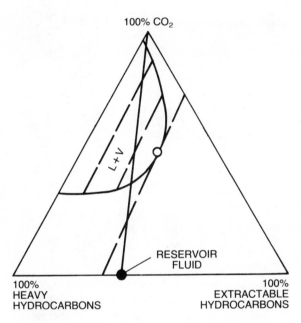

c) PSEUDO TERNARY DIAGRAM AT PRESSURE B

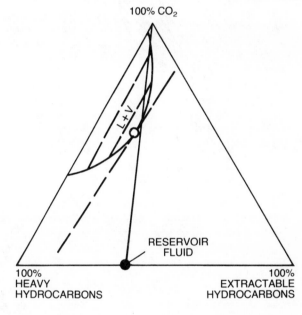

d) PSEUDO TERNARY DIAGRAM AT PRESSURE C

Fig. 2.22—Hypothetical phase behavior for Type I systems (temperature greater than approximately 120°F).

hypothetical phase behavior for three pressures. Pressure A (Fig. 2.22b) is too low for dynamic miscibility. The reservoir fluid lies to the left of the limiting tie line through the plait point. The pressure/composition diagram shows that as CO_2 is added to reservoir fluid at this pressure, mixture compositions at first lie in a single-phase region and then enter the two-phase region by crossing the bubble-point curve. This behavior is consistent with that shown by the hypothetical pseudoternary diagram.

At the higher pressures, B and C (Figs. 2.22c and 2.22d), dynamic miscibility can be achieved because the reservoir fluid lies to the right of the limiting tie line. At

Pressure B, CO_2/oil mixtures still enter the two-phase region by crossing the bubble-point curve as CO_2 is added to oil; but at the higher pressure, C, the two-phase region is entered by crossing the dewpoint curve as CO_2 is added to reservoir oil.

This conceptual picture of Type I phase behavior suggests that the MMP should be no higher than the pressure at which the plait point occurs on the P-X diagram.

Type IIa Phase Behavior

Fig. 2.23 shows an example of Type II phase behavior. For convenience, Type II behavior with the features

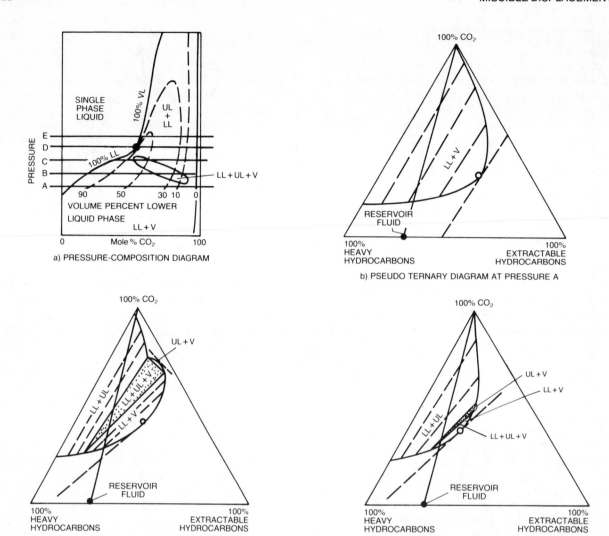

a) PRESSURE-COMPOSITION DIAGRAM

b) PSEUDO TERNARY DIAGRAM AT PRESSURE A

c) PSEUDO TERNARY DIAGRAM AT PRESSURE B

d) PSEUDO TERNARY DIAGRAM AT PRESSURE C

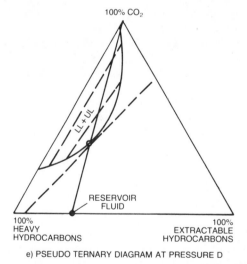

e) PSEUDO TERNARY DIAGRAM AT PRESSURE D

Fig. 2.23—Hypothetical phase behavior for Type IIa systems (temperature less than approximately 120°F).

shown in Fig. 2.23 is labeled Type IIa in this monograph.

The pressure/composition diagram (Fig. 2.23a) has a region where two liquids are in equilibrium. These are an oil-rich liquid, designated LL because it is usually the denser, lower-liquid phase in a static cell, and a CO_2-rich liquid, designated UL for upper liquid. There is also a two-phase region where vapor and oil-rich liquid are in equilibrium (V+LL), and there is a three-phase region of vapor, oil-rich liquid, and CO_2-rich liquid (V+LL+UL). Type IIa behavior has these distinguishing characteristics: (1) the three-phase region is found at progressively lower pressures as CO_2 is added to reservoir fluid (i.e., the region slopes downward to the right on the P-X diagram) and (2) retrograde behavior is observed in the liquid/liquid region. Orr *et al.*[13] speculate that Type IIa phase behavior occurs when the reservoir fluid has some critical concentration of light ends (unspecified). Conceivably, light hydrocarbons or nitrogen in the injected CO_2 also would have a similar effect on phase behavior.

The phase behavior that would be observed in a PVT cell for this hypothetical system as CO_2 is added to reservoir fluid depends on pressure level in the following manner. Pressure A (Fig. 2.23a) is above but not too far removed from the reservoir fluid bubble point. At this pressure, CO_2/oil mixtures at first enter a single-phase liquid region but then pass into a two-phase vapor/liquid region (V+LL) as more CO_2 is added by crossing a locus of mixture bubble points. At the higher pressure, B, the same behavior is found initially; but at high CO_2 concentrations, mixtures pass into the three-phase region. At still higher CO_2 concentrations, mixtures pass from the three-phase region into a two-phase liquid region (LL+UL). Phase behavior at Pressure C is similar to that at Pressure B except the three-phase region is smaller and is encountered at lower CO_2 concentrations. At Pressure D, mixtures pass through a single-phase liquid region and then through a critical point before entering the liquid/liquid region. At Pressure E, mixtures pass from the single-phase liquid region into the retrograde liquid/liquid two-phase region. As the CO_2-rich liquid enters the two-phase region, the volume percent of oil-rich second phase at first increases and then decreases as overall CO_2 concentration increases.

Fig. 2.23a shows the three-phase region disappearing before it intersects the bubble-point curve. Another possible configuration of this diagram would be for the three-phase region to terminate at the bubble-point curve. The particular form of the P-X diagram probably depends on oil composition.[48]

Figs. 2.23b through 2.23e show hypothetical pseudoternary diagram phase behavior at these pressures and illustrate how dynamic miscibility could be achieved for reservoir fluids that show Type IIa phase behavior. At Pressure A, Fig. 2.23b, there is only a vapor/liquid multiphase region. At this pressure, the two-phase region might also occur as a band across the diagram, similar to Fig. 2.21d, depending on the critical pressure of the CO_2/extractable-hydrocarbon pseudobinary. As pressure increases, a three-phase (LL+UL+V) region and a two-phase (LL+UL) region appear, as shown in Fig. 2.23c. Phase compositions in the LL+UL+V

region are defined by the corners of the three-phase triangle. With further increases in pressure, the three-phase region at first expands in size but then begins to shrink (Fig. 2.23d) until it ultimately disappears (Fig. 2.23e). The two-phase V+LL region continuously decreases in size with increasing pressure. Finally, a pressure is reached where only the two liquid phases, LL+UL, are present, as shown in Fig. 2.23e. There also should be a small UL+V two-phase region (Fig. 2.23c), but this region has not been reported to date in published phase behavior experiments.

Dynamic miscibility can occur at pressures where the three-phase region is present.[5,11,12,47] For this to happen, pressure must be sufficiently high that the limiting tie line of the LL+V region passes to the left of the reservoir fluid composition. The limiting tie line of the UL+V region would also have to pass to the left of the 100% CO_2 apex of the triangle (Fig. 2.23c) for in-situ mass transfer to establish a transition zone of contiguously miscible compositions from CO_2 to reservoir oil. If this latter condition is not fulfilled, true dynamic miscibility may not occur until pressure is sufficiently high that the three-phase region (and therefore the UL+V region) is not present. Dynamic miscibility is achieved primarily by a vaporizing-gas drive type mechanism, except that extractable hydrocarbons are both vaporized into a vapor phase (LL+V region) and extracted into a CO_2-rich liquid phase (LL+UL region). Displacement mechanism in that part of the transition zone through the UL+V region would be similar to the condensing-gas drive mechanism.

Stalkup,[12] Gardner *et al.*,[11] and Orr and Taber[48] have published experimental pressure/composition diagrams for CO_2 and reservoir fluids that show Type IIa behavior. Fig. 2.6 shows the diagram of Gardner *et al.*[11,*] These latter authors also published two experimental pseudoternary diagrams, Figs. 2.24 and 2.25, which were determined from single-contact and multicontact experiments in a PVT cell.[**] Fig. 2.24 shows pseudoternary diagram phase behavior at a pressure where a three-phase region was found on the P-X diagram. Fig. 2.25 shows pseudoternary diagram behavior at higher pressure where only two liquid phases coexist in the multiphase region. According to the concepts discussed in the preceding paragraphs, dynamic miscibility should be possible for the phase behavior shown in Figs. 2.24 and 2.25; and this was, in fact, observed in slim-tube displacements conducted by Gardner *et al.*

Although Figs. 2.24 and 2.25 represent phase behavior with C_6^- as a pseudocomponent, this does not indicate that no higher molecular weight hydrocarbons were extracted by the CO_2. Gardner *et al.*[11] found that oil produced near the end of their flooding experiments contained almost no C_{35}^- material, and as discussed by these authors, the choice of the C_6^- fraction for representing pseudoternary phase behavior was dictated by practical convenience. It should be kept in mind that hydrocarbons heavier in molecular weight than the C_6^-

---

*More recent work with Wasson crude did not show the second critical point indicated in Fig. 2.6 at high CO_2 concentrations (F.M. Orr Jr., personal communication, New Mexico Inst. of Mining and Technology, 1982).

**Note that these diagrams are plotted in volume fractions, whereas Fig. 2.6 is plotted in mole fractions.

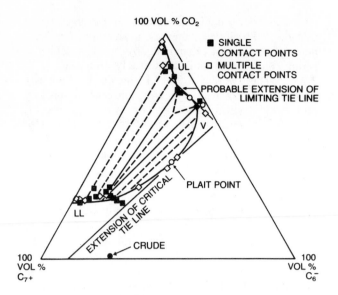

Fig. 2.24—Composite ternary diagram for the CO_2/Wasson crude system at 1,350 psia and 105°F; lines indicate phase behavior representation used in simulations (after Ref. 11).

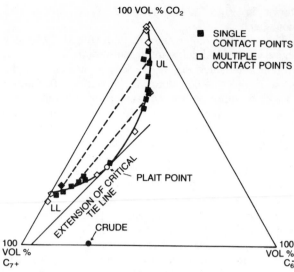

Fig. 2.25—Composite ternary diagram for the CO_2/Wasson crude system at 2,000 psia and 105°F; lines indicate phase behavior representation used in simulations (after Ref. 11).

fraction are extracted by CO_2 and contribute importantly to the development of dynamic miscibility in CO_2 miscible flooding.

Type IIb Phase Behavior

Another example of Type II phase behavior is shown in Fig. 2.26. Pressure/composition diagrams with the features of Fig. 2.26a are called "Type IIb" phase behavior in this monograph. Type IIb behavior has these distinguishing characteristics: (1) the three-phase region is found at progressively higher pressures as CO_2 is added to reservoir fluid (i.e., the region slopes upward to the right on a P-X diagram), (2) retrograde behavior is not observed in the liquid/liquid region, at least not for pressures of interest, and (3) the critical point on the 100% LL line occurs at very high pressures (not shown in Fig. 2.26a). Orr *et al.* [13] speculate that this type phase behavior may occur when the reservoir fluid is essentially devoid of light ends.

Figs. 2.26b through 2.26f show hypothetical pseudoternary diagram phase behavior. As pressure increases, a three-phase region forms in the vicinity of the plait point of the LL+V two-phase region and moves toward the left side of the diagram. The liquid/liquid LL+UL region grows with increasing pressure, while the LL+V region shrinks. Finally, at sufficiently high pressures, only UL and LL phases coexist in the multiphase region. For dynamic miscibility to be possible, the limiting tie line of the LL+UL region must pass to the left of the reservoir fluid composition, and the limiting tie line of the UL+V region (if this region still exists at the pressure where the first condition is fulfilled) must pass to the left of the 100%-CO_2 apex of the triangle.

Fig. 2.7 is an example of actual Type IIb pressure/composition phase behavior from experiments published by Orr *et al.* [13]

Type IIc Phase Behavior

Fig. 2.8 shows a third example of Type II phase behavior, called "Type IIc" in this monograph. Pressure/composition diagrams for Type IIc behavior have these distinguishing characteristics: (1) the three-phase region slopes downward and to the right and (2) retrograde behavior is not observed in the liquid/liquid region for pressures of interest.

Conceptual pseudoternary diagrams for a CO_2/oil system with this phase behavior show that dynamic miscibility will be attained at a pressure sufficiently high that the limiting tie line of the LL+V region passes to the left of the reservoir fluid composition and the limiting tie line of the UL+V region passes to the left of the CO_2 apex of the triangle (see Fig. 2.27a). If these conditions are not met before the three-phase region disappears, then dynamic miscibility will not occur until the pressure is high enough for the limiting tie line of the LL+UL region to pass to the left of the reservoir fluid composition (Fig. 2.27b).

2.8 Methods for Determining Miscibility Conditions

Review of Methods

The pressure required for dynamic miscibility with natural gas, flue gas, or CO_2, as well as the gas composition and pressure necessary for miscibility with enriched hydrocarbon gases, are best determined from displacement experiments.

Equilibrium gas and liquid compositions necessary for constructing pseudoternary diagrams are difficult and time consuming to obtain experimentally.* Moreover,

*A recently introduced method for obtaining equilibrium phase compositions by continuously contacting equilibrium fluids with injection gas in a static cell shows promise for obtaining data suitable for constructing pseudoternary diagrams more rapidly and with less difficulty than has been possible in the past with batch experiments. [4,49]

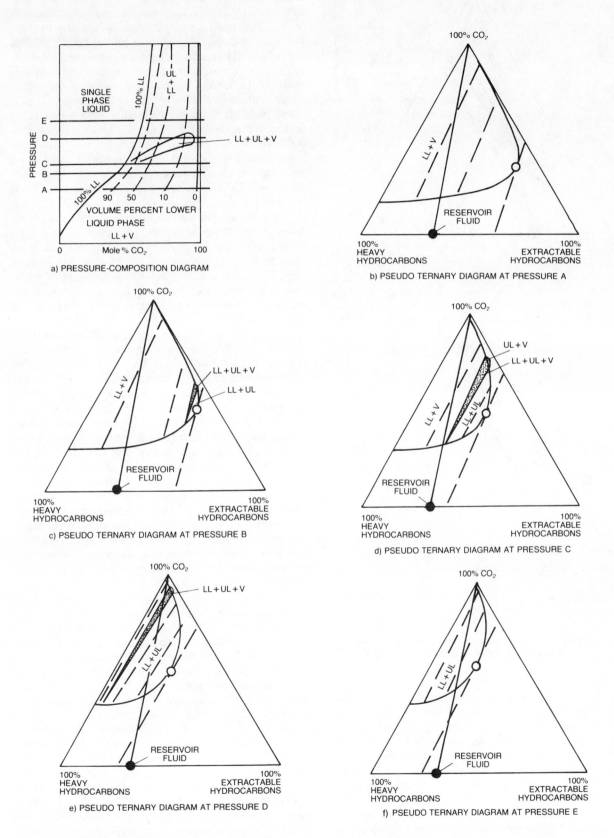

Fig. 2.26—Hypothetical phase behavior for Type IIb systems (temperature less than approximately 120°F).

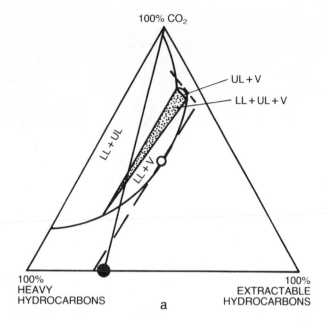

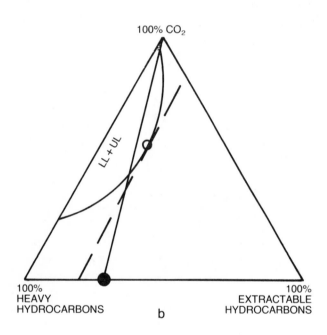

Fig. 2.27—Hypothetical phase behavior for Type IIc systems (temperature less than approximately 120°F).

the slopes of the tie lines determined under batch and flow conditions may be different,[2] since the exact composition path followed by gas/liquid mixtures will be different, depending on whether the fluids are flowing or whether they are mixed together batchwise in an equilibrium cell. The composition path actually followed during displacement conceivably may influence the pressure and solvent composition required for miscibility.[17]

Flow experiments also are preferred over calculations as a method for determining miscibility conditions. In principle, the pressure or injected gas composition necessary for dynamic miscibility can be determined by calculating vapor/liquid equilibria with appropriate equations of state or K-value correlations while concurrently simulating the multiple contacting and in-situ

mass transfer of components with a mathematical flow model. Several recent papers[17,50-52] have described techniques for calculating the phase behavior of reservoir fluid and injection gas systems, including CO_2. In some instances, miscibility conditions can be calculated that are sufficiently accurate for design purposes. However, calculation of miscibility conditions has several disadvantages. Generally, equations of state or K-value correlations are not sufficiently accurate for a priori calculations, particularly in the plait-point region that is so critical for achieving dynamic miscibility. Often it is necessary to calibrate the correlations or equations of state with experimental phase behavior data—i.e., parameters in the correlation or equation of state are adjusted to achieve a prediction of the known data. Even when this is done, predictions in the plait-point region may not be accurate unless the experimental data were taken in the plait-point region, which usually is not the case. In any event, more effort can be expended to determine miscibility conditions in this manner than by flow tests. Combined phase behavior/flow calculations can be a valuable technique, however, for calculating overall reservoir performance of the dynamic miscible processes (see Sec. 4.2).

Correlations of experimental miscibility data have been proposed for estimating dynamic miscibility conditions with rich gases[57] and CO_2,[27,29-31,53-56] and a previously unpublished correlation for vaporizing-gas drive miscibility pressure with lean hydrocarbon gas is presented in Chap. 7. The discussion in Chaps. 6, 7, and 8 shows that on some occasions predictions of miscibility pressure with the better correlations can be close to the actual values, perhaps within 50 to 100 psi; but sometimes the correlations may be seriously in error, perhaps by 1,000 psi or more. Unfortunately, one is never sure how accurately a correlation is predicting. Correlations are useful for screening reservoirs for suitability for miscible processes, but unless there is sufficient margin in the operating pressure to allow for the potential error in correlation estimates, miscibility pressure should be determined experimentally for project designs.

A number of displacement techniques have been used in the past by different investigators for determining miscibility conditions, and several criteria have been used for interpreting the displacements. Some miscibility experiments have been conducted in horizontal or vertical consolidated porous media of various lengths and diameters and at various flow rates.[10,27,29] Other experiments were conducted in long, coiled sections of small-diameter tubing packed with sand or glass beads.[27,30,54,55] In still another method, displacements were conducted in vertical sandpacks under gravity-stable conditions.[58] Criteria for interpreting the displacements have included breakthrough and ultimate recoveries at a given volume of solvent injection,[27] visual observations of the core effluent,[10,30] composition of produced gases,[10] shape of the breakthrough and ultimate recovery curves vs. pressure, or combinations of these criteria. The different experimental techniques and interpretation criteria can lead to different conclusions for miscibility pressure and have caused considerable confusion in trying to compare data taken by different investigators. Thus, the value of many of the available published miscibility data is questionable for correlation pruposes. The remainder of this section describes recommended methods for conducting and interpreting displacement experiments.

Important Factors in the Design of Displacement Experiments

Two considerations are especially important in the design of flow experiments to determine miscibility pressure: (1) dynamic miscibility should be attained over a negligible fraction of the displacement path length and (2) experiments should be designed to have essentially complete sweepout after injection of approximately 1.2 PV of solvent.

Dynamic miscibility between oils and solvents that are not miscible on first contact is achieved by in-situ mass transfer of components between phases. For gases such as CO_2, natural gas, nitrogen, and flue gas, some period of immiscible displacement must occur before mass transfer between reservoir oil and the advancing gas front establishes dynamic miscibility at the front. A quantitative calculation of the distance required to develop miscibility from solution of appropriate differential equations describing the displacement process is not available. However, an approach that is useful in chemical engineering for describing mass transfer in packed absorption or solvent extraction towers is useful for illustrating factors to consider in designing miscibility experiments to minimize the length of column displaced immiscibly.[59]

Continuous mass transfer in an absorption tower is represented by equivalent multiple-batch contacts, and the tower is visualized as having equilibrium stages representing the equivalent number of batch contacts. The length of tower containing a given number of equilibrium stages is

$$L = \frac{NUd_p{}^2}{C \cdot D}, \qquad \qquad \qquad (2.1)$$

where

C = constant,
N = number of equilibrium stages,
U = velocity of gas flow,
d_p = diameter of packing material, and
D = effective molecular diffusion coefficient.

In miscibility experiments, N can be viewed as the number of multiple contacts required to establish dynamic miscibility and L as the length over which immiscible displacement occurs. Eq. 2.1 is not useful for quantitative calculation of L, because the constant is not known. However, this equation is useful for illustrating the qualitative effect of important design parameters on L. The equation shows that the length of the immiscible phase of the displacement relative to total displacement path length can be minimized by a combination of suitably small flow velocity and particle size of porous medium in the miscibility apparatus and by proper selection of total length of the column. Guidelines for these values are given later in this section.

Displacement experiments to determine miscibility pressure or composition should be designed such that viscous fingering and gravity tonguing do not confuse interpretation of the experiments. Viscous fingering and gravity tonguing can occur because miscible solvents generally are less viscous and less dense than the oils they displace. In displacements where either viscous fingering or gravity tonguing is a significant factor, sweepout at pressures above the miscibility pressure may be incomplete even after several pore volumes of injection, and a relatively high unrecovered average oil saturation may be left in the test column. Such a result, of course, clouds the interpretation of miscibility.

Gravity-Stable Experiments

Gravity-stable experiments are one method for preventing viscous fingering or gravity tonguing. In this technique, the density difference between solvent and oil is exploited by injecting the less dense solvent at the top of the test column and conducting the displacement vertically downward at a rate low enough for the density difference between solvent and oil to overcome the tendency for solvent fingers to protrude into the oil. If the solvent is more dense than the oil, it is injected at the bottom of the test column.

The maximum flow velocity for a completely stable displacement is[60]

$$u_{st} = 0.0438k \left(\frac{d\rho}{d\mu} \right)_{min}, \qquad \qquad (2.2)$$

where

u_{st} = Darcy velocity, ft/D,
k = permeability, darcies, and

$\left(\dfrac{d\rho}{d\mu} \right)_{min}$ = minimum derivative of solvent/oil mixture density with respect to solvent/oil mixture viscosity as a function of solvent concentration, lbm/cu ft-cp.

When displacement velocity is below u_{st} in miscibility experiments, the entire transition zone is completely stable, and both breakthrough and ultimate recoveries are useful for judging the pressure at which dynamic miscibility first occurred.

The velocity required by Eq. 2.2 may be impractically low. Eq. 2.3 gives a less restrictive criterion for a displacement that will be gravity stable except for a portion of the transition zone.[61] The critical rate is defined as

$$u_c = 0.0438 \frac{(\rho_o - \rho_s)k}{(\mu_o - \mu_s)}, \qquad \qquad (2.3)$$

where

ρ_o, ρ_s = densities of oil and solvent, lbm/cu ft, and
μ_o, μ_s = viscosities of oil and solvent, cp.

For displacement velocities between u_{st} and u_c, transition-zone instability is possible, causing earlier solvent breakthrough than would occur for a completely stable displacement. However, ultimate recovery data are still useful for judging miscibility because there will be no gross fingering of solvent into oil, and good volumetric sweepout will be achieved after about 1 PV of solvent injection. (For a more detailed discussion of gravity-stable displacement, see Sec. 3.6, Chap. 3.)

Fig. 2.19 shows ultimate and breakthrough recoveries for the displacement of a reservoir fluid by a lean hydrocarbon gas at a velocity below the critical rate but above the stable rate. The displacements were conducted in a 6-ft-long Boise sandstone core. In this particular example, both ultimate recovery and breakthrough recovery give useful information and result in the same interpretation for miscibility pressure.

Slim-Tube Experiments

When the density difference between oil and solvent is small, as usually is the case for CO_2 displacements, the velocity required for a gravity-stable displacement may be impractically low. In this case, the importance of fingering can be minimized by designing experiments such that mixing by transverse dispersion either prevents or at least severely retards fingering before the solvent front has travelled a significant fraction of the test column length.

It is shown in Sec. 3.4 that viscous fingers below a critical width will be suppressed by transverse dispersion and will not grow initially. Eq. 3.19 gives an approximate formula for estimating this critical width.[62] If a miscibility displacement experiment is conducted under conditions where the critical finger width exceeds the diameter of the flow tube, all fingers that are initiated by flow perturbations will be suppressed by transverse dispersion and will not grow. Eq. 3.19 shows that very small diameters and low flow velocities are required to do this.

If the flow velocity required to prevent the initial growth of viscous fingers is impractically low, experiments may still be designed such that finger growth is suppressed within a small fraction of the flow-tube length by transverse dispersion. For example, the significance of transverse dispersion in suppressing the growth of fingers above the critical finger width is characterized by the dimensionless transverse dispersion group, $K_t L/vW^2$, where

K_t = coefficient of transverse dispersion, cm^2/sec,[63,64]
L = path length of displacement, cm,
W = width of flow model, cm, and
v = interstitial velocity, cm/sec.

In slim-tube experiments, a sufficiently large value of the dimensionless transverse dispersion group is achieved to suppress fingering by making the diameter of the flow tube sufficiently small and the length sufficiently great.

An approximate guideline for designing experiments can be obtained by using Eq. 3.8 in Chap. 3 for the coefficient of transverse dispersion and by selecting a minimum value for the dimensionless transverse dispersion group from the experiments of Pozzi and Blackwell[65]—e.g.

$$\frac{\left[\left(\frac{D_o}{F\phi}\right)+0.0157v\sigma d_p\right]L}{vW^2} > 10. \qquad \ldots \ldots \ldots (2.4)$$

In this equation, F is the formation factor, ϕ is porosity, D_o is the effective molecular diffusion coefficient (cm^2/sec), and d_p is particle diameter of the column packing (cm). The inhomogeneity factor, σ, is given in Fig. 3.5 for various size beads or sand grains. Eq. 2.4 shows that fingering can be minimized in miscibility experiments by a suitable combination of long column length, small column diameter, and low flow velocity. Eq. 2.4 shows that increasing the particle size increases transverse dispersion, which helps to minimize fingering, but Eq. 2.1 shows that decreasing the particle size is desirable for minimizing the path length of immiscible displacement before dynamic miscibility is achieved. A balance must be struck between these two conflicting demands in the design of miscibility flow experiments.

Recommended Apparatus and Procedure

Fig. 2.28 is a schematic of a slim-tube apparatus recommended by Yellig and Metcalfe[30] for miscibility experiments. The apparatus consists of a 40-ft-long, ¼-in.-diameter coiled stainless steel tube packed with 160- to 200-mesh Ottawa sand. For each displacement test, the same pack initially is saturated with oil at the desired test temperature and pressure. The supply cylinder is filled with solvent and allowed to equilibrate to the test temperature and pressure before the solvent is injected into the sandpack with a positive displacement pump. The supply cylinder contains a movable piston so water from the positive displacement pump can be injected on one side of the piston to displace solvent on the other side. Yellig and Metcalfe recommend a displacement velocity no greater than 40 ft/D until 70% PV is injected, at which point they increase velocity to 80 ft/D. Effluent from the sandpack flows at displacement temperature and pressure through a high-pressure capillary-tube sight glass for visual observation and then passes through a backpressure regulator that maintains the desired test pressure. Liquids are measured in a burette, while gases are measured in a wet test meter. The coiled sandpack, the solvent supply cylinder along with associated valves for introducing fluids to the sandpack, the capillary sight glass, and the backpressure regulator all are contained in a constant-temperature air bath.

Yellig and Metcalfe[30] found that this combination of length, diameter, particle size, and displacement velocity gave oil recovery and visual cell data from which miscibility pressure for CO_2 floods could be determined within ±100 psi for a wide variety of test conditions and within ±50 psi on many occasions. Apparently, dynamic miscibility is established within a sufficiently small fraction of the total tube length and viscous fingering is retarded or suppressed sufficiently by transverse

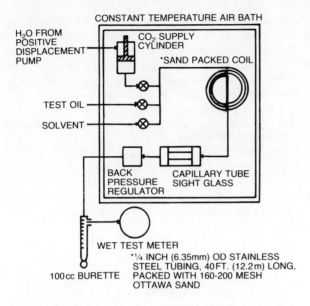

Fig. 2.28—Flow diagram for slim-tube apparatus (after Ref. 30).

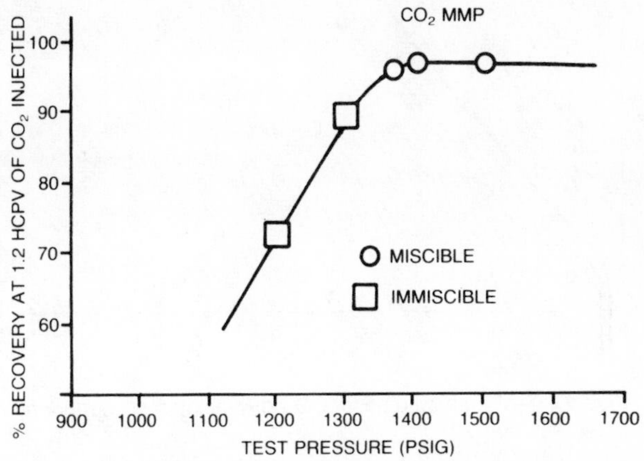

Fig. 2.29—Test results for fixed oil composition and fixed temperature (after Ref. 30).

dispersion that these factors do not confuse interpretation of the data.

Other combinations of tube dimension, velocity, and particle diameter can give good results also as long as the slim-tube design follows the principles discussed in this section. The values given are not intended to represent limits and may be conservative for many situations.

Both oil recovery and visual cell observations are used to interpret slim-tube miscibility experiments. The following paragraphs describe a recommended method for interpretation.

Oil Recovery. Oil recovery from a single displacement is not a sufficient criterion for determining MMP or minimum enrichment required for rich injection gases. In the vaporizing-gas drive process or CO_2-flood process, complete recovery of oil is not expected, even for simple hydrocarbon systems (see Sec. 3.6). For the condensing-gas drive process, the pseudoternary diagram concept of dynamic miscibility indicates that complete oil recovery should be achieved for simple hydrocarbon systems in sandpacks, and, in fact, this is observed for such systems and for some reservoir fluids. Complete recovery of oil by condensing-gas drive may not be achieved with other reservoir fluids because of phase behavior that is more complicated than that represented in Secs. 2.2 and 2.3 by the simple ternary and pseudoternary diagram concept. Some precipitation of solid and liquid phases may occur.[7,58] For these reasons, it could be wrong simply to set a minimum recovery level as a criterion for miscibility.

The breakover point in the ultimate recovery curve from a series of displacements is recommended as a criterion to determine the pressure or composition where dynamic miscibility first occurs.[30] Fig. 2.29 shows oil recovery vs. pressure for a series of CO_2 flood experiments in a slim-tube apparatus. Oil recovery shown on the figure is the value after 1.2 PV of injection, which was equal to or very near the maximum final recovery

obtained in the series of tests. Oil recovery was determined both by direct measurement of produced oil in a burette and by solvent extraction of the sandpack after the experiment with subsequent analysis of the solvent. In this example, a sharp break in the recovery curve signifies a change in the displacement mechanism from immiscible to miscible displacement as pressure is increased. Ultimate recoveries in slim-tube displacements typically are 90 to 95% or higher at pressures above the breakover point.

Yellig and Metcalfe[30] reported that the sharpness with which the recovery curve breaks over and stabilizes at increasing pressures depends on temperature. At higher temperatures, recovery approached a maximum more gradually in their experiments. The sharpness of break in the recovery curve as well as the level of ultimate recovery achieved for miscible displacements also depends on oil and injected gas compositions and on the slim-tube dimensions and operating conditions.[54]

Orr *et al.*[13] speculate that for CO_2 flooding a breakover point in the oil-recovery curve may not always signify a change from immiscible displacement to a displacement that attains dynamic miscibility in the strict sense of the phase behavior concepts presented in this chapter—i.e., no tie line passing through the oil composition. Instead, the breakover point might signify a condition where the two-phase region is not growing appreciably smaller with further pressure increases but where the transition zone nevertheless penetrates the two-phase region close to the plait point. Under these conditions, the displacement would be "near-miscible." Ultimate recovery would be somewhat lower than if true dynamic miscibility had been attained, but still would be relatively high because of effective hydrocarbon extraction and reduced IFT in the plait-point regions.

Both Ref. 54 and Fig. 2.30 give examples of actual experimental slim-tube oil recovery data that showed a sharp breakover of the recovery curve but ultimate recoveries lower than what are typically found in slim-

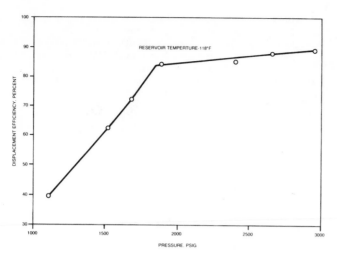

Fig. 2.30—Slim-tube data for CO_2 displacing a midcontinent oil.

tube tests. There is no way of knowing whether strict dynamic miscibility was achieved in these examples without detailed phase behavior data, although the displacements at pressures above the breakover point were judged to be miscible from visual cell observations. In any event, Orr *et al.*[13] concluded that the breakover criterion is "sufficient to insure that phase behavior will be favorable enough to produce an efficient displacement, which is, of course, what matters, whether or not it is strictly miscible."

Visual Cell Observations. Visual cell observations can be useful for judging miscibility in vaporizing-gas drive and CO_2-flood experiments. According to the concept of how vaporizing-gas drive miscibility is achieved with methane, nitrogen, or flue gas by multiple contacts between injected gas and reservoir oil, only single phase effluent should be observed as the transition zone changes from displaced oil to injected gas; and, in practice, single-phase effluents usually are observed above the miscibility pressure with these injection fluids. Occasionally, the observation is not clear-cut—turbidity, fine bubbles dispersed in the flowing stream, or similar phenomena may be seen.

Interpretation of visual cell observations for CO_2 floods is more complicated and requires an experienced observer. Several authors have described these phenomena.[30,47,54] It should be kept in mind, however, that no single description of visual cell observations, like, for example, those discussed in the remainder of this section, may describe all situations.

In general, immiscible CO_2 displacements show clear-cut two-phase flow with sharp interfaces in the visual cell after CO_2 breakthrough. The color of fluids in the visual cell may vary from clear vapor displacing a black oil to clear or straw-colored fluid displacing a dark red oil. These distinctions may be less clear-cut for pressures close to the breakover point of the ultimate recovery curve.

For CO_2 displacements that achieve dynamic miscibility, a gradual color change from dark reservoir fluid through red, orange, and, finally, to a yellow or almost clear fluid have been reported as the transition zone is produced.[30,47] Some limited multiphase flow

also has been reported at pressures above but near the MMP, although the interface between phases is not as sharply defined as for an immiscible displacement and has a large surface area, probably indicative of near-critical fluids. Flow finally changes to single phase with increasing amounts of CO_2. Small gas bubbles as well as two liquid phases have been reported flowing concurrently in the visual cell for a short time at pressures where a three-phase region exists.[47,52]

The following description taken from Henry and Metcalfe's paper[47] gives an idea of what these authors observed in the visual cell for displacements that achieved dynamic miscibility as well as what they observed for displacements that were immiscible.

...shows the typical changes that were observed within a sight glass during flow under conditions in which multiple phases are generated.* The sight glass first contained black oil. As the CO_2 content of the oil increased in the transition zone, the color changed from black to red to orange. At this point in time, the oil gradually separated into two liquid phases, a light orange or amber phase over a dark orange or red phase. Generally, the upper phase continued to become lighter in color eventually becoming clear. Shortly after two liquid phases appeared, tiny, clear, spherical gas bubbles moved across the top of the sight glass. These bubbles grew larger in size, became canoe-shaped and eventually filled the top third of the sight glass. The clear vapor phase generally grew with time at the expense of the two liquid phases. However, in some cases, the vapor phase appeared for a short while and then disappeared again. In every case, as the sight glass became clearer, asphaltenes would precipitate out leaving a black residue behind. After the fluids within the sight glass had become almost totally clear, dark black residue moved through the sight glass intermittently separated by slugs of clear vapor.

At pressures above the multiple phase region, the sequence of events was the same except that gas bubbles were not observed at any time during the displacements. The upper liquid phase would become clearer and clearer and would fill the sight glass at the expense of the lower liquid phase. Again, a residue would be left behind and dark black slugs would move through the sight glass long after the CO_2 front had passed.

At pressures below the multiple phase region, the displacement was generally immiscible with gas bubbles showing up in the black oil.

The mechanism of the condensing-gas drive process is such that a zone of distinct two-phase flow may be observed in the visual cell for miscible displacements. Because of this, visual observations of the slim-tube effluent may not be helpful for judging minimum miscibility conditions.

Summary

The slim-tube method gives reproducible, accurate results that are fairly easy and convenient to obtain experimentally. It is the preferred method for determining miscibility pressure. However, it does not simulate many important aspects of reservoir flooding, and levels of

*That is, for pressures where a three-phase region was present.

ultimate recovery, both for immiscible and for miscible tests, should not be considered as indicative of the unit displacement efficiency to expect in reservoir rocks. If slim-tube experiments do not give recovery and visual cell data from which relatively clear-cut and accurate determinations of miscibility pressure are possible, this may be the result of viscous fingering and/or failing to establish dynamic miscibility in a relatively short distance, and longer tube lengths, lower velocities, and smaller particle sizes should be investigated.

References

1. Glasstone, S: *Textbook of Physical Chemistry* (second edition), Van Nostrum, New York City (1946) 791.
2. Hutchinson, C.A. Jr. and Braun, P.H.: "Phase Relations of Miscible Displacement in Oil Recovery," *AIChE J.* (1961) **7**, 64–72.
3. Tiffin, D.L.: "Effects of Mobile Water on Multiple Contact Miscible Gas Displacement," paper SPE 10687 presented at the 1982 SPE/DOE Enhanced Oil Recovery Symposium, Tulsa, April 4–7.
4. Orr, F.M. Jr. and Silva, M.K.: "Equilibrium Phase Compositions of CO_2-Hydrocarbon Mixtures: Measurement by a Continuous Multiple Contact Experiment," paper SPE 10726 presented at the 1982 SPE/DOE Enhanced Oil Recovery Symposium, Tulsa, April 4–7.
5. Yellig, W.F.: "Carbon Dioxide Displacement of a West Texas Reservoir Oil," paper SPE 9785 presented at the 1981 SPE/DOE Enhanced Oil Recovery Symposium, Tulsa, April 5–8.
6. Wilson, J.F.: "Miscible Displacement—Flow Behavior and Phase Relationships for a Partially Depleted Reservoir," *Trans.*, AIME (1960) **219**, 223–28.
7. Shelton, J.L. and Yarborough, L.: "Multiple Phase Behavior in Porous Media During CO_2 or Rich Gas Flooding," *J. Pet. Tech.* (Sept. 1977) 1171–78.
8. Bossler, R.B. and Crawford, P.B.: "Miscible-Phase Floods May Precipitate Asphalt," *Oil and Gas J.* (Feb. 23, 1959) **57**, 137–45.
9. Huang, E.T.S. and Tracht, J.H.: "The Displacement of Residual Oil by Carbon Dioxide," paper SPE 4735 presented at the 1974 SPE Improved Oil Recovery Symposium, Tulsa, April 22–24.
10. Rathmell, J.J., Stalkup, F.I., and Hassinger, R.C.: "A Laboratory Investigation of Miscible Displacement by Carbon Dioxide," paper SPE 3483 presented at the 1971 SPE Annual Meeting, New Orleans, Oct. 3–6.
11. Gardner, J.W., Orr, F.M. Jr., and Patel, P.D.: "The Effect of Phase Behavior on CO_2 Flood Displacement Efficiency," *J. Pet. Tech.* (Nov. 1981) 2067–81.
12. Stalkup, F.I.: "Carbon Dioxide Miscible Flooding: Past, Present, and Outlook for the Future," *J. Pet. Tech.* (Aug. 1978) 1102–12.
13. Orr, F.M. Jr., Yu, A.D., and Lein, C.L.: "Phase Behavior of CO_2 and Crude Oil in Low Temperature Reservoirs," *Soc. Pet. Eng. J.* (Aug. 1981) 480–92.
14. Simon, R., Rosman, A., and Zana, E.: "Phase Behavior Properties of CO_2 Reservoir Oil Systems," *Soc. Pet. Eng. J.* (Feb. 1978) 20–26.
15. Graue, D.J. and Zana, E.: "Study of a Possible CO_2 Flood in the Rangely Field, Colorado," *J. Pet. Tech.* (July 1981) 1312–18.
16. Turek, E.A. *et al.*: "Phase Equilibria in Carbon Dioxide-Multicomponent Hydrocarbon Systems: Experimental Data and an Improved Prediction Technique," paper SPE 9231 presented at the 1980 SPE Annual Technical Conference and Exhibition, Dallas, Sept. 21–24.
17. Metcalfe, R.S., Fussell, D.D., and Shelton, J.L.: "A Multicell Equilibrium Separation Model for the Study of Multiple Contact Miscibility in Rich-Gas Drives," *Soc. Pet. Eng. J.* (June 1973) 147–55.
18. Helferich, F.G.: "Theory of Multicomponent, Multiphase Displacement in Porous Media," *Soc. Pet. Eng. J.* (Feb. 1981) 51–62.
19. Shelton, J.L. and Schneider, F.N.: "The Effects of Water Injection on Miscible Flooding Methods Using Hydrocarbons and Carbon Dioxide," *Soc. Pet. Eng. J.* (June 1975) 217–26.
20. Slobod, R.L. and Koch, H.A. Jr.: "High-Pressure Gas Injection—Mechanism of Recovery Increase," *Drill. and Proc. Prac.*, API (1953) 82.
21. Koch, H.A. Jr. and Hutchinson, C.A.: "Miscible Displacements of Reservoir Oil Using Flue Gas," *Trans.*, AIME (1958) **213**, 7–19.
22. McNeese, C.R.: "The High Pressure Gas Process and the Use of Flue Gas," paper presented at the 1963 ACS Symposium on Production and Exploration Chemistry, Los Angeles, March 31–April 5, A-67.
23. Rushing, M.D. *et al.*: "Miscible Displacement with Nitrogen," *Pet. Eng.* (Nov. 1977) 26–30.
24. Rushing, M.D. *et al.*: "Nitrogen May Be Used for Miscible Displacement in Oil Reservoirs," *J. Pet. Tech.* (Dec. 1978) 1715–16.
25. Yarborough, L.: "Vapor-Liquid Equilibrium Data for Multicomponent Mixtures Containing Hydrocarbon and Nonhydrocarbon Components," *J. Chem. Eng. Data* (1972), **17**, No. 2, 129.
26. Jacoby, R.H. and Rzasa, M.J.: "Equilibrium Vaporization Ratios for Nitrogen, Methane, Carbon Dioxide, Ethane, and Hydrogen Sulphide in Absorber Oil-Natural Gas and Crude Oil-Natural Gas Systems," *Trans.*, AIME (1952) **195**, 99–110.
27. Holm, L.W. and Josendal, V.A.: "Mechanisms of Oil Displacement by Carbon Dioxide," *J. Pet. Tech.* (Dec. 1974) 1427–36; *Trans.*, AIME, **257**.
28. Metcalfe, R.S. and Yarborough, L.: "Effect of Phase Equilibria on the CO_2 Displacement Mechanism," *Soc. Pet. Eng. J.* (Aug. 1979) 242–52.
29. Holm, L.W. and Josendal, V.A.: "Effect of Oil Composition on Miscible-Type Displacement by Carbon Dioxide," *Soc. Pet. Eng. J.* (Feb. 1982) 87–98.
30. Yellig, W.F. and Metcalfe, R.S.: "Determination and Prediction of CO_2 Minimum Miscibility Pressures," *J. Pet. Tech.* (Jan. 1980) 160–68.
31. Christian, L.D. *et al.*: "Planning a Tertiary Oil Recovery Project for Jay-Little Escambia Creek Fields Unit," *J. Pet. Tech.* (Aug. 1981) 1535–44.
32. Peterson, A.V.: "Optimal Recovery Experiments with N_2 and CO_2," *Pet. Eng.* (Nov. 1978) 40–50.
33. Perry, G.E. *et al.*: "Weeks Island 'S' Sand Reservoir B Gravity Stable Miscible CO_2 Displacement, Iberia Parish, Lousiana," *Proc.*, 4th Annual U.S. DOE Symposium on Enhanced Oil and Gas Recovery and Improved Drilling Methods, Tulsa (1978) **1B**.
34. Meldrum, A.H. and Nielsen, R.F.: "A Study of Three-Phase Equilibria for Carbon Dioxide-Hydrocarbon Mixtures," *Prod. Monthly* (Aug. 1955) 22–35.
35. Stewart, W.C. and Nielsen, R.F.: "Phase Equilibria for Mixtures of Carbon Dioxide and Several Normal Saturated Hydrocarbons," *Prod. Monthly* (Jan. 1954) 27–32.
36. Reamer, H.H. and Sage, B.H.: "Phase Equilibria in Hydrocarbon Systems—Volumetric and Phase Behavior of the n-Decane-CO_2 System," *J. Chem. Eng. Data* (1963) **8**, 508–13.
37. Schneider, G. *et al.*: "Phasengleichgewichte und kritische Erscheinungen in binaren Mischsystemen bis 1500 bar: CO_2 mit n-Octan, n-Undecan, n-Tridecan und n-Hexadecan," *Chem. Ing. Tech.* 39 (1967) 649–56.
38. Zarah, B.Y., Luks, K.D., and Kohn, J.P.: "Phase Equilibria Behavior of Carbon Dioxide in Binary and Ternary Systems with Several Hydrocarbon Components," *AIChE Symposium Series* (1974) **70**, No. 140, 91–101.
39. Kulkarni, A.A. *et al.*: "Phase Equilibria Behavior of Carbon Dioxide-n-Decane at Low Temperatures," *J. Chem. Eng. Data* 19 (1974) 92–94.
40. Snedeker, R.A.: "Phase Equilibria in Systems with Supercritical Carbon Dioxide," PhD dissertation, Princeton U. (1955).
41. Schneider, G.: "Phase Equilibria in Binary Fluid Systems of Hydrocarbons with Carbon Dioxide, Water, and Methane," *AIChE Symposium Series* (1968) **64**, 9–15.
42. Liphard, K.G. and Schneider, G.M.: "Phase Equilibria and Critical Phenomena in Fluid Mixtures of Carbon Dioxide plus 2, 6, 10, 15, 19, 23-hexamethyltetracosane up to 423°K and 100 MPa," *J. Chem. Thermodynamics* (1975) **7**, 805–14.
43. Huie, N.C., Luks, K.D., and Kohn, J.P.: "Phase Equilibria

Behavior of Systems Carbon Dioxide-n-Eicosane and Carbon Dioxide-n-Decane-n-Eicosane,'' *J. Chem. Eng. Data* (1973) **18,** 311–13.

44. Yang, H.W., Luks, K.D., and Kohn, J.P.: "Phase Equilibria Behavior of the System Carbon Dioxide-n-Butylbenzene-2 Methylnaphthalene," *J. Chem. Eng. Data* (1976) **21,** 330–35.

45. Gupta, V.S.: "Measurement and Prediction of Phase Equilibria in Liquid Carbon Dioxide-Hydrocarbon Ternary Systems," PhD dissertation, Pennsylvania State U. (1969).

46. Francis, A.W.: "Ternary Systems of Liquid Carbon Dioxide," *J. Phys. Chem.* (1954) **58,** 1099–1114.

47. Henry, R.L. and Metcalfe, R.S.: "Multiple Phase Generation During CO_2 Flooding," paper SPE 8812 presented at the 1980 SPE/DOE Enhanced Oil Recovery Symposium, Tulsa, April 20–23.

48. Orr, F.M. Jr. and Taber, J.J.: *Displacement of Oil by Carbon Dioxide,* Annual Report for Oct. 1980-Sept. 1981 prepared for the U.S. DOE under Contract No. DE-AS19–80BC10331, Bartlesville, OK (1980).

49. Orr, F.M. Jr., Silva, M.K., and Lien, C.L.: "Equilibrium Phase Compositions of CO_2-Crude Oil Mixtures: Comparison of Continuous Multiple Contact and Slim-Tube Displacement Tests," paper SPE 10725 presented at the 1982 SPE/DOE Enhanced Oil Recovery Symposium, Tulsa, April 4–7.

50. Fussell, L.T.: "A Technique for Calculating Multiphase Equilibria," *Soc. Pet. Eng. J.* (Aug. 1979) 203–08.

51. Fussell, D.D. and Yanosik, J.L.: "An Iterative Sequence for Phase Equilibrium Calculations Incorporating the Redlich-Kwong Equation of State," *Soc. Pet. Eng. J.* (June 1978) 173–82.

52. Rowe, A.M. Jr.: "Internally Consistent Correlations for Predicting Phase Compositions for Use in Reservoir Composition Simulators," paper SPE 7475 presented at the 1978 SPE Annual Technical Conference and Exhibition, Houston, Oct. 2–4.

53. Cronquist, C.: "Carbon Dioxide Dynamic Miscibility with Light Reservoir Oils," *Proc.,* Fourth Annual U.S. DOE Symposium on Enhanced Oil and Gas Recovery and Improved Drilling Methods, Tulsa (1978) **1B;** also private communication.

54. Johnson, J.P. and Pollin, J.S.: "Measurement and Correlation of CO_2 Miscibility Pressures," paper SPE 9790 presented at the 1981 SPE/DOE Enhanced Oil Recovery Symposium, Tulsa, April 5–8.

55. Lee, J.: "Effectiveness of Carbon Dioxide Displacement Under Miscible and Immiscible Conditions," Research Report RR-40, Petroleum Recovery Inst., Calgary (March 1979).

56. Dumyuskin, I.I. and Namiot, A.Y.: "Mixing Conditions of Oils with Carbon Dioxide," *Neft. Khozyaistvo* (March 1978) 59–61.

57. Benham, A.L., Dowden, W.E., and Kunzman, W.J.: "Miscible Fluid Displacement—Prediction of Miscibility," *Trans.,* AIME (1960) **219,** 229–37; *Miscible Processes,* Reprint Series, SPE, Dallas (1965) **8,** 123–131.

58. Rutherford, W.M.: "Miscibility Relationships in the Displacement of Oil by Light Hydrocarbons," *Soc. Pet. Eng. J.* (Dec. 1962) 340–46; *Trans.,* AIME, **225.**

59. Sherwood, T.K. and Pigford, R.L.: *Adsorption and Extraction,* McGraw-Hill Book Co. Inc., New York City (1952) 52–53.

60. Dumoré, J.M.: "Stability Considerations in Downward Miscible Displacement," *Soc. Pet. Eng. J.* (Dec. 1964) 356–62, *Trans.,* AIME, **231.**

61. Hill, S.: "Génie Chemique," *Chem. Eng. Sci.* (1952) **1,** No. 6, 246.

62. Gardner, J.W. and Ypma, J.G.J.: "An Investigation of Phase Behavior-Macroscopic Bypassing Interaction in CO_2 Flooding," paper SPE 10686 presented at the 1982 SPE/DOE Enhanced Oil Recovery Symposium, Tulsa, April 4–7.

63. Perkins, T.K. and Johnston, O.C.: "A Review of Diffusion and Dispersion in Porous Media," *Soc. Pet. Eng. J.* (March 1963) 70–84; *Trans.,* AIME, **228;** *Miscible Processes,* Reprint Series, SPE, Dallas (1965) **8,** 77–91.

64. Blackwell, R.J.: "Laboratory Studies of Microscopic Dispersion Phenomena," *Soc. Pet. Eng. J.* (March 1962) 1–8; *Trans.,* AIME, **225;** *Miscible Processes,* Reprint Series, SPE, Dallas (1965) **8,** 69–76.

65. Pozzi, A.L. and Blackwell, R.J.: "Design of Laboratory Models for Study of Miscible Displacement," *Soc. Pet. Eng. J.* (March 1963) 28–40.

66. Stalkup, F.I.: "The Use of Phase Surfaces to Describe Condensing Gas Drive Experiments," *Soc. Pet. Eng. J.* (Sept. 1965) 184–88; *Trans.,* AIME, **234.**

Chapter 3
Factors Affecting Displacement Behavior

3.1 Mobility and Mobility Ratio

The Darcy equation, which describes the flow of fluids in a porous medium, relates the velocity of a fluid to the pressure gradient by a proportionality factor as illustrated by Eq. 3.1 for horizontal flow in one dimension.

$$u_i = \frac{k_i}{\mu_i} \frac{dp}{dx} , \dots\dots\dots\dots\dots\dots (3.1)$$

where

u = superficial velocity, cm/s,
k = permeability, md,
p = pressure, atm,
x = length, cm, and
μ = viscosity, cp.

The subscript i represents a particular fluid. The proportionality factor is termed the mobility of the fluid and is a measure of the facility with which the fluid flows through the rock — i.e., how mobile it is. From Eq. 3.1, the mobility of fluid i, λ_i, is defined as the effective permeability of the rock to Fluid i divided by its viscosity, k_i/μ_i. For example, the mobility of oil is k_o/μ_o and the mobility of a miscible solvent is k_s/μ_s.

When one fluid displaces another, the mobility ratio, M, of the displacement is defined as the mobility of the displacing fluid divided by the mobility of the displaced fluid. Mobility ratio is one of the most important parameters of a miscible displacement and has a profound influence on volumetric sweepout of the solvent and on the integrity of solvent slugs, as is shown in the remainder of this chapter.

Consider an idealized situation where solvent displaces oil at the irreducible water saturation and where mixing of solvent with the oil is negligible. No water is flowing and the permeabilities to oil and solvent are equal. Mobility ratio in this case is simply the ratio of oil and solvent viscosities, μ_o/μ_s. Viscosity ratios for

displacements with the low-viscosity solvents discussed in this monograph are always greater than unity (unfavorable). In practice, mixing of solvent and oil does occur during the course of the displacement, which can result in an effective viscosity ratio that is less than the ratio of pure component viscosities, although the effective ratio still remains greater than one and unfavorable.

Definition of effective mobility ratio becomes more complicated and uncertain when mobile water is present as, for example, in tertiary recovery floods or in secondary or tertiary recovery floods when water is injected with the solvent. Solvent/water injection is a technique to reduce solvent mobility by reducing solvent relative permeability.

Consider, for example, the idealized situation where water and solvent are injected simultaneously at a fixed ratio to displace oil in a secondary recovery flood. In practice, the solvent and water usually are injected in alternate, small slugs; but to simplify the illustration, simultaneous injection is assumed here. In this example, the injection ratio was selected such that both water and solvent fronts travel at the same velocity. Experimental data and theoretical analyses are lacking to define the mobility that best characterizes movement of the solvent front in this situation. Possible choices are (1) the average total mobility of the solvent/water region, (2) the total mobility at the backside of the solvent/water shock front, or (3) solvent mobility alone at either the backside of the solvent front or at the average solvent saturation. In tertiary recovery floods, some mobile water is also present in the tertiary oil bank.

As a working assumption until data are available that indicate otherwise, the following definition is recommended for characterizing mobility ratio between an oil bank and the solvent displacing the oil bank when mobile water is present in either region.

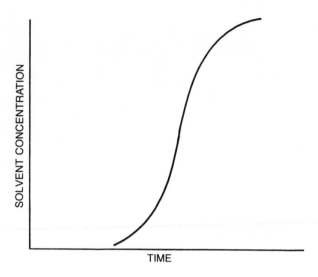

Fig. 3.1—Profile of the solvent effluent concentration produced from a sandpacked tube in an equal-viscosity, equal-density experiment.

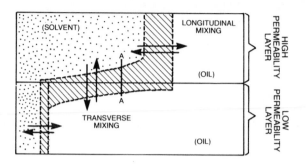

Fig. 3.2—Mixing of solvent and oil by longitudinal and transverse dispersion.

$$M = \frac{\left(\dfrac{k_s}{\mu_s} + \dfrac{k_w}{\mu_w}\right)_{sw\ avg}}{\left(\dfrac{k_o}{\mu_o} + \dfrac{k_w}{\mu_w}\right)_{ow\ avg}} \qquad \ldots \ldots \ldots \ldots (3.2)$$

This is consistent with the definition of mobility ratio adopted by Craig for waterflooding.[1,2]

In most miscible floods there will be more than one displacing front, as, for example, in tertiary recovery floods and in floods where solvent slugs are injected and driven by other fluids. Motion of any particular front is affected not only by the mobility ratio across that front but by the mobilities of other regions behind and ahead of the front and by the relative sizes of the different flow regions present. So far, there is no method to define rigorously the mobility ratio that characterizes solvent sweepout in such situations, and, in fact, the effective ratio probably changes during the course of a displacement as relative sizes of the various regions change. Sec. 3.4, Areal Sweepout, contains more discussion of this subject.

3.2 Mixing of Fluids by Dispersion

Suppose a first-contact miscible solvent is injected into a linear tube packed with sand to displace an oil that has the same density and viscosity as the solvent. This is one of the simplest types of miscible displacement. Viscosity and density differences do not influence the flow behavior. Fig. 3.1 illustrates how the effluent solvent concentration profile from such an experiment would look. Mixing between solvent and oil in the sandpack results in an S-shaped concentration profile. At first, solvent is produced at low concentration. This is followed by a period of steeply rising concentration and, finally, by a period where effluent concentration gradually approaches injected concentration. Because of this mixing, a transition zone of solvent/oil mixtures separates 100% solvent from 100% oil. This mixing in the direction of

flow is called "longitudinal dispersion."

Suppose another experiment is conducted in a model such as the one shown in Fig. 3.2. This two-dimensional model contains two layers of sand. One of the layers is much more permeable than the other such that solvent, which is injected across the left face of the model, mostly enters only the permeable layer. Once again, the solvent and oil have equal density and equal viscosity. In this experiment, the solvent not only mixes with the oil by longitudinal dispersion in the direction of flow; it also mixes with oil transverse to the direction of flow in the less permeable layer as indicated in the figure. If concentrations were measured in situ through the section marked AA, an S-shaped concentration profile again would be found. This mixing transverse to the direction of flow is called "transverse dispersion."

Longitudinal and transverse dispersion can influence miscible displacement profoundly. In the solvent-slug process, the slug may be diluted by mixing to such an extent that some compositions lie within the two-phase region. When this happens, the displacement is no longer miscible. Mixing by dispersion also moderates the viscosity and density differences between solvent and oil. In some situations, this moderation of viscosity and density contrasts may be sufficient to alter the flow behavior significantly.

Three mechanisms contribute to the mixing of miscible fluids: molecular diffusion, microscopic convective dispersion, and macroscopic convective dispersion. Molecular diffusion is a result of random thermal motion of molecules. Microscopic convective dispersion results from flow paths caused by rock inhomogeneities that are small compared with dimensions of laboratory cores; for example, flow through connecting pores of unequal length with subsequent mixing of fluids by molecular diffusion within pores. Macroscopic dispersion results from the flow paths caused by permeability heterogeneities that are large compared with the dimensions of laboratory cores but are smaller than gross correlatable reservoir features such as strata.

Molecular Diffusion and Microscopic Convective Dispersion

If two miscible fluids are brought into contact with an initially sharp interface, the subsequent mixing caused by molecular diffusion is represented by the well-known Fick[3] diffusion equation:

$$\frac{dG_i}{dt} = -D_{o_i} A \frac{\partial C_i}{\partial x} , \quad \dots\dots\dots\dots\dots (3.3)$$

where

G = quantity of material that has diffused across the plane represented by the original sharp interface, moles,

t = time,

D_o = molecular diffusion coefficient, L^2/t,

A = cross sectional area for diffusion, L^2,

C = concentration, moles/L^3, and

x = distance, L.

Eq. 3.3 as written applies if there is no change in volume upon mixing the two fluids and describes the net transport of each molecular species in the mixture.

Usually the diffusion coefficient, D_o, depends on mixture composition, and a rigorous solution of Eq. 3.3 would take this dependency on concentration into account. However, the mathematics for this are quite complicated and data are generally lacking. Often, for reservoir problems, an adequate representation of the mixing of solvent caused by diffusion may be achieved by selecting an average diffusion coefficient at the 50% solvent concentration.[4] For a discussion of variable diffusion coefficient, see Crank.[5]

For Eq. 3.3 to apply to diffusion in a porous medium, the molecular diffusion coefficient must be adjusted to account for the tortuous path for diffusion in the pores of the rock. An *effective* diffusion coefficient for use in Eq. 3.3 may be calculated from Eq. 3.4[6-8]:

$$\frac{D}{D_o} = \frac{1}{F\phi} , \quad \dots\dots\dots\dots\dots\dots\dots\dots (3.4)$$

where

F = formation electrical resistivity factor and

ϕ = fractional porosity.

This equation recognizes the analogy between diffusion and electrical conductivity in a porous medium and has been verified by several investigators.[9,10]

When fluids flow through a porous medium, more mixing takes place in the direction of flow than would be expected from molecular diffusion alone. This additional mixing caused by flow or convection appears to be explained by a "mixing cell" theory as illustrated by Fig. 3.3. As shown in this figure, the individual Streamlines 1, 2, and 3 follow a tortuous path through the porous medium, although the average direction of each streamline must be in the direction of mean flow. Suppose that different solvent concentrations initially are traveling along each streamline. The solvent concentrations associated with Streamlines 1 and 2 enter Pore A through small pore connections or pore throats. Within

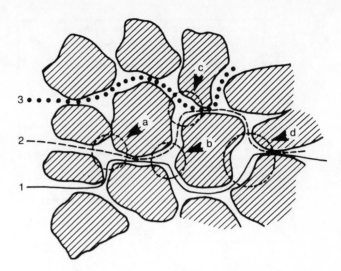

Fig. 3.3—Microscopic convective dispersion.

Pore A the solvent concentration is equalized by molecular diffusion such that a uniform concentration emerges from Pore A. The solvent of altered composition associated with Streamline 2 then mixes in Pore C with the solvent of composition associated with Streamline 3. In Pore C, diffusion again equalizes concentrations such that solvent of twice-altered composition emerges from Pore C along Streamline 2. Mixing between Streamlines 1 and 2 then occurs in Pore D, etc.

Fig. 3.3 also illustrates how fluids are mixed by convective dispersion transverse to the direction of flow. Again, consider the three Streamlines 1, 2, and 3, but suppose that Streamline 1 initially carries only solvent molecules, whereas Streamlines 2 and 3 initially carry only oil molecules. In Pore A, the fluids from Streamlines 1 and 2 are mixed, and Streamline 2 leaves Pore A carrying some solvent. At Pore C, fluid from Streamline 2 is mixed with solvent-free fluid from Streamline 3. Streamline 3 now leaves Pore C carrying some solvent. In this manner, the solvent becomes progressively dispersed normal to the direction of flow.

Microscopic Longitudinal and Transverse Dispersion Coefficients

The diffusion-convection equation, given below as Eq. 3.5, describes the overall transport and mixing of fluids flowing through a porous medium.[11] Terms here show the relation "convective flow" plus "dispersive flow" equals "accumulation."

$$-(v \cdot \nabla C) + \nabla \cdot (\tilde{K} \cdot \nabla C) = \frac{\partial C}{\partial t} , \quad \dots\dots\dots (3.5)$$

where

∇ = Laplacian operator,

v = interstitial velocity, and

$\tilde{K}$ = tensor formulation of dispersion coefficient.

The dispersion coefficient has a molecular diffusion contribution and a convective dispersion contribution such that $K = D + E$, where D is the effective molecular diffusion coefficient (Eq. 3.4), and E is the convective

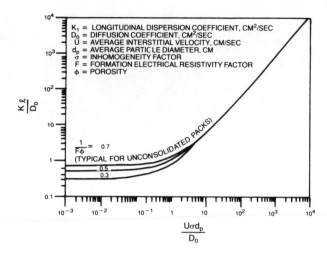

Fig. 3.4—Longitudinal dispersion coefficients for porous media (after Ref. 4).

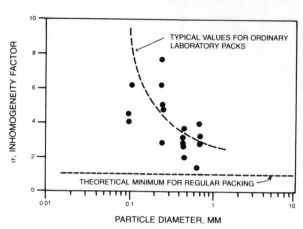

Fig. 3.5—Inhomogeneity factors for random packs of spheres (after Ref. 4).

TABLE 3.1—VALUES OF σd_p FOR OUTCROP SANDSTONES[4]

Source	Dispersion	Rock	σd_p (cm)
Crane and Gardner[7]	Transverse	Berea	0.25
Brigham et al.[6]	Longitudinal	Berea	0.39
	Longitudinal	Torpedo	0.17
Raimondi et al.[143]	Longitudinal	Berea	0.46
Handy[54]	Longitudinal	Boise	0.55
		average	0.36

dispersion coefficient. Convective dispersion is nonisotropic because the longitudinal and transverse convective dispersion coefficients, E_ℓ and E_t, are not equal. This necessitates a tensor formulation of the dispersion relations in Eq. 3.5 when convective dispersion is important.

For a constant velocity in the x direction only, Eq. 3.5 reduces to the form shown in Eq. 3.6:

$$K_\ell \frac{\partial^2 C}{\partial x^2} + K_t \left(\frac{\partial^2 C}{\partial y^2} + \frac{\partial^2 C}{\partial z^2} \right) - v \frac{\partial C}{\partial x} = \frac{\partial C}{\partial t} , \quad \dots (3.6)$$

where

$K_\ell = D + E$, the coefficient of longitudinal dispersion, and

$K_t =$ the coefficient of transverse dispersion.

The first term in Eq. 3.6 accounts for longitudinal dispersion in the x direction, and the second and third terms account for transverse dispersion in the y and z directions.

After reviewing published data on longitudinal dispersion, Perkins and Johnston[4] proposed the following equation for the longitudinal dispersion coefficient that describes mixing between fluids of equal density and equal viscosity.

$$\frac{K_\ell}{D_o} = \frac{1}{F\phi} + 0.5 \frac{v\sigma d_p}{D_o} , \quad \frac{v\sigma d_p}{D_o} < 50, \quad \dots (3.7)$$

where

$v =$ interstitial velocity, cm/s,
$K_\ell =$ coefficient of longitudinal dispersion, cm^2/s,
$\sigma =$ inhomogeneity factor, and
$d_p =$ average particle diameter, cm.

A graphical representation of this equation is shown in Fig. 3.4. At low flow rates ($v\sigma d_p/D_o < 0.1$), molecular diffusion dominates longitudinal mixing, whereas at higher flow rates ($4 < v\sigma d_p/D_o < 50$), convective dispersion dominates longitudinal mixing and the dispersion coefficient is approximately proportional to velocity. For values of $v\sigma d_p/D_o$ between 0.1 and 4, both molecular diffusion and convective dispersion are important mechanisms of longitudinal mixing. At values of $v\sigma d_p/D_o > 50$, the dispersion is greater than would be explained by the mixing-cell theory and presumably shows that diffusion is not equalizing the concentration entirely within each pore space. Most data indicate that K_ℓ/D_o is roughly proportional to $(v\sigma d_p/D_o)^{1.2}$ in this region.[4]

The inhomogeneity factor, σ, accounts for microscopic, or at least very small-scale, permeability heterogeneities. Fig. 3.5 is a correlation of σ vs. particle diameter, d_p, developed by Perkins and Johnston[4] for random packs of various-size glass beads. Presumably, the increase in σ as d_p decreases reflects increased bridging by the beads at smaller bead diameters. Packing and cementing irregularities and possibly larger-scale heterogeneities occur in laboratory-size samples of cemented reservoir rocks. Table 3.1 presents values of $d_p\sigma$ published by Perkins and Johnston[4] for several cemented rocks.

Reservoir flow rates, except in the vicinity of wells, often are in the range 0.1 to 1 ft/D. If an average value of 0.36 cm is assumed for $d_p\sigma$ and if molecular diffusion coefficients of 2×10^{-5} and 10^{-3} cm^2/s are assumed for liquids and gases, respectively, at reservoir conditions, reservoir values for $v\sigma d_p/D_o$ may lie in the range

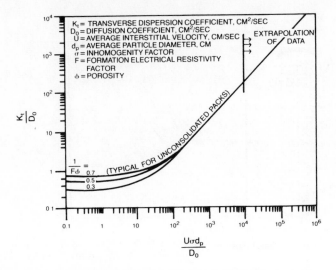

Fig. 3.6—Transverse dispersion coefficients for porous media (after Ref. 4).

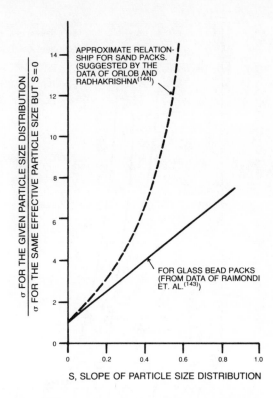

Fig. 3.7—Effect of particle-size distribution on dispersion (after Ref. 4).

of 0.6 to 6 for liquids and 0.01 to 0.1 for gases. According to Fig. 3.4, both microscopic convective dispersion and molecular diffusion may be important in the longitudinal mixing of liquids at reservoir conditions, although molecular diffusion dominates in the longitudinal mixing of gases. Laboratory flow rates may be much higher, sometimes in the range 1 to 50 ft/D. Convective dispersion can dominate the longitudinal mixing of liquids in laboratory experiments and in the reservoir near wells, and may dominate in the mixing of gases, depending on flow rate.

After reviewing transverse dispersion data, Perkins and Johnston[4] recommended the following equation for calculating the transverse dispersion coefficient for fluids of equal density and equal viscosity.

$$\frac{K_t}{D_o} = \frac{1}{F\phi} + 0.0157 \frac{v\sigma d_p}{D_o}, \quad \frac{v\sigma d_p}{D_o} < 10^4. \quad \dots (3.8)$$

This equation is illustrated graphically in Fig. 3.6. Fig. 3.5 and Table 3.1 may be used with Eq. 3.8 and Fig. 3.6. For values of $v\sigma d_p/D_o < 50$, molecular diffusion dominates transverse mixing. For values of this parameter greater than approximately 300, convective dispersion dominates. Thus, at field displacement rates, microscopic transverse mixing is controlled by molecular diffusion, although in many laboratory experiments both molecular diffusion and convective dispersion may play an important role.

Methods for determining longitudinal and transverse dispersion coefficients from laboratory displacements are discussed by various authors.[4,6,12-14] These methods generally involve fitting the solvent effluent concentration profile from a laboratory displacement with an appropriate solution of the diffusion-convection equation and determining the value of dispersion coefficient that results in the best agreement of experimental and calculated concentration profiles.

Additional Variables Influencing Longitudinal and Transverse Microscopic Dispersion Coefficients

Eqs. 3.7 and 3.8 along with Figs. 3.4 and 3.6 for estimating microscopic longitudinal and transverse dispersion coefficients give only approximate values, since other variables such as particle-size distribution of the porous medium, fluid saturations, mobility ratio, density ratio, ratio of particle diameter to column diameter in laboratory cores, and particle shape all affect the magnitude of the dispersion coefficients to some degree.[4] Fig. 3.7 depicts a correction to be made to the inhomogeneity factor for particle-size distribution. Often for a given particle-size distribution, when particle size is plotted on logarithmic-probability paper vs. the cumulative percent of particles in the sample that are smaller than the stated size, a straight line can be drawn through the data. The slope, S, to enter in Fig. 3.7 is the slope of the log-normal particle-size distribution and is calculated from

$$S = \log d_p]_{84} - \log d_p]_{16} = \log \frac{d_p]_{84}}{d_p]_{16}}, \quad \dots (3.9)$$

where $d_p]_{84}$ = particle diameter of the 84% cumulative fraction, mm.

Fig. 3.8 shows how saturation of the hydrocarbon phase affected the longitudinal dispersion coefficient of a solvent/oil displacement for several strongly water-wet laboratory sandstones. For these experiments, as the saturation of the hydrocarbon phase decreased below its saturation at irreducible water (i.e., the water saturation increased), the longitudinal dispersion coefficient began to increase markedly. Probably, one effect of fluid saturation is to alter the pore-size distribution available for miscible displacement. The effect of saturation on dispersion coefficient may not be nearly as dramatic for weakly water- or oil-wet rocks, but published ex-

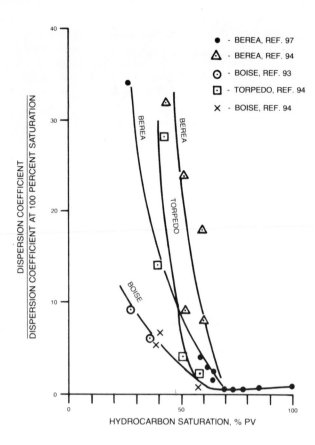

Fig. 3.8—Influence of saturation on hydrocarbon dispersion coefficient.

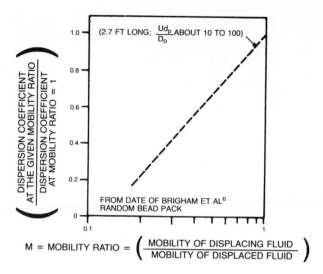

Fig. 3.9—Suppression of dispersion resulting from favorable mobility ratios (after Ref. 4).

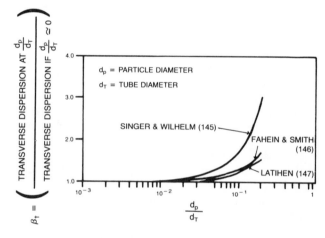

Fig. 3.10—Effect of d_p/d_T on transverse dispersion, turbulent flow (after Ref. 4).

perimental data are lacking either to confirm or to disprove this hypothesis.

Mobility ratio also affects the dispersion coefficient. For oil/solvent mobility ratios greater than one, viscous instabilities develop between solvent and oil, and the displacement is no longer one of the simple miscible displacement with mixing by dispersion. However, if the solvent is less mobile than the oil, the usual dispersion mechanisms will continue to operate, but the favorable mobility ratio will suppress to some degree the effects of packing or permeability heterogeneities. The effect of a favorable mobility ratio is shown in Fig. 3.9.[4]

Gravity also may suppress dispersion to some degree for displacements where the less dense solvent overrides the denser oil or displaces the denser oil downward. Although quantitative corrections are not available, in such a displacement any unevenness of the front caused by packing or permeability heterogeneities should be reduced. Perkins[4] speculates that the dispersion coefficient might be reduced by a factor as large as σ, the inhomogeneity factor.

Pozzi and Blackwell[15] empirically determined a correlation for the effective transverse dispersion coefficient in packs of unconsolidated sand and glass beads for displacements with fluids of unequal densities and unequal viscosities. Ref. 15 should be consulted for details.

In laboratory experiments, the ratio of particle diameter to core diameter may influence the magnitude of the measured transverse dispersion coefficient. Fig. 3.10 shows the approximate magnitude of this effect as a function of the particle-to-core-diameter ratio.

Particle shape also may influence dispersion. This variable may be of more concern in packed beds of irregularly shaped particles than in reservoir rocks, although it may be a factor in reservoir rocks in some situations.

Macroscopic Dispersion

Mixing of fluids also can be caused by permeability heterogeneities that are large compared with reservoir core dimensions. Warren and Skiba[16] used a Monte Carlo technique to show that (1) an increased level of reservoir dispersion would result solely from variations in the permeability within strata and (2) the magnitude of this macroscopic dispersion was related both to the scale of the permeability heterogeneities as well as to the distribution function of the permeabilities. They concluded that the mixing caused by permeability variations of this nature also could be modeled as a diffusional type process using the conventional diffusion-convection equation if an effective macroscopic dispersion coefficient was used in the equation rather than the molecular diffusion coefficient. Warren and Skiba hypothesized that experiments performed in the laboratory probably do not yield a valid measure of macroscopic dispersion, since the scale of heterogeneity that is significant in the reservoir is larger than can be defined by laboratory-size core samples.

Several recent reservoir dispersion tests appear to support the concept that mixing in some reservoirs can be significantly greater than that expected from molecular diffusion and microscopic convective dispersion alone.[17-20] In these experiments, a nonreactive tracer was injected into the reservoir in a slug of water and then was produced back from the same well. Such a procedure is part of the tracer method for determining waterflood residual oil saturation and is intended specifically to determine the effective mixing coefficient around the wellbore to aid in interpretation of the residual oil test.[20,21] ("Mixing coefficient" is defined as the effective dispersion coefficient divided by the interstitial velocity and is measured in units of length.) This type of test gives a measurement of the mixing that occurs within the average radius of the tracer slug. If the reservoir is stratified, the tracer slug advances farthest into the more permeable layers and therefore tests the effect of heterogeneities in these strata to a larger degree than in the less permeable strata. However, when the well used for tracer injection then is put on production, the tracer that entered the more permeable strata also flows back to the well faster than the tracer that invaded the less permeable strata, with the result that the midpoints of the tracer concentration profiles in all strata return to the producing well more or less simultaneously. Such an experiment measures the combined influence on mixing of the microscopic and macroscopic dispersion within and between strata. Drift of fluids away from the well during the test, however, can cause some deviation from a simultaneous return of tracer from different directions during backflow, complicating measurement of macroscopic dispersion. Mixing coefficients reported from reservoir experiments of this nature have been on the order of 0.1 to 2 ft. This compares with a mixing coefficient in laboratory sandstone cores attributable to microscopic dispersion only that is usually on the order of 0.005 to 0.02 ft.

Significance of Longitudinal and Transverse Dispersion

In most miscible displacements not stabilized by gravity, transverse dispersion plays a much more important role than does longitudinal dispersion. This is because in most displacements solvent fingers penetrate into the oil for a variety of reasons, exposing a large surface area along the sides of the fingers over which transverse dispersion can occur. In contrast, longitudinal dispersion occurs over a much smaller area of the solvent/oil interface at the finger tips. Transverse dispersion can affect the growth of viscous fingers (especially in laboratory models) and thus have an influence on sweepout. Solvent contained in the exposed fingers can be diluted below miscible composition by transverse dispersion. More is said about these effects in Secs. 3.4 and 3.5. Transverse dispersion also can moderate the growth of a gravity tongue of solvent, and it can severely dilute solvent slugs in communicating strata. This is discussed further in Sec. 3.6.

Longitudinal dispersion assumes a more important role when the miscible flood is gravity stable in a dipping reservoir. When the less dense solvent displaces oil downdip below a critical displacement rate, gravity acts to keep the solvent and oil segregated and prevents protrusions of solvent fingers into the oil. Eq. 3.10 approximates the mixing that would occur from longitudinal dispersion in a gravity-stable displacement vertically downward in a pinnacle reef. The solution describes the concentration profile created by longitudinal dispersion for one-dimensional flow in an infinitely long system, initially containing no solvent, but into which a constant solvent concentration is continuously injected beginning at time zero.[12] Implicit in this solution are the assumptions of no viscous fingering of the solvent and no volume change on mixing.

$$C = \frac{1}{2}\, \text{erfc}\left(\frac{x-vt}{2\sqrt{Kt}}\right) + \frac{1}{2\sqrt{\pi}} e^{-[(x-vt)/(2\sqrt{Kt})]^2}$$

$$\cdot \left[\frac{2\sqrt{Kt}}{x+vt} - \frac{1}{2}\left(\frac{2\sqrt{Kt}}{x+vt}\right)^3 + \ldots\right], \quad \ldots \ldots (3.10)$$

where

x = distance from injection end, ft,
v = interstitial velocity, ft/D
t = time, days,
K = longitudinal dispersion coefficient, sq ft/D, and
$\text{erfc}(\xi) = 1 - \text{erf}(\xi)$.

Erf(ξ) is the error function, which can be evaluated from tables of the error function.[22]

According to Brigham,[12] Eq. 3.10 defines the solvent concentration flowing across a plane rather than the in-situ concentration in the plane. For reservoir velocities and dimensions, only the first term is important. All terms may be important for one-dimensional displacements in short laboratory cores, a situation discussed in detail by Brigham.

Brigham also discusses methods for calculating mixing in other flow geometries for displacements where the mixing is caused by longitudinal dispersion only in the absence of viscous fingering or gravity tonguing.[23]

3.3 Flow Regimes in Miscible Displacement

From flow experiments in a vertical cross-sectional laboratory model packed with glass beads,* Crane *et al.*[24] found that four flow regimes are possible at unfavorable mobility ratios, depending on the value of the dimensionless group characterizing the ratio of viscous and gravity forces. Fig. 3.11 illustrates conceptually the different flow regimes observed by these authors. At very low values of the viscous-to-gravity ratio (Region I, Fig. 3.11a), the displacement is characterized by a single gravity tongue overriding the oil. The geometry of this tongue and vertical sweepout both depend on the particular viscous/gravity ratio of the displacement. At

*This model had a small width compared with its height and length, and flow was essentially two dimensional in the vertical and longitudinal directions.

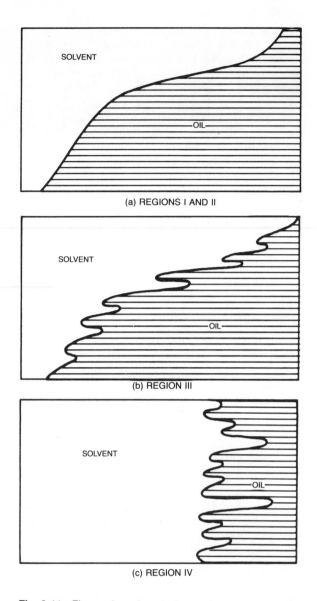

Fig. 3.11—Flow regimes for miscible displacement in a vertical cross section.

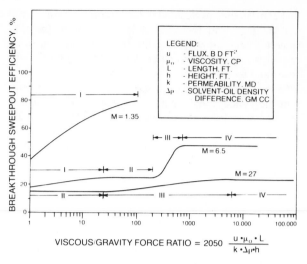

Fig. 3.12—Flow regimes in a two-dimensional, uniform linear system—schematic.

$$R_{v/g} = R_1 \left(\frac{L}{h} \right) = \left(\frac{u\mu_o}{kg\Delta\rho} \right) \left(\frac{L}{h} \right), \quad \dots \dots \quad (3.11)$$

where

u = Darcy velocity,

L = distance between wells,

h = height of reservoir,

k = permeability to oil,

μ_o = oil viscosity,

g = gravitational acceleration, and

$\Delta\rho$ = difference in oil and solvent densities.

Expressed in oilfield units, the viscous/gravity ratio is

$$R_{v/g} = \frac{2{,}050 \; u(\text{B/D} - \text{sq ft}) \; \mu_o(\text{cp}) \; L(\text{ft})}{k(\text{md}) \; \Delta\rho(\text{g/cm}^3) \; h(\text{ft})} \quad \dots \quad (3.12)$$

higher values of the viscous/gravity ratio, the displacement is still characterized by a single gravity tongue (Region II, Fig. 3.11a), but vertical sweepout becomes independent of the particular value of the viscous/gravity ratio until a critical value is exceeded. Beyond this critical value, a transition region is encountered (Region III, Fig. 3.11b) where secondary fingers form beneath the main gravity tongue.[8] In this region, sweepout for a given value of pore volumes injected increases sharply with increasing values of the viscous/gravity ratio. Finally, a value of viscous/gravity ratio is reached where the displacement is entirely dominated by multiple fingering in the cross section, and vertical sweepout again becomes independent of the particular value of the viscous/gravity ratio (Region IV, Fig. 3.11c).

Fig. 3.12 further illustrates the different flow regimes in miscible displacement and shows how sweepout at solvent breakthrough in a vertical cross section is affected both by the flow regime and by mobility ratio. In this figure the ratio of viscous and gravity forces is calculated by the following expression.

The sweepout values of this figure and the values of viscous/gravity ratio marking the transition from one flow regime to another were estimated from several sources[15,24-26] and should be regarded as semi-quantitative.

Eqs. 3.11 and 3.12 are for displacements at irreducible water—i.e., only the solvent/oil phase is flowing (for waterflooded conditions see the discussion in Sec. 3.6). These equations also assume that vertical permeability is the same as horizontal permeability. When this is not the case, a suggested approximation is to substitute $\sqrt{k_v k_H}$ for k in these equations, where k_v and k_H are the permeabilities in the vertical and horizontal directions, respectively. (See also the discussion in Sec. 3.6.)

The value of viscous/gravity ratio at which a transition occurs from one flow region to another depends on the mobility ratio.[24,26] Crane et al.[24] found the transition from Region I to Region II flow occurred at an $R_{v/g}$ of about 20 when the mobility ratio was 6.5, whereas the transition occurred at an $R_{v/g}$ of about unity at a mobility ratio of 27. Other examples of the Region I/II transition can be seen in the data of Craig et al.[1,26]

For these same mobility ratios, Crane *et al.*[24] also reported R_1 values at which the change from Region II to Region III flow occurred and at which the change from Region III to Region IV flow occurred ($R_1 = 2,050$ $u\mu_o/k\Delta\rho$).* The change from Region II to Region III flow occurred at an R_1 value of roughly 30 when the mobility ratio was 6.5 and at an R_1 value of approximately 10 when the mobility ratio was 27. At these mobility ratios, they found multiple fingering to be developed fairly completely (Region IV) for R_1 values of 70 and 600. In the Crane *et al.* experiments, the value for L/h was approximately 10, and the $R_{v/g}$ values at which these changes in flow regime occurred were approximately 10 times the R_1 values.

Pozzi and Blackwell[15] also studied the change from Region II to Region III flow. For mobility ratios of 1.85, 16.3, and 69, these authors reported the change occurred at $R_{v/g}$ values of approximately $0.4(L/h)^2$, which is only in rough agreement with the data of Crane *et al.*[24]

Although there are few published data for three-dimensional miscible displacements, the available published data[1,26] as well as several unpublished sources** support the qualitative description of flow behavior outlined in the previous discussion for flow in a vertical cross section. Unfortunately, there are very few data relating sweepout to viscous/gravity ratio for experiments where three-dimensional geometry has been modeled.[26] Most sweepout data have been taken either in two-dimensional areal flow models, in which a very small model thickness and a high injection rate caused flow to be well within the Region IV flow regime and dominated by multiple fingering, or in two-dimensional vertical cross-section models in which displacements were dominated by gravity tonguing in Flow Regimes I and II.

When two-dimensional model data are used to estimate the flow regime that will prevail in three-dimensional pattern flooding, a problem arises in selecting the appropriate linear velocity for calculating the viscosity/gravity ratio using the expression given above for two-dimensional flow. For example, flow velocities in a five-spot pattern vary significantly from near the wellbore to midway between wells, and there are no published guidelines for calculating the viscous/gravity ratio that best characterizes vertical sweepout. Until better guidelines are available, a suggested working assumption for five-spot flow is†

$$u = \frac{1.25i}{hL}, \quad \dots\dots\dots\dots\dots\dots\dots\dots (3.13)$$

where i is the injection rate in B/D per well. For line-drive flow, a suggested assumption is‡

*Units same as in Fig. 3.12.

**Blackwell, R.J.: personal communication, Exxon Production Research Co. (1982) and White, G.L.: personal communication, ARCO Oil & Gas Co. (1982).

†This formula was derived by comparing breakthrough sweepout calculated from two-dimensional areal and vertical sweepout data with limited five-spot sweepout data. In effect, it *assumes* that the proper width for calculating an average effective linear velocity is one-fifth the distance between wells.

‡This formula was derived by *assuming* that the proper width for calculating an average effective linear velocity for a 1:1 line drive on regular spacing is one-half the distance between wells.

$$u = \frac{i}{hL}. \quad \dots\dots\dots\dots\dots\dots\dots\dots\dots\dots (3.14)$$

The reader is cautioned that these assumptions could be substantially in error.

Subsequent sections of this chapter discuss areal and vertical sweepout and the estimation of volumetric sweepout from areal and vertical sweepout data.

Example Calculation 1

Problem. Calculate the viscous/gravity ratio for vaporizing-gas drive flooding in a reservoir that has not been waterflooded previously. Assume these data: 40-acre five-spot pattern, $i = 2,000$ B/D (gas injection at reservoir conditions), $\mu_o = 0.4$ cp, $k = 75$ md, $h = 35$ ft, $L = 933$ ft, $\Delta\rho = 0.4$ g/cm^3, $M = 25$, and $k_v/k_h = 1$.

Solution.

$$u \cong \frac{1.25(2,000)}{35(933)} = 0.0766 \text{ B/D-sq ft}$$

and

$$R_{v/g} = \frac{2,050(0.0766)(0.4)(933)}{75(0.4)(35)} = 56.$$

Therefore, from Fig. 3.12, flow is dominated by gravity tonguing.

Example Calculation 2

Problem. Calculate the viscous/gravity ratio for CO_2 flooding in a reservoir that has not been waterflooded previously. Assume these data: 40-acre five-spot pattern, $i = 500$ B/D (CO_2 injection at reservoir conditions), $\mu_o = 1.3$ cp, $k = 4$ md, $h = 25$ ft, $L = 933$ ft, $\Delta\rho = 0.1$ g/cm^3, $M = 25$, and $k_v/k_H = 1$.

Solution.

$$u \cong \frac{1.25(500)}{25(933)} = 0.0286 \text{ B/D-sq ft}$$

and

$$R_{v/g} = \frac{2,050(0.0268)(1.3)(933)}{4(0.1)(25)} = 6,660.$$

Therefore, from Fig. 3.12, flow is dominated by viscous fingering.

3.4 Viscous Fingering

The displacement of oil by first-contact miscible solvents in homogeneous porous media is mechanistically simple when the solvent/oil mobility ratio is less than or equal to one and when gravity does not influence the displacement by segregating the two fluids. For these conditions, oil is displaced efficiently ahead of the solvent, and the

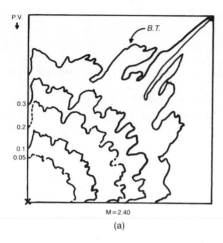

(a)

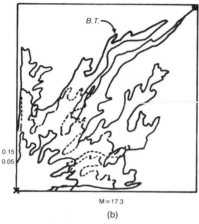

M = 17.3

(b)

Fig. 3.13—Displacement fronts for different mobility ratios and injected pore volumes until breakthrough, quarter of a five-spot (after Ref. 28).

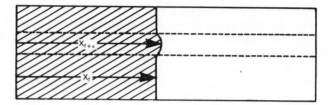

Fig. 3.14—Simplified model of frontal instability (after Ref. 3).

solvent does not penetrate into the oil except as dictated by dispersion. The displacement front is stable, and a mixed zone develops and grows according to the principles discussed in Sec. 3.2.

For mobility ratios greater than one, the displacement has a quite different character. The solvent front becomes unstable, and numerous fingers of solvent develop and penetrate into the oil in an irregular fashion. These viscous fingers result in earlier solvent breakthrough and poorer oil recovery after breakthrough for a given volume of solvent injected than would be the case if the displacing front remained stable.

Figs. 3.13a and 3.13b illustrate the viscous fingering observed in areal laboratory five-spot models of consolidated sand for displacements at various mobility ratios.[28] Gravity segregation was not a factor in these experiments, and flow was in the Region IV flow regime (see Sec. 3.3). Because of the small thickness of the models, flow was essentially two-dimensional, and fingering probably did not occur in the vertical cross section. In these illustrations, the severity of the areal fingering increases as mobility ratio becomes more unfavorable, resulting in earlier solvent breakthroughs.

The reason for displacement front instability when the mobility ratio is unfavorable can be visualized with the simple illustration shown in Fig. 3.14.[3] In this illustra-

tion, solvent displaces oil linearly from a porous medium that initially is fully saturated with oil. The mobility ratio in this case is just the ratio of oil and solvent viscosities. Longitudinal dispersion is assumed negligible. In the absence of heterogeneity, the front should remain a plane surface throughout the displacement. Suppose, however, that a small region encountered by the front is more permeable than the surrounding region. The front entering the region of higher permeability momentarily travels faster than the rest of the front, causing a small bump to protrude a distance, ϵ, from the otherwise plane interface.

The subsequent history of the perturbation can be examined by writing equations for the velocity of both the perturbation and the main front. To do this, the portion of the porous medium within the dashed lines in Fig. 3.14 is treated as an isolated system. If x_f is the distance from the inflow end to the undisturbed front, the distance to the front of the bump is $x_f + \epsilon$. From Darcy's equation for linear flow, the velocity of the undisturbed front is

$$\frac{dx_f}{dt} = \frac{k\Delta p}{\phi\mu_s[M L + (1-M)x_f]}, \quad \dots\dots\dots\dots (3.15)$$

where $M = \mu_o/\mu_s$ and velocity of the leading edge of the perturbation is

$$\frac{d(x_f + \epsilon)}{dt} = \frac{k\Delta p}{\phi\mu_s[M L + (1-M)(x_f + \epsilon)]}. \quad \dots\dots (3.16)$$

It follows that

$$\frac{d\epsilon}{dt} = \frac{-k\Delta p(1-M)\epsilon}{\phi\mu_s[M L + (1-M)x_f]^2}, \quad \dots\dots\dots\dots (3.17)$$

provided $\epsilon \ll x_f$. Therefore, $\epsilon = e^{Ct}$, where

$$C = \frac{-k\Delta p(1-M)}{\phi\mu_s[M L + (1-M)x_f]^2}. \quad \dots\dots\dots\dots (3.18)$$

Thus, ϵ initially grows exponentially with time immediately after formation of the perturbation if $M > 1$ but decays exponentially with time if $M < 1$. This example is highly over simplified for representing actual finger initiation and growth in a porous medium, but it serves to illustrate that the viscous force causes any irregularity of the front arising from permeability heterogeneity to grow if $M > 1$. Longitudinal and transverse dispersion, which were not taken into account in this simple example, act to oppose this growth by moderating the viscosity contrast. Whether or not the finger propagates after being initiated and, if so, at what rate depends on the importance of dispersion in the displacement.

There have been many experimental[8,16,25,28-36] and mathematical[33,37-45] studies of viscous fingering both for miscible and for immiscible displacement. As a result, the following description of the viscous fingering phenomenon has emerged.

Finger Initiation

The exact process of finger initiation is still obscure although it generally is attributed to the presence of permeability heterogeneities. Finger initiation is easy to visualize in porous media since pore structure is microscopically random even in glass-bead packs that are macroscopically or superficially homogeneous. These small microscopic variations are sufficient to initiate the fingers. Viscous fingers are observed even under appropriate displacement conditions in laboratory Hele-Shaw models, which are models constructed of two parallel plates with the gap between the models filled with liquid. Apparently, extremely small variations in the plate surface and gap width are sufficient to initiate fingers.

There have been several attempts to investigate finger initiation and propagation mathematically by the frontal perturbation method.[40,44-47] In this method, a spectrum of wavelengths of perturbations of the front is assumed—e.g., a Fourier series describing variations of the frontal position about an average smooth line. The resulting analyses showed that for any given set of displacement conditions, perturbations below a critical wavelength will be eliminated by dispersion and only those perturbations above the critical wavelength will continue to grow at an unfavorable mobility ratio. Gardner and Ypma[47] published the following approximate formula for critical wavelength for the case of an initially sharp solvent/oil interface.

$$\lambda_c = 2^{5/2} \cdot \pi \cdot \frac{\mu_o + \mu_s}{\mu_o - \mu_s} \cdot \frac{K_t}{v}, \quad \dots\dots\dots\dots (3.19)$$

where

$$\lambda_c = \text{finger width, cm,}$$
$$\mu_o, \mu_s = \text{oil and solvent viscosities, cp,}$$
$$K_t = \text{transverse dispersion coefficient,}$$
$$\text{cm}^2/\text{s, and}$$
$$v = \text{average interstitial velocity, cm/s.}$$

It can be inferred from the perturbation analyses that dispersion is insufficient to damp out all the flow perturbations in systems of reservoir dimensions and most laboratory dimensions. This conclusion has been supported by numerous experiments.[25,28-31,33,35,37]

In their experimental study of viscous fingering in linear and radial models, Perkins et al.[37] observed a small region devoid of fingers at the inlet of their models. These researchers postulated that although fingers might be initiated at the inlet by permeability variations, the rate of growth in length of the finger would have to exceed the initial rate of growth of the longitudinal mixed zone before fingers could be propagated. Perkins et al. developed a simplified expression for the rate of growth in finger length and developed an approximate calculation for the linear and radial distances at which viscous fingering first would be observed. Other authors have derived different expressions for the rate of growth in finger length,[41,43] but regardless of which expression for growth of finger length is used with the Perkins et al. finger initiation calculations, the calculations show that the initial region devoid of fingers is entirely negligible on a reservoir scale and in most laboratory systems.

Finger Growth

Once fingers above the critical wavelength are initiated, they begin to grow in length. Additional solvent fingers may be initiated from the ends of already growing fingers, with the result of a fingering pattern that looks like the limbs and twigs of a tree (Figs. 3.13a and 3.13b). Increasing the mobility ratio increases the viscous instability and consequently increases the rate at which the fingers grow.

While the fingers grow in length, they also grow in average width. This is accomplished through a combination of spreading by transverse dispersion as well as by suppression of smaller fingers by transverse dispersion and by merging and coalescence of smaller fingers into larger ones.[37] Examples of merging and coalescence of fingers can be seen in Fig. 3.13b, and as a result of this mechanism, oil trapped between two merging fingers is mixed into the larger finger that forms.[37] The result is that fewer and fewer fingers grow larger and larger in size as the displacement progresses.

These mechanisms work to moderate growth in finger length by mixing oil and solvent, which moderates the effective viscosity ratio of the displacement. Longitudinal dispersion is a relatively unimportant factor in growth of finger length.[15,25,29] A very high rate of transverse dispersion, on the other hand, can stabilize the displacement by wiping out fingers or at least reducing them to one or two large ones.[8,25,35]

The relative importance of transverse dispersion in a miscible displacement is characterized by the dimensionless transverse dispersion group, $K_t L/v W^2$, where L is the total displacement length and W is the reservoir width or thickness over which fingering is occurring. The greater the value of this dimensionless group, the more significance transverse dispersion has on the displacement. Laboratory experiments in two-dimensional models where flow is predominantly linear, although limited in number, indicate that for a given mobility ratio, sweepout remains relatively constant as long as the dimensionless transverse dispersion group is below a value of roughly 0.01.[8,15,25] Above this value, transverse dispersion damps finger growth (viscous or gravity) to the extent that sweepout is affected.

There is no similar semiquantitative criterion for judging the importance of transverse dispersion in other areal geometries such as five-spot patterns. The value of the dimensionless transverse dispersion group in laboratory pattern flood experiments typically is much larger than the value for this group for field conditions, sometimes by a factor of 100 or more. A high value of the dimensionless transverse dispersion group probably has caused partial suppression and blurring of fingers in some areal models [14,25,48,49] and slowing of finger growth in some core floods. [50,51] The value of the macroscopic mixing coefficient* in reservoir floods would have to be unusually large, perhaps several feet or greater, for the dimensionless transverse dispersion group for a reservoir flood to be as great as the value of this dimensionless group in those laboratory experiments where fingering appears to have been partially retarded by transverse dispersion.

There have been many mathematical treatments of miscible displacement with viscous fingering. [37,38,40-43,52] In 1962, Peaceman and Rachford [42] solved the diffusion-convection and continuity equations numerically by a finite-difference technique. These authors used the resulting mathematical model to calculate the behavior of several two-dimensional laboratory experiments, one at a mobility ratio of 5:1, the other at 86:1. Small, random variations of permeability with position were sufficient to initiate finger growth in the mathematical model. Although finger development observed in the laboratory model was more complex than that calculated by the mathematical model, the rate of propagation of fingers calculated by the mathematical model was in close agreement with the experimental observations. The length of the fingered region was observed to increase in a near linear manner with pore volumes injected. Also, the calculated and experimental dependence of oil recovery on the quantity of solvent injected were in good agreement. At that time, the method was impractical for reservoir problems because the number of grids and the computing time required for such problems was excessive for the computing systems available. Since then, the computing speed and storage capacity of computers have improved, and they continue to improve steadily as does the efficiency of numerical solution techniques. This could be a valuable future approach to account for viscous fingering on displacement behavior, at least in simulations of limited extent such as repeating pattern elements. Conceivably, at some future time, viscous fingering could be accounted for in large reservoir simulations by such a technique.

Other mathematical treatments and experimental studies also show that the length of the fingered region in linear and diverging radial systems increases linearly with pore volumes injected as long as transverse dispersion is not excessive. [25,37,41,43] The different theories, however, have resulted in different relationships for finger growth.

Koval [43] developed a mathematical treatment of viscous fingering that is analogous to the Buckley-Leverett [53] calculation method for immiscible displacement. The Koval method has had relatively good success in predicting the results of experimental laboratory miscible displacements that were conducted under conditions where the growth of multiple viscous fingers was not affected appreciably by transverse dispersion. For example, Koval found good agreement between the predictions of his method and experimental data of

Blackwell et al., [25] Brigham et al., [6] and Handy [54] as well as with displacements in heterogeneous cores published in his paper; Claridge reported that the method worked well for radial data, [55] and Kyle and Perrine found fair agreement between the Koval theory and experiments. [33] The Koval method, however, does not adequately predict all available data. [37]

One technique discussed in Sec. 3.5 for estimating areal sweepout in miscible floods uses the original Koval method to account for viscous fingering, [55] and an empirical variation of the method is commonly used to approximate the effect of viscous fingering in reservoir simulations [56] (see Sec. 4.2). Because this method or variations of it have been useful in practical applications, it is discussed in some detail in the following paragraphs with particular attention given to its assumptions and limitations.

In the Koval method, a given solvent saturation, S_s, is assumed to travel at a constant, characteristic velocity. For linear flow,

$$\left(\frac{dx}{dt}\right)_{S_s} = V_{\hat{S}_s} = \frac{q_t}{A\phi}\left(\frac{df_s}{dS_s}\right)_{S_s}, \qquad \ldots\ldots\ldots (3.20)$$

where f_s is the volume fraction of solvent in the flowing stream at a distance x from the inlet. The solvent saturation is the volume fraction of solvent in a given differential volume element regardless of whether the solvent and oil are homogeneously mixed or macroscopically segregated. An assumption of linear volume blending is implicit in the use of S_s as a volume fraction.

The Buckley-Leverett expression for fractional flow in an immiscible water/oil displacement with negligible gravity and capillary pressure influence is

$$f_w = \frac{1}{1 + \frac{k_o}{k_w}\frac{\mu_w}{\mu_o}}, \qquad \ldots\ldots\ldots\ldots\ldots (3.21)$$

where k_o and k_w are permeabilities to oil and water. Koval assumed that fractional flow could be expressed by a similar equation for segregated miscible displacement. He reasoned that permeability to either solvent or oil could be expressed as the total permeability multiplied by the average saturation of each fluid and that solvent fractional flow could be calculated as follows when viscous fingering predominates.

$$f_s = \frac{1}{1 + \frac{(1-S_s)}{S_s}\cdot\left(\frac{\mu_{s_{eff}}}{\mu_{o_{eff}}}\right)\left(\frac{1}{H}\right)}$$

$$= \frac{1}{1 + \frac{(1-S_s)}{S_s}\cdot\frac{1}{EH}}, \qquad \ldots\ldots\ldots\ldots (3.22)$$

where

---

*Mixing coefficient is the ratio of effective dispersion coefficient to interstitial velocity.

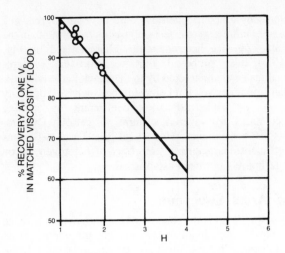

Fig. 3.15—Defining curve for heterogeneity factor (after Ref. 43).

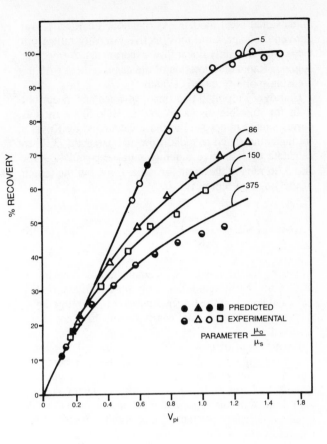

Fig. 3.16—Comparison of Blackwell's experimental data with predictions based on *K*-factor method (after Ref. 43).

$\mu_{s_{eff}}$, $\mu_{o_{eff}}$ = the effective viscosities in the solvent and oil fingers, cp,

E = effective viscosity ratio between the forward-projecting solvent fingers and backward-projecting oil fingers, and

H = a heterogeneity factor characterizing the heterogeneity of a given rock sample.

The effective viscosity ratio is different from the nominal viscosity ratio of pure oil and solvent because of solvent/oil mixing in the fingered region. From laboratory miscible displacements in homogeneous porous media,[25] Koval derived the following expression for the effective viscosity ratio.

$$E = \left[0.78 + 0.22 \left(\frac{\mu_o}{\mu_s} \right)^{\frac{1}{4}} \right]^4, \quad \dots\dots\dots\dots (3.23)$$

where μ_o and μ_s are the viscosities of pure oil and solvent.

Koval characterized rock heterogeneity by the percent oil recovery at 1 PV of solvent injected in a matched viscosity flood. He defined a homogeneous system to be one in which recovery at these displacement conditions was 99%. In such a system the heterogeneity factor is one. From displacements in a variety of heterogeneous sandstone cores, the relationship shown in Fig. 3.15 was derived for heterogeneity factor. Koval found the heterogeneity for a given rock sample to be constant regardless of the mobility ratio of the displacement.

The solution to Eqs. 3.20 and 3.22 for the pore volumes of solvent injected at solvent breakthrough is

$$V_{pD\,BT} = \frac{1}{K}, \quad \dots\dots\dots\dots\dots\dots (3.24)$$

where $K = EH$. Oil recovery after breakthrough and frac-

tion of solvent in the effluent are given by Eqs. 3.25 and 3.26:

$$N_{pv} = \frac{2(KV_{pDi})^{\frac{1}{2}} - 1 - V_{pDi}}{K - 1} \quad \dots\dots\dots (3.25)$$

and

$$f_{se} = \frac{K - (K/V_{pDi})^{\frac{1}{2}}}{K - 1}, \quad \dots\dots\dots\dots (3.26)$$

where

f_{se} = solvent fractional flow in the effluent,
N_{pv} = oil recovery in fractional pore volume, and
V_{pDi} = pore volumes injected.

The length of the fingered region is

$$\Delta\ell = \left(K - \frac{1}{K} \right) x_m, \quad \dots\dots\dots\dots (3.27)$$

where x_m is the mean displacement distance—i.e., the distance to the solvent front calculated as if the displacement had been piston-like.

Fig. 3.16 compares predictions by the Koval method with oil recovery from Blackwell *et al*.'s[25] two-dimensional areal models. As mentioned earlier, Koval derived the expression for effective viscosity ratio from these predominantly linear flow experiments. Agreement is good. Rapid deterioration of the displacement with increasing mobility ratio is evident.

Claridge[55] reported that finger-invaded and swept-out radii for miscible displacements in Hele-Shaw models corresponded rather well to those calculated by Koval's formulas translated to radial flow and using Eq. 3.23 to calculate the effective mobility ratio. Eqs. 3.24, 3.25, and 3.26 remain the same for radial flow, but the length of the fingered region is

$$\Delta r = r_m \left(\sqrt{K} - \frac{1}{\sqrt{K}} \right), \quad \ldots \ldots \ldots \ldots \ldots (3.28)$$

where r_m is the mean displacement radius. Perkins et al.[37] found a different relationship for effective mobility ratio and length of the fingered region in diverging radial displacements, and this contradiction has not been resolved.

The problem of how best to utilize these laboratory-derived results in field applications has not been resolved. The limitations of Eq. 3.23 for effective viscosity ratio and Fig. 3.15 for heterogeneity factor are not well defined at this time, and more investigation in this regard is certainly warranted. One might anticipate that the effective viscosity ratio in a miscible flood dominated by viscous fingering would be affected both by solvent/oil mixing caused by micro- and macroscopic transverse dispersion and by solvent/oil mixing caused by hydrodynamic mechanisms such as crossflow, merging, and coalescence of fingers. If this is true, Eq. 3.23 may predict too favorable an effective viscosity ratio at reservoir conditions, because the dimensionless transverse dispersion group, $K_t L / v h^2$, generally will not be as great at reservoir conditions as in the laboratory experiments that Koval used to estimate the coefficients of Eq. 3.23. Claridge,[58] however, after examining laboratory displacement data that were available in the literature, concluded that Eq. 3.23 gave adequate predictions over a range of the dimensionless transverse dispersion group from 10^{-4} to 1. Although this range is higher than the values of dimensionless transverse dispersion that typically prevail at reservoir conditions, Claridge's finding suggests that the effective viscosity ratio in a displacement dominated by viscous fingering may be influenced more by the hydrodynamics of finger growth than by transverse dispersion. At this time, the factors affecting viscosity ratio moderation are not well understood, and the overall adequacy of Eq. 3.23 for describing reservoir floods needs further investigation and evaluation.

Koval, in Fig. 3.15, tried to relate core-scale heterogeneity to the efficiency of unit viscosity ratio displacements in small laboratory cores. Only the effect of small-scale (less than core dimensions) heterogeneities can be accounted for by this test, and the degree to which the test characterizes core-scale heterogeneities is not well established. If there is a level of macroscopic heterogeneity in the reservoir that is too large in scale to be observed in small core tests but is smaller than gross correlatable features, these macroscopic heterogeneities could have a different and possibly more severe effect on viscous fingering than the small, core-size heterogeneities characterized by Fig. 3.15. Koval proposed that reservoir-scale heterogeneities be characterized by the Dykstra-Parsons coefficient[57] and derived a relationship between heterogeneity factor and Dykstra-Parsons coefficient by making calculations for layered systems of various Dykstra-Parsons coefficients. How well the Dykstra-Parsons coefficient characterizes the effect of heterogeneity on fingering has not been established.

3.5 Areal Sweepout

Areal sweepout is the fraction of the pattern area invaded by pure solvent. In an areal sense, it is the fraction of initial displaceable volume that is displaced by the solvent. This is different from areal coverage, which is the fraction of the pattern area invaded by the leading edge of the solvent/oil mixed zone. Areal coverage is greater than areal sweepout for a given volume of solvent injected because of oil mixed into the solvent fingers by the mechanisms discussed in Sec. 3.4.

Areal sweepout and areal coverage are affected by factors such as pattern geometry, mobility ratio, areal heterogeneity, and degree of mixing between solvent and oil. They are functions of the volume of solvent injected, and increase with increasing solvent injection.

Pattern geometry, mobility ratio, and areal heterogeneity all affect potential and streamline distribution. Pressure gradient is largest and velocity the greatest along the shorter streamlines connecting wells, and more rapid solvent invasion into the reservoir and breakthrough into producing wells occur along these streamlines. Solvent invades the reservoir more slowly along the longer streamlines, and areal invasion increases with increasing cumulative solvent injection.

The remainder of this section reviews areal sweepout data that were taken under conditions where gravity segregation was not a factor in the experiments. Because of the small thickness of the models, flow was essentially two-dimensional, and fingering probably did not occur in the vertical dimension. Although these data were not taken under conditions where three-dimensional geometry was modeled rigorously, they customarily are used to represent volumetric sweepout at high values of the viscous to gravity force ratio where Region IV flow behavior prevails (Sec. 3.3).

Continuous Solvent Injection, Secondary Recovery

At unfavorable mobility ratios and large values of the viscous/gravity ratio, areal sweepout and areal coverage are affected greatly by viscous fingering, as seen from the tracings of the leading edge of the solvent front shown in Figs. 3.13a and 3.13b for the experiments of Habermann.[28] Factors affecting finger growth were discussed in Sec. 3.4; in mathematical calculations or laboratory measurements of areal sweepout, it is important that these factors be reasonably well scaled to represent field conditions. Otherwise, computed or experimentally measured sweepout is likely to be too favorable.

There have been numerous attempts to measure areal

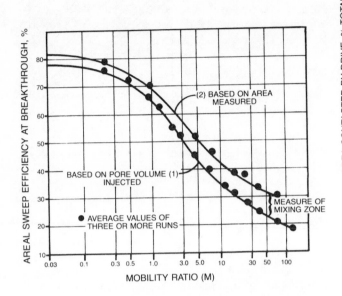

Fig. 3.17—Breakthrough sweep efficiency and a measure of mixing zone for two-zone displacements, five-spot (after Ref. 28).

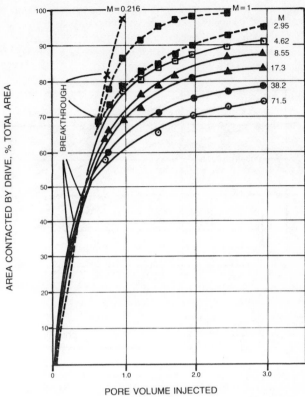

Fig. 3.18—Area contacted by drive after breakthrough, quarter of a five-spot (after Ref. 28).

sweepout and areal coverage in the laboratory. An extensive review of these data was given in the waterflooding monograph by Craig.[1] Areal data have been obtained by such diverse methods as electrolytic models,[59,60] blotter-type models,[61] gelatin models,[62] potentiometric models,[63] resistance-type models,[64] Hele-Shaw models,[29,65] and porous media flow models.[2,28,49,66] Attempts also have been made to calculate areal sweepout.[67-71] Unfortunately, most of the data either are invalid for miscible displacement or at best questionable.

At unfavorable mobility ratios, the data from methods that do not involve fluid displacement are not valid for either miscible or immiscible displacement because these methods are incapable of showing viscous fingering. As a result, the data are too optimistic. Even many of the flow model data appear to be taken under conditions where an unrealistically high rate of transverse dispersion partially retarded the growth of viscous fingers. This blunting of fingers, combined in some instances with experimental inaccuracy in resolving the exact location of the solvent front, also appears to have caused optimistically high areal coverage for a given displaceable volume of solvent injection in some experiments.[2,49,66] There have been relatively few experiments in which these factors do not appear to have caused somewhat optimistic sweepout.[28,29,31,72]

Many of the early areal sweep experiments in porous media measured areal coverage instead of areal sweepout. The area invaded by the solvent front was defined by colored solvents in transparent models[28] or

by adding an X-ray-absorbing material to the solvent and taking X-ray shadow graphs.[2,49,66] Habermann's results for a homogeneous five-spot appear to be the best-scaled and most accurate data of this type.[28] Habermann measured both areal coverage and areal sweepout at solvent breakthrough and areal coverage after breakthrough for various mobility ratios. These data are shown in Figs. 3.17 and 3.18. Solvent injection is expressed in terms of displaceable hydrocarbon pore volumes—i.e., the product of total pore volume and the mobile solvent saturation. These data are valid only for a single-front displacement, where solvent is injected continuously and where oil is the only mobile fluids initially.

The Habermann data of Fig. 3.17 show that areal coverage and areal sweepout at solvent breakthrough decrease continuously with increasing mobility ratio. Whereas areal sweepout at breakthrough is 0.67 for a mobility ratio of unity, it decreases to 0.36 for a mobility ratio of 10 and to 0.2 for a mobility ratio of 70.

The difference between the areal coverage and areal sweepout curves, which ranges from 4 to 10% sweep depending on mobility ratio, is a rough measure of the leading half of the mixed zone under the conditions of Habermann's experiments. Although the Habermann data were better scaled than those of most other sweep experiments, transverse dispersion most likely was still large relative to that expected from molecular diffusion at field rates and dimensions, and the difference between areal coverage and sweepout in Fig. 3.17 is an upper bound for field displacements where dispersion is

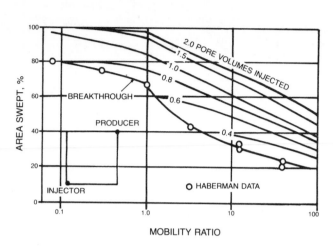

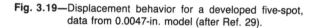

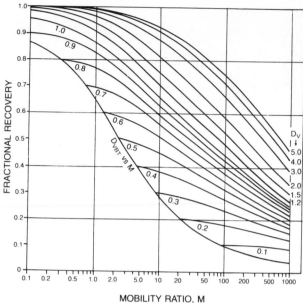

Fig. 3.19—Displacement behavior for a developed five-spot, data from 0.0047-in. model (after Ref. 29).

Fig. 3.20—Oil recovery in miscible flooding for five-spot well patterns (after Ref. 55).

primarily the result of molecular diffusion.*

Fig. 3.18 shows that areal coverage after solvent breakthrough continues to increase with increasing solvent injection, although the areal coverage achieved for a given displaceable volume of injection is smaller the higher the mobility ratio. For example, after injection of one displaceable volume of solvent, areal coverage reaches 0.75 for a mobility ratio of 10 but only 0.62 for a mobility ratio of 70.

Areal sweepout for a five-spot pattern was measured by Mahaffey et al.[29] In their Hele-Shaw model experiments, longitudinal and transverse dispersion were scaled to represent mixing caused in the reservoir by molecular diffusion only. The Mahaffey et al. data therefore should give a lower limit for sweepout in the reservoir as affected by dispersion.

Fig. 3.19 shows the Mahaffey et al. areal sweepout data for various mobility ratios and displaceable volumes injected. The dominant influence of an unfavorable mobility ratio and viscous fingering is again evident. Even after two displaceable volumes of injection, areal sweepout is only 0.75 for a mobility ratio of 10 and 0.52 for a mobility ratio of 100.

For comparison, the breakthrough sweepout data of Habermann also are shown on this figure. Agreement between the two sets of data is good, indicating that the higher level of dispersion in Habermann's experiments did not affect breakthrough results materially.

Areal coverage data for five-spot, direct line-drive, staggered line-drive, and nine-spot patterns were reported by Dyes et al.[49] and Kimbler et al.[73] Swept area was defined by the X-ray shadowgraph technique. The areal coverage observed for the five-spot experiments was higher than that reported by Habermann,

particularly at the higher mobility ratios, probably because of unscaled dispersion and inaccuracy in resolving the solvent front by the shadowgraph technique. Presumably the areal coverage data for the other patterns are equally optimistic.

A method for adjusting the Dyes et al. data is suggested by some work published by Claridge.[55] He observed that the high areal coverage in the Dyes et al. five-spot model seemed to agree with the fractional area lying inside a curve drawn through the tips of fingers in photographs of the displacements in the better-scaled work. Claridge reasoned that areal sweepout could be estimated from the Dyes et al. data by applying the Koval equations for linear displacement efficiency of an unstable displacement (Sec. 3.4) to the reported values for area contacted. Fig. 3.20 shows the final areal recovery efficiency correlation for a normal five-spot reported by Claridge after applying this procedure and smoothing results. Agreement is fair with the scaled five-spot data of Mahaffey et al. in Fig. 3.19. Breakthrough sweepout predicted by the Claridge method at unfavorable mobility ratio is less favorable than the Mahaffey data, but sweepout for solvent injection greater than 0.5 displaceable volumes is more favorable than the Mahaffey data. Presumably the same procedure would be as valid for estimating areal sweepout from the line-drive and nine-spot data of Dyes et al. and Kimbler et al. as from the five-spot data.

Solvent Slugs, Secondary Recovery

All the areal sweep data discussed so far apply to a single-front displacement. In many field applications of miscible displacement, a slug process is practiced in which a solvent slug is driven with another fluid such as gas or water. There are only a few instances where sweep data of this kind have been reported.[28,29,72,74]

*These data probably are most appropriate for representing sweepout for field conditions where the macroscopic dispersion is large—i.e., where the mixing coefficient is on the order of several feet or larger.

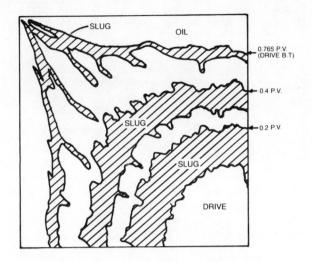

Fig. 3.21—Distribution of miscible slug at low mobility ratio; Run 142, slug size = 10% PV, $\mu_o/\mu_s = 1.59$, $\mu_s/\mu_o = 1.25$, $\mu_o/\mu_d = 1.99$ (after Ref. 28).

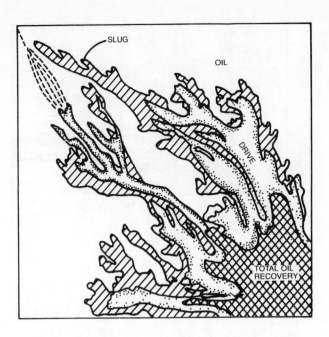

Fig. 3.22—Distribution of miscible slug at high mobility ratio; Run 141, slug size = 10% PV, $\mu_o/\mu_s = 27.1$, $\mu_s/\mu_o = 1.72$, $\mu_o/\mu_d = 46.6$ (after Ref. 28).

Severe viscous fingering has a profound effect on the sweepout and oil recovery efficiency of small solvent slugs. Figs. 3.21 and 3.22 illustrate the movement of two 10% PV solvent slugs in the laboratory five-spot model of Habermann.[28] In both of these tests the mobility ratio was unfavorable at both the leading and trailing edges of the slug. For the displacement shown in Fig. 3.21, the mobility ratios between solvent/oil, drive fluid/solvent, and drive fluid/oil were 1.59, 1.25, and 1.99, respectively. For the displacement shown in Fig. 3.22, the mobility ratios between solvent/oil, drive fluid/solvent, and drive fluid/oil were 27.1, 1.72, and 46.6, respectively. For the slightly unfavorable mobility ratio experiment, the leading and trailing edges of the slug develop irregular shapes because of fingering, but the slug appears to remain intact, although in the region of converging radial flow, it becomes highly ''strung out'' in the backward-projecting fingers. At the higher mobility ratio, the slug does not remain intact as a well-defined band, but is concentrated in the tips of the fingers. In fact, for this example, the drive fluid has penetrated completely through the slug in several locations, and breakthrough of the slug and drive fluid will occur almost simultaneously. At the higher mobility ratio, the surface area of the slug in contact with both oil and drive fluid has increased greatly, and the width of individual fingers of solvent is smaller than the width of the more or less intact slug of solvent in Fig. 3.21. Presumably, these conditions should lead to a more rapid dilution of the slug and subsequent loss of miscibility from dispersion. Mobility ratios that are very unfavorable cannot be tolerated in miscible slug displacements without viscous fingering destroying the integrity of the solvent bank. Even when the mobility ratio is favorable at either the leading or trailing edge of the slug, fingering will occur at the other edge if an un-

favorable mobility ratio prevails. Similar behavior was reported by Mahaffey *et al.*[29] for their five-spot experiments scaled for molecular diffusion.

Although the sweep data for solvent slugs are too limited to permit generalized correlation, the following behavior is observed. A solvent slug with viscosity intermediate between that of oil and the drive fluid increases the areal sweepout compared with the sweepout that would be achieved by drive fluid alone. Similarly, a solvent slug with a lower viscosity than that of the drive fluid decreases the areal sweepout of the drive fluid. Thus, the effective mobility ratio of a displacement depends on both the slug size and solvent viscosity.

In addition to viscous fingering, miscible sweepout in a slug process is affected by (1) breaching of the slug by immiscible drive gas and (2) dilution of the slug by dispersion, if this results in a local loss of miscibility. Lacey *et al.*[72] published areal sweepout results for a high-pressure five-spot model where propane slugs displaced a methane-saturated refined oil at 1,550 psig and the propane slugs in turn were driven by methane. Although the phase behavior wasn't reported, Lacey *et al.* stated that it was similar to that of some reservoir oils. If so, the slug could have been diluted to approximately 60 to 80% propane before miscibility was lost. In these experiments, the viscosity ratio was approximately 10 at the leading edge of the slugs and 8.5 at the trailing edge—viscosity ratios that are realistic for hydrocarbon solvents or for CO_2 displacing many potential miscible-flood candidate oils and in turn displaced by a drive gas.

Fig. 3.23 shows sweepout for slug sizes of 2.5, 7, and 17% PV. Sweepout is also shown on this figure for two continuous solvent injection displacements at mobility ratios characteristic of the leading and trailing edges of the solvent banks. In addition, sweepout is given for an

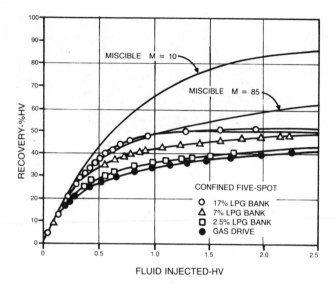

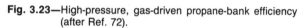

Fig. 3.23—High-pressure, gas-driven propane-bank efficiency (after Ref. 72).

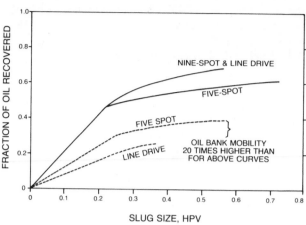

Fig. 3.24—Oil recovery up to 98% water cut as a function of well pattern, oil bank mobility, and slug size (after Ref. 74).

immiscible gas drive at a mobility ratio of 85.

Although the dispersion was not scaled exactly, the experiments are probably sufficiently scaled to reflect in a semiquantitative manner the influence of viscous fingering and slug dilution on displacements with small solvent slugs in homogeneous porous media with negligible gravity override. Sweepout ultimately reached a level as high as 85% for continuous solvent injection at $M=10$ and reached a value as high as 62% for continuous solvent injection at $M=85$. However, the highest sweepout in the slug displacements was only slightly higher than 50% for the 17% PV slug and was as low as 43% for the 2.5% PV slug, illustrating the effect of loss of miscibility both by slug dilution and by immiscible drive gas breaching the slug.

At 1 HCPV of total fluid injection for all slugs, the incremental oil recovery over that achieved by immiscible gas drive was approximately equal to the volume of the slug injected. For greater values of total fluid injected, although the 17% PV slug recovered more incremental oil than the 7% PV slug, it appears to be less efficient than the 7% PV slug in terms of incremental recovery per volume of solvent injected.

Tertiary Recovery

The only tertiary recovery sweepout data were published by Claridge[74] for experiments in a trapping Hele-Shaw model. He examined the influence of well pattern and slug size on recovery of tertiary oil when the solvent slug was driven immiscibly with water and a highly favorable mobility ratio existed at the water/slug front. Two situations were examined: one where the mobility ratio of tertiary oil bank to waterflooded region was 2.2 (unfavorable) and the mobility ratio of solvent slug to oil bank was 49 (unfavorable), the other where the mobility ratio of tertiary oil bank to waterflooded region was 0.12 (favorable) and the mobility ratio of solvent slug to oil bank was 25 (unfavorable).

Sweepout for the Claridge experiments is shown in

Fig. 3.24. Although strictly speaking, results are valid only for the conditions studied, a number of important observations can be made. The mobility of the solvent and crude oil relative to that of the floodwater strongly affected both the degree of fingering of oil bank through the waterflooded region and the resulting areal sweepout. When the tertiary oil bank mobility was higher than the mobility of the waterflooded region, the oil bank fingered into the water while, at the same time, solvent fingered into the oil bank. The well pattern had a pronounced effect. Nine-spot and line-drive patterns were better than a five-spot pattern when mobility ratio at the leading edge of the oil bank was favorable, but the five-spot pattern performed better than a line drive when mobility ratio was unfavorable at the leading edge of the oil bank.

For this single-layer flood with no gravity segregation, recovery of waterflood residual oil increased with increasing solvent slug size, reaching about 57% with a 50% HCPV solvent slug when oil bank mobility in the five-spot was favorable, but reaching only 38% for the same slug size when oil bank mobility was unfavorable. The ratio of slug injected/oil recovered was about constant up to a slug size of approximately 23% HCPV, after which the ratio increased continuously with increasing slug size.

Photographs showed that the drive water advanced with a much smoother and more continuous front than that of the solvent, although the shape of the water front was affected to some extent by the unfavorable mobility-ratio solvent/oil displacement that preceded it. Some water cusped with solvent into the producing well along the shortest streamlines between injector and producer, resulting in relatively early water breakthrough, but subsequent water production was moderate until the main drive-water front arrived.

Experiments in the nine-spot model showed that ultimate recovery efficiency was increased by shutting in the well where solvent breakthrough occurred first until solvent had broken through in all wells. When the initial breakthrough well continued producing and was not shut in, the mobile solvent tended to flow preferentially to this well, resulting in a slowdown or stopping of oil bank movement to the other wells.

3.6 Vertical Sweepout

Vertical sweepout in miscible displacement is influenced primarily by gravity and by permeability stratification. Solvents usually are less dense than either oil or brine, although CO_2 may be more dense than some oils, depending on reservoir temperature and pressure. Drive gases such as hydrocarbon gas or flue gas are less dense than either the miscible solvent, reservoir oil, or brine. Because of these density differences, solvents and drive gases may segregate and override the other reservoir fluids. In horizontal floods, gravity segregation is usually detrimental, decreasing vertical sweepout and oil recovery. Exceptions may occur in some stratified reservoirs. Occasionally, gravity can be used to advantage in dipping reservoirs to improve sweepout and displacement efficiency.

The effect of permeability stratification on vertical sweepout is well known from waterflooding.[1] The unfavorable mobility ratio in miscible flooding, however, aggravates the normal tendency of injected fluids to enter the most permeable strata. Transverse dispersion of solvent slugs in communicating strata, causing slug dilution and loss of miscibility, is another factor affecting vertical sweepout, as is crossflow of solvent and oil bank between strata.

In this section, the factors affecting vertical sweepout in both horizontal and dipping reservoirs are discussed.

Influence of Gravity Forces
in Horizontal Reservoirs, Secondary Recovery

Craig *et al.*[1,26] measured vertical sweepout at breakthrough in homogeneous and isotropic cross-sectional laboratory models for the flow regimes defined as Regions I and II in Sec. 3.3. These experiments simulated secondary recovery by continuous solvent injection. Craig *et al.* reported that the experiments were carried out under conditions where the effects of transverse dispersion were unimportant. Results are correlated in Fig. 3.25 with viscous/gravity force ratio and mobility ratio. These experiments show that when vertical permeability is not restricted, gravity segregation can cause low vertical sweepout at breakthrough, 30% or less, for typical field values of viscous/gravity ratio and for mobility ratios typical of miscible displacement. Unfavorable mobility ratios act to accelerate the growth of the gravity tongue and to reduce breakthrough sweepout, while larger viscous forces result in improved sweepout at breakthrough.

When vertical permeability is restricted—i.e., when the ratio of vertical to horizontal permeability is less than one, vertical sweepout is higher than would be the case for isotropic permeability since restricted vertical permeability retards solvent segregation. Vertical sweepout for this type situation has been studied by several investigators with mathematical models.[75-77] Spivak[77] published a correlation for vertical sweepout at breakthrough for immiscible displacement and accounted for

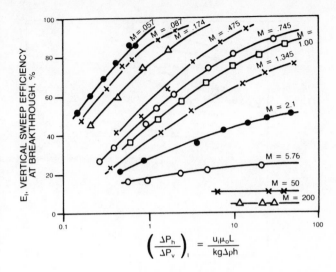

Fig. 3.25—Vertical sweep efficiency at breakthrough, linear uniform systems (after Ref. 1).

$$\left(\frac{\Delta p_h}{\Delta p_v}\right)_1 = 2,050 \frac{u_i(\text{B/D-sq ft})\mu_o(\text{cp})L(\text{ft})}{k(\text{md})\Delta\rho(\text{g/cm}^3)h(\text{ft})}.$$

the influence of vertical permeability with a $\sqrt{k_v k_H}$ term. Other authors used $\sqrt{k_v/k_H}$[76] and k_v/k_H[75] to characterize the effect of vertical permeability. As a first approximation, when vertical permeability is not equal to horizontal permeability, its effect on vertical sweepout can be estimated by substituting $\sqrt{k_v k_H}$ for k_x in calculating the viscous/gravity ratio for Fig. 3.25.

There have been several theoretical treatments of vertical sweepout in homogeneous and isotropic linear cross-sectional flow.[78-80] Both Hawthorne[79] and Dietz[78] derived equations for immiscible gas and water drives characterized by a single gravity tongue that showed that vertical sweepout generally was related to the viscous/gravity ratio and to mobility ratio. With suitable modifications, the derivations are also appropriate for miscible displacement in the limiting case of no transverse dispersion. These equations show that when the viscous/gravity ratio becomes sufficiently large, vertical sweepout depends on mobility ratio only (Region II). In these circumstances, each point on the solvent/oil interface has its own constant velocity. Solvent arrives at the outlet when $1/M$ displaceable volumes have been injected (M=mobility ratio), and all of the displaced fluid is produced after injection of M displaceable volumes of solvent. Between these limits, the number of displaceable solvent volumes, D_v, required to achieve a given sweepout E_I is related to mobility ratio by[24]

$$E_I = \frac{1}{M-1}[2(M \cdot D_v)^{1/2} - 1 - D_v]. \qquad \ldots\ldots\ldots (3.29)$$

Equation 3.29 is useful for rough estimates of vertical sweepout vs. solvent throughput when the flow regime is in Region II. However, the equation predicts a breakthrough sweepout that is low compared with Fig. 3.25 and compared with Pozzi and Blackwell's

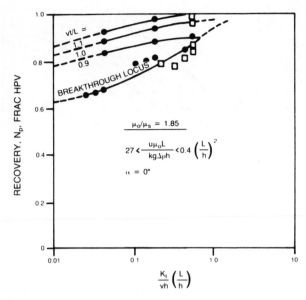

Fig. 3.26—Vertical sweep efficiency (N_P) correlation for a viscosity ratio of 1.85 when transverse mixing is by molecular diffusion (after Ref. 15).

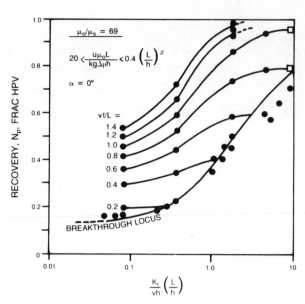

Fig. 3.28—Vertical sweep efficiency (N_P) correlation for a viscosity ratio of 69 when transverse mixing is by molecular diffusion (after Ref. 15).

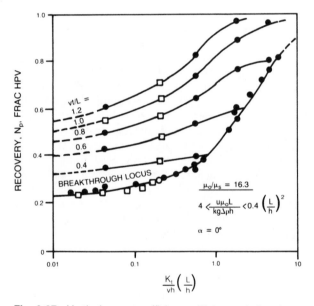

Fig. 3.27—Vertical sweep efficiency (N_P) correlation for a viscosity ratio of 16.3 when transverse mixing is by molecular diffusion (after Ref. 15).

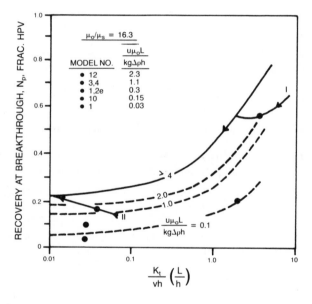

Fig. 3.29—Vertical sweep efficiency at breakthrough depends on viscous/gravity ratio when its value is less than C_L; transverse mixing is by molecular diffusion; recovery can either decrease (arrowed Path I) or decrease (arrowed Path II) with increasing injection rate in specific models (after Ref. 15).

breakthrough data at low values of dimensionless transverse dispersion.[15] For a given value of D_v, it also predicts lower sweepout after breakthrough than observed in experiments.[15,24,*]

Transverse dispersion also can be a factor affecting vertical sweepout. Solvent/oil mixing caused by transverse dispersion moderates density and viscosity contrasts, and, if sufficiently great, can increase sweepout. This is illustrated in Figs. 3.26 through 3.29, which show results of experiments conducted by Pozzi

and Blackwell for linear cross-sectional flow in homogeneous and isotropic models.[15] Figs. 3.26 through 3.28 are valid when the flow regime is in Region II (E_I is independent of viscous/gravity ratio) and show vertical sweepout at breakthrough at various values of displaceable volumes injected as a function of the dimensionless transverse dispersion group $(K_tL)/(vh^2)$. The figures are for mobility ratios of 1.85, 16.3, and 69. Fig. 3.29 is valid for Region I flow and shows breakthrough sweepout vs. dimensionless transverse dispersion for a mobility ratio of 16.3 and at several values of the viscous/gravity ratio.

*White, G.L.: unpublished report, ARCO Oil & Gas Co. (1979).

These figures show that vertical sweepout increases with increasing values of the dimensionless transverse dispersion group. However, for most well spacings, transverse dispersion does not influence vertical sweepout appreciably unless the reservoir or stratum being flooded is relatively thin. For example, if $K_t = 5 \times 10^{-3}$ sq ft/D (molecular diffusion), $v = 0.5$ ft/D, $M = 16.3$, and $L = 1,320$ ft, the reservoir has to be less than about 10 ft thick for transverse dispersion to begin to have a pronounced effect on vertical sweepout.

Craig *et al.* also reported limited data for breakthrough sweepout in five-spot models.[1,26] These are shown in Fig. 3.30 for mobility ratios up to 1.85. For these low mobility ratios, Fig. 3.30 shows that five-spot volumetric sweepout at breakthrough increases with increasing viscous/gravity ratio up to a value of at least 10. Region I flow in the vertical cross section is indicated over this range of viscous/gravity ratio, which is consistent with the findings of Fig. 3.25.

Five-spot sweepout at higher mobility ratios than reported in Fig. 3.30 and for various displaceable volumes of injection after breakthrough can be estimated from the cross-sectional vertical sweepout data in Figs. 3.25 through 3.29 and from five-spot areal sweepout data by applying the formula $E_v = E_A E_I$. Some unpublished five-spot model experiments at adverse mobility ratios up to 10 and where cross-sectional flow was in Regions I and II showed favorable areal sweepout with little or no areal viscous fingering.* Areal sweepout increased to near 100% after only 0.2 to 0.4 displaceable volumes of injection after breakthrough. The segregating and solvent spreading effect of gravity apparently prevented appreciable viscous fingering areally. These unpublished experiments indicate that areal sweepout data dominated by viscous fingering, such as shown in Fig. 3.19, are too severe for calculating volumetric sweepout under gravity-dominated flow. Instead, areal sweepout in these three-dimensional model experiments seemed to be described approximately by unit mobility ratio data for areal models.

In estimating volumetric sweepout by the $E_v = E_A E_I$ formula, a problem arises in selecting the appropriate linear velocity for calculating E_I from Figs. 3.25 through 3.29. Flow velocities in a five-spot vary significantly from near the wellbore to midway between wells, and there are no published guidelines for calculating a viscous/gravity ratio that best characterizes vertical sweepout. Until better guidelines are available, a suggested working assumption is to multiply the five-spot viscous/gravity ratio calculated by the formula of Fig. 3.30 by a factor of five. This will give a very approximate estimate of the viscous/gravity ratio to use with Figs. 3.25 through 3.29. The factor of five was estimated by comparing breakthrough sweepout calculated by the $E_v = E_A E_I$ method with experimental breakthrough sweepout from Fig. 3.30.** The reader is cautioned that this approximation could be substantially in error.

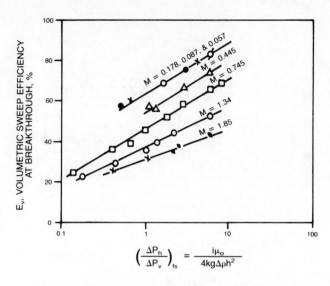

Fig. 3.30—Volumetric sweep efficiency at breakthrough, five-spot uniform systems (after Ref. 1).

$$\left(\frac{\Delta p_h}{\Delta p_v}\right)_{fs} = 512 \frac{i(\text{B/D-well})\mu_o(\text{cp})}{k(\text{md})\Delta\rho(\text{g/cm}^3)h^2(\text{sq ft})}.$$

Influence of Gravity Forces in Horizontal Reservoirs, Tertiary Recovery

There are no published vertical sweepout data for tertiary recovery miscible flooding similar to Figs. 3.25 through 3.29. Miller[76] and Warner[75] have studied gravity segregation for this condition with reservoir simulators, and Miller published a correlation identifying conditions where vertical sweepout was dominated by a gravity tongue or by viscous fingering.

In tertiary recovery, water is mobile at the start of solvent injection and affects the viscous/gravity ratio as well as the overall mobility ratio. A rough estimate of vertical sweepout can be made from Figs. 3.25 through 3.29 by substituting

$$\frac{1}{\sqrt{k_H k_v}\left(\dfrac{k_{roi}}{\mu_o} + \dfrac{k_{rwi}}{\mu_w}\right)}$$

for μ_o/k_x in calculating the viscous/gravity ratio and by taking $\Delta\rho$ to be the density difference between solvent and water.[76] In the above expression, k_{roi} and k_{rwi} are the oil and water relative permeabilities at the start of solvent injection.

Overall mobility ratio is affected both by the mobility of the tertiary oil bank and by the initial fluid mobilities at the start of solvent injection. A rough approximation for effective mobility ratio of the displacement neglects the tertiary oil bank mobility and assumes mobility ratio to be $(k_{rs}/\mu_s) \div [(k_{roi}/\mu_o) + (k_{rwi}/\mu_w)]$. The procedure just discussed for estimating vertical sweepout in tertiary recovery flooding should be used for *rough estimates* only. It is unsubstantiated by experimental data and could be considerably off the mark.

In tertiary recovery flooding, both oil and water are displaced from the reservoir volume swept by solvent. Actual oil recovery can be less than the oil displaced from the swept volume. Reservoir simulator calculations

---

*White, G.L.: unpublished report, ARCO Oil & Gas Co. (1979).

**This procedure assumes that E_A is estimated for a unit mobility ratio.

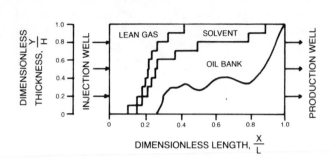

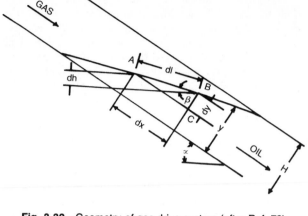

Fig. 3.31—Computed distribution of miscible fluids and tertiary oil bank in a gravity-dominated displacement (after Ref. 76).

Fig. 3.32—Geometry of gas-drive system (after Ref. 79).

by several investigators[75,76,81] show that when the displacement is strongly gravity-controlled, much of the oil displaced by the advancing gravity tongue of solvent is pushed down into the center of the cross section, except for a small tertiary oil bank immediately ahead of the solvent tongue. This situation is shown in Fig. 3.31. Solvent breakthrough occurs shortly after breakthrough of the tertiary oil bank, and the bulk of miscibly displaced oil is produced concurrently with solvent. As vertical sweepout continues to increase with continued solvent injection, oil that crossflowed toward the middle and bottom of the reservoir cross section is produced gradually. Stopping solvent injection and switching to water injection forces some of the remaining crossflowed oil bank back into the solvent-swept region and resaturates this previously miscibly swept volume.

The discussion thus far has been limited to situations where solvent only is injected. Miller[76] also made reservoir simulator calculations of situations where water is injected with the solvent to reduce solvent mobility (see discussion in Sec. 3.8). He found that gravity affected vertical sweepout in two ways. Density differences caused override of the oil and water by solvent. In addition, gravity counterflow segregation occurred between injected water and the hydrocarbon phase behind the solvent front. Miller found that if the gravity tongue were caused primarily by counterflow segregation, an increase in reservoir thickness led to better performance since the fluids had a longer distance over which to segregate relative to the well spacing. This is contrasted with displacements, such as those shown in Fig. 3.25, where the gravity tongue is formed primarily by the density difference between two miscible fluids rather than by segregation from water. In this latter situation, an increased reservoir thickness has a detrimental effect on the displacement since there is a greater potential gradient between injection and producing wells at the top of the reservoir than at the bottom.

Influence of Gravity in Dipping Reservoirs

In some reservoirs with dip, gravity can be used to advantage to improve sweepout and oil recovery. This is achieved by injecting the solvent updip and producing

the reservoir at a rate low enough for gravity to keep the less dense solvent segregated from the oil, suppressing fingers of solvent as they try to form. For such a gravity-stabilized displacement the interface assumes an angle of tilt, β, relative to the horizontal as shown in Fig. 3.32. The angle of tilt can be calculated from the following equation.[79]

$$\frac{\sin\beta}{\cos(\alpha-\beta)} = \frac{22.8u\left(\dfrac{\mu_o}{k_o} - \dfrac{\mu_s}{k_s}\right)}{(\rho_o - \rho_s)}, \quad \dots\dots\dots (3.30)$$

where

$$
\begin{aligned}
u &= \text{Darcy velocity, ft/day,} \\
k_o, k_s &= \text{oil and solvent permeabilities, darcies,} \\
\rho_o, \rho_s &= \text{oil and solvent densities, lbm/cu ft,} \\
\mu_o, \mu_s &= \text{oil and solvent viscosities, cp, and} \\
\alpha &= \text{angle of dip relative to the horizontal.}
\end{aligned}
$$

The maximum flow velocity for a completely stable displacement is given by the following equation according to Dumoré.[82]

$$u_{st} = 0.0439\left(\frac{d\rho}{d\mu}\right)_{min} k\sin\alpha, \quad \dots\dots\dots\dots (3.31)$$

where $(d\rho/d\mu)_{min}$ is the minimum as a function of solvent concentration of the derivative of solvent/oil mixture density with respect to solvent/oil mixture viscosity. For dynamic miscibility, the viscosities established in the solvent/oil transition zone by the multicontact mechanism are the appropriate viscosities for Eq. 3.31. In this equation, solvent and oil permeabilities are assumed equal.

The velocity, u_{st}, calculated by Eq. 3.31 is called the "stable rate." Below this velocity, the miscible displacement is completely stable throughout the transition zone of compositions ranging from pure oil to pure solvent composition.

Another criterion used for the design of gravity-stable miscible floods is the critical rate, u_c, given by Eq. 3.32.[78,83]

$$u_c = \frac{0.0439(\rho_o - \rho_s)\sin\alpha}{\left(\dfrac{\mu_o}{k_o} - \dfrac{\mu_s}{k_s}\right)} . \quad\ldots\ldots\ldots\ldots (3.32)$$

This equation is derived by assuming a sharp interface between oil and solvent—i.e., a negligibly small transition zone.

Eq. 3.31 is the more stringent criterion. If the actual displacement velocity, u, is greater than u_c—i.e., $u > u_c$—the displacement is unstable and solvent fingers into the oil. If $u < u_{st}$, the displacement is completely stable. However, if $u_{st} < u < u_c$, a portion of the transition zone is unstable and fingers into the oil. At this latter condition there is no instability causing the pure solvent to finger into the oil because the displacement is below the critical rate, and an efficient displacement of oil by the undiluted solvent still results for continuous solvent injection. However, the fingering within a portion of the solvent/oil transition zone causes more rapid dilution of a solvent slug by mixing than would be the case with a completely stable displacement.

If the one-quarter-power blending rule describes solvent/oil viscosities* and if linear blending describes solvent/oil densities,** the ratio of stable rate to critical rate is given by the following equation.

$$\frac{u_{st}}{u_c} = \frac{\left(1 - \dfrac{1}{M}\right)}{4(M^{1/4} - 1)} , \quad\ldots\ldots\ldots\ldots\ldots (3.33)$$

where M is the nominal mobility ratio. For a mobility ratio of five, this equation shows the stable rate to be 0.4 times the critical rate, whereas for a mobility ratio of 100 the stable rate is only 0.115 times the critical rate.

If water is used to drive a solvent slug in a dipping reservoir, being the denser phase it underruns the solvent and may trap oil by flowing ahead of the solvent. To prevent this, a stable solvent/water front can be maintained by injecting water at a rate that is greater than the rate calculated by Eq. 3.32 when water density, viscosity, and permeability are substituted for these properties of oil.[78,84] Since water immiscibly displaces solvent, the appropriate permeability to water is evaluated at the residual solvent saturation. If oil viscosity is more than two or three times water viscosity, the minimum rate calculated for the water/solvent front may exceed the critical rate calculated for the solvent/oil front.

The effectiveness of gravity segregation in improving displacement efficiency decreases rapidly after the displacement rate exceeds the critical rate. In reservoirs with low permeability and dip, the critical rate often is too low to be practical.

Hawthorne proposed a method for estimating vertical sweepout in dipping reservoirs when the displacement is not gravity stabilized.[79] His theory is limited to situations where the displacement is characterized by a single gravity tongue and where transverse dispersion is negligible resulting in a sharp solvent/oil interface. The reservoir was assumed to be homogeneous and isotropic, at irreducible water saturation, and with the interface initially horizontal. In this situation the angle β (Fig. 3.32) varies from point to point along the interface. The expression $\sin\beta/\cos(\alpha - \beta)$ is equivalent to $[\sin\alpha - (dy/dx)\cos\alpha]$, where x is in the direction of flow and y is normal to the direction of flow. Solvent fractional flow through any plane perpendicular to the direction of flow is calculated from

$$f_s = \frac{\left[1 - 0.0439\dfrac{k_o A}{\mu_o q_t}(\rho_o - \rho_s)(\sin\alpha - \dfrac{dy}{dx}\cos\alpha)\right]}{1 + \left(\dfrac{\mu_s}{\mu_o}\right)\left(\dfrac{k_o}{k_s}\right)\left(\dfrac{y}{1-y}\right)} ,$$

$$\ldots\ldots\ldots\ldots\ldots\ldots\ldots (3.34)$$

where A is the area normal to flow (sq ft), and q_t is the total flow rate (cu ft/D). The advance of any point on the interface is calculated from

$$(\Delta x)_y = \frac{q_t}{A}\left(\frac{df_s}{dS_s}\right)_y \Delta t, \quad\ldots\ldots\ldots\ldots (3.35)$$

where

$$\left(\frac{df_s}{dS_s}\right)_y = \frac{df_s}{dy} \cdot \frac{1}{\left(\dfrac{dS_s}{dy}\right)_y} \quad\ldots\ldots\ldots\ldots (3.36)$$

and

$$S_s = \left(1 - \frac{y}{H}\right). \quad\ldots\ldots\ldots\ldots\ldots (3.37)$$

There is no analytical solution available for Eqs. 3.34 through 3.37. The advance of the fluid interface at any instant of time depends on the shape of the interface at that time. Hawthorne used a finite-difference scheme for evaluating dy/dx and df_s/dy at a large number of points on the interface and calculated the frontal advance for small time increments.

Influence of Permeability Stratification

In stratified reservoirs, injection fluids preferentially enter the strata of higher permeability. As a result, breakthrough into producing wells from these strata will occur before other strata of lower permeability are swept completely. If sufficient injection fluid is injected to sweep the less permeable strata, some fraction must flow

---

$*\mu = \left[\dfrac{C_s}{\mu_s^{1/4}} + \dfrac{(1-C_s)}{\mu_o^{1/4}}\right]^{-4}$, where C_s = solvent volume fraction.

$**\rho = C_s\rho_s + (1-c_s)\rho_o$, where C_s = solvent volume fraction.

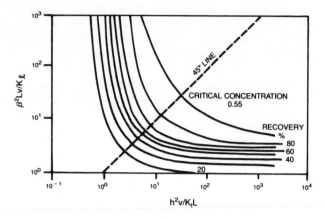

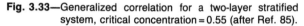

Fig. 3.33—Generalized correlation for a two-layer stratified system, critical concentration = 0.55 (after Ref. 85).

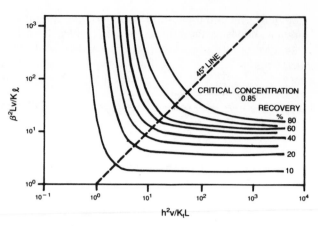

Fig. 3.34—Generalized correlation for a two-layer stratified system, critical concentration = 0.85 (after Ref. 85).

nonproductively through the higher permeability strata that already have been swept. The effect of stratification is compounded in miscible flooding, where unfavorable mobility ratios usually prevail and small solvent slugs may be injected. At an unfavorable mobility ratio, injection fluids flow through the higher permeability strata more disproportionately than would be indicated by the permeability-thickness product alone. In this situation, the technique of water injection with the solvent, discussed in Sec. 3.8, is useful to promote more uniform vertical entry of the solvent.

The effect of permeability stratification on the vertical sweepout of injection fluids has been studied extensively, and methods for characterizing stratification and accounting for its influence in waterflooding are summarized by Craig.[1] Predictive methods for miscible floods are discussed in the next chapter.

Another effect of permeability stratification is that much larger slugs than the average value enter the more permeable strata and much smaller slugs than the average value enter the least permeable layers. The small slugs may be diluted below miscible composition so quickly by transverse and longitudinal dispersion that little benefit of miscible displacement in these strata may be realized.

Calculations by Koonce and Blackwell[85] show the potentially significant effect that transverse dispersion may have on solvent slug dilution, loss of miscibility, and vertical miscible sweepout in stratified reservoirs. They solved the diffusion-convection equation numerically for a two-layer reservoir using the method of characteristics.[86] One layer had 10 times the permeability of the other layer, so the solvent slug entered the more permeable layer predominantly. Figs. 3.33 and 3.34 illustrate some of the results of these computations. In these figures, β is the slug size expressed as a fraction of the pore volume of the most permeable layer, H is the thickness of this layer, v is the interstitial velocity in this layer, and K_ℓ and K_t are the longitudinal and transverse dispersion coefficients. The critical concentration is the concentration to which the slug can be diluted before entering the two-phase region, and the recovery refers to the permeable layer of length L. Koonce and Blackwell reported that Figs. 3.33 and 3.34 are valid as long as the

permeability ratio of the layers is greater than three. Calculations for other critical concentrations are given in Ref. 86.

These two figures reveal that fairly large slug sizes could be required in stratified reservoirs with communication between strata when well spacing is large and thicknesses of the permeable layers are small. Although initially the slug size (expressed as fractional pore volume) required to achieve a given recovery decreases with system length, eventually the slug size requirement begins to increase with system length because of transverse dispersion. For example, for a critical concentration of 0.85, a 7% PV slug was needed for 75% recovery in a 5-ft stratum, 100 ft in length. For a 200-ft length, a 5.3% PV slug was required for the same recovery level; but for a 660-ft length, an 11% PV slug was necessary.

Although these calculations were made for stratified flow, they also indicate the importance of transverse dispersion as a mechanism causing loss of slug miscibility when solvent flow is in viscous fingers or a gravity tongue.

Crossflow of solvent and oil between strata, caused by differences in flow potential, is another aspect of miscible flooding in stratified reservoirs that can affect the sweepout of injection fluids as well as affect the recovery of oil that has been miscibly displaced. The importance of location of the permeable strata on vertical sweepout and oil recovery in reservoirs where crossflow between strata can occur was illustrated by calculations made by Warner[75] with a numerical miscible-flood reservoir simulation model.

In Warner's study, miscible flooding in a waterflooded reservoir was simulated. Gravity segregation between the solvent slug and water was an important mechanism. Higher oil recovery was computed for a stratified reservoir with communicating layers than for a homogeneous reservoir, when a high-permeability layer was located at the middle or bottom of the reservoir model. Decreased recovery was observed with the permeable layer at the top. Results of these calculations are shown in Table 3.2.

Other studies have been published, both experimental and mathematical, concerning the effect of crossflow between layers on vertical sweepout in waterflooding and miscible flooding.[87-90,148] Several of these investigations found that at unfavorable mobility ratios crossflow between communicating layers decreased vertical sweep efficiency compared with a noncommunicating layer case, whereas for favorable mobility ratios, the reverse was true.

Still another effect of stratification that must be taken into account is its influence on capture efficiency—i.e., recovery of oil that is displaced by the miscible solvent. In secondary recovery floods where a free gas saturation is present, oil displaced in permeable strata may crossflow into and resaturate less permeable strata. A similar phenomenon may occur in tertiary recovery flooding when the less permeable strata have been swept by prolonged water injection. In this case, crossflow of the tertiary oil bank into these strata may result in a loss of miscibly displaced oil if a sufficiently large solvent slug is not injected to flood out these less permeable strata.

3.7 Unit Displacement Efficiency

Under some conditions all the oil may not be displaced from a given volume of rock even though the solvent composition and pressure are sufficient for miscibility and even though the rock has been swept by solvent. This undisplaced oil left in the solvent-swept rock is analogous to the microscopic residual oil left after waterflooding. In addition to subtracting from total oil recovery, the residual phase may decrease solvent injectivity below the rate that would be anticipated for complete oil displacement away from the injection well. A miscible flood residual oil saturation also may reduce solvent relative permeability and cause a less adverse mobility ratio than would have resulted from complete oil displacement by the solvent.

The primary causes for a miscible-flood residual oil saturation are (1) trapping of oil by mobile water at water saturations above the irreducible value, (2) bypassing of oil located in dead-end pores and low-permeability occlusions that are not flushed by solvent, and (3) precipitation of hydrocarbons as a result of mixing into multiphase regions. These mechanisms and the magnitude of residual oil saturation resulting from them are discussed in this section.

Effect of a Mobile Water Saturation

Water saturations that are greater than the initial irreducible water saturation may be present at the solvent/oil front for two reasons: (1) the tertiary oil bank* that is formed when a previously waterflooded reservoir is miscibly flooded will not drive the water saturation completely back to its irreducible value and (2) water injected with the solvent for mobility reduction may overtake the solvent/oil front. Aspects of miscible displacement in the presence of mobile water have been reported in a number of publications.[92-98] From these research findings it is apparent that under some conditions part of the oil saturation can be immobilized or trapped by mobile water even though the remaining oil is miscibly

*As discussed in Sec. 3.8, a significant solvent-free tertiary oil bank may not form because of viscous fingering and dispersion, which may cause solvent to penetrate appreciably into the displaced waterflood residual oil. In this sentence "tertiary oil bank" is used in a general sense to describe the displaced oil, whether solvent-free or mixed with solvent.

TABLE 3.2—COMPUTED INFLUENCE OF RESERVOIR DESCRIPTION ON OIL RECOVERY IN A DISPLACEMENT WHERE GRAVITY SEGREGATION IS IMPORTANT
(Process: 25% HCPV CO_2 followed by continuous water injection; calculations for contrast in layer permeabilities = 4:1; $k_v/k_H = 0.1$)

Reservoir Description	Oil Recovery in Homogeneous Reservoir
homogeneous	1.0
high-permeability layer at top	0.9
high-permeability layer in middle	1.9
high-permeability layer at bottom	2.2

displaced in a more or less normal manner. The degree of trapping is very dependent on rock wettability.

Figs. 3.35 and 3.36 show the magnitude of oil trapped by mobile water in several strongly water-wet laboratory sandstones. Mobile water saturations in these sandstones were established by simultaneous injection of laboratory oils and water, and first-contact miscible displacements were conducted by the simultaneous injection of solvent and water at the same ratio. Fig. 3.35 correlates the trapped oil with water saturation and expresses trapped oil as a fraction of the irreducible waterflood residual oil saturation.[94] The fraction of waterflood residual oil left undisplaced in these Boise, Torpedo, and Berea sandstones ranged from 0.2 at a 40% PV water saturation to 0.8 at a water saturation close to 60% PV. The irreducible water saturation in these cores ranged from 16 to 21% PV.

Fig. 3.36 correlates the fraction of undisplaced waterflood residual oil with the fractional permeability of water.[97] The data for consolidated Berea sandstone show an almost linear increase in trapped oil as the fractional water permeability increases from zero (irreducible water saturation) to one (waterflood residual oil saturation). The correlation in Fig. 3.36 was found to hold whether water saturation was increased from the irreducible value or was decreased from its value at waterflood residual oil.

Shelton and Schneider[99] proposed a theory relating the degree of oil trapping to the relative permeability characteristics of the rock. This theory postulates that the magnitude of the trapped oil saturation for a given oil relative permeability is approximated by the saturation difference between the imbibition oil relative permeability curve and the drainage oil relative permeability curve starting at 100% water saturation. Consider, for example, the relative permeability curves shown in Fig. 3.37. Oil relative permeability, Curve I, is a drainage curve and was measured as oil saturation was increased from a starting point at which the core was completely water-filled. Curve II is an imbibition curve and was measured as oil saturation was decreased from a starting point at the irreducible water saturation. At a constant oil relative permeability, the oil saturation on the imbibition curve is greater than the saturation on the drainage curve. Yet both saturations have the same relative permeability. According to Shelton and Schneider, the saturation difference between these two curves approximates the oil saturation that is immobile and does not contribute to flow.

If oil saturation were increased from the waterflood

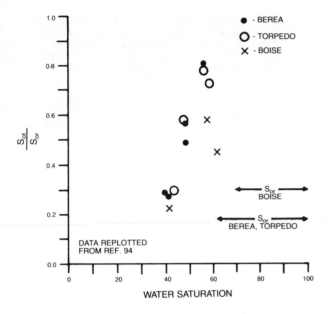

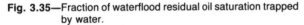

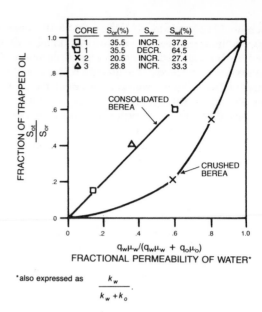

Fig. 3.35—Fraction of waterflood residual oil saturation trapped by water.

Fig. 3.36—Correlation of oil trapped on imbibition with the fractional permeability of water (after Ref. 97).

residual oil value, as would occur in tertiary recovery flooding, there could be some hysteresis in oil relative permeabilities such that the imbibition Curve II was not exactly retraced. However, the resulting new drainage curve still would differ appreciably from the original Drainage Curve I, and the trapped oil saturation according to the Shelton and Schneider theory would be the difference in saturation between these two curves.

There are only limited published data that can be used to test the theory of Shelton and Schneider, but the available data indicate that the oil trapped under imbibition conditions is about the same as or some less than the saturation difference between imbibition and drainage relative permeability curves. The trapped oil saturations measured by Shelton and Schneider in strongly water-wet Berea sandstone were slightly less than the saturation difference between imbibition and drainage curves. Trapped saturations measured by Raimondi and Torcaso[97] were about equal to the saturation difference between the imbibition and drainage curves when oil saturation was only 5 to 6% PV above waterflood residual saturation; however, the trapped oil saturation was less than half of the difference between the relative permeability curves when the oil saturation was as much as 10% PV higher than waterflood residual oil saturation.

Although most of the data on oil trapping have been taken for water-wet rocks, the theory of Shelton and Schneider and some experiments[94,96,98] suggest that oil trapping may not be as severe for weakly water- or oil-wet rocks. For oil-wet rocks, there is much less hysteresis between the oil drainage and imbibition relative permeability curves. The hysteresis occurs instead in the water relative permeability, since water is the nonwetting phase. Weakly water-wet rocks exhibit

behavior intermediate between the strongly water- and oil-wet cases. Most reservoir rocks are expected to be much less strongly water-wet than the laboratory sandstones for which data are given in Figs. 3.35 and 3.36. Several studies are reported for preserved reservoir rocks that have found little or no trapping of oil caused by mobile water, at least for the water saturations investigated, perhaps because of the mixed wettability of these systems.[94,100-103]

In evaluating the severity of oil trapping for any particular reservoir rock, both the wettability and the pore geometry of the rock need to be accounted for, since both of these factors influence the shape and separation of the imbibition and drainage relative permeability curves. The potential severity of oil trapping is best determined from displacement experiments in samples of preserved wettability reservoir rock. If preserved wettability rock is not available, oil trapping should be determined in reservoir rock in which an attempt has been made to restore reservoir wettability by aging with crude oil.[104] However, aging with crude oil may not completely restore the rock to its original wettability.

Effect of Dead-End Pore Volume and Low-Permeability Occlusions

Oil can be bypassed by miscible solvents because of dead-end pore structure and because of microscopic-to-macroscopic permeability heterogeneities. Several publications have shown the effect of laboratory core-scale heterogeneities on oil bypassing in carbonate cores.[102,105,106] There are no data published so far showing a significant effect of this sort in sandstones.

Bypassed oil saturations ranging from 0.13 to 25% PV were reported for a series of cores from one carbonate reservoir.[102] This study also reported that the amount of bypassed oil appeared to be related qualitatively to pore-size distribution. As the pore-size distribution on a volume basis broadened, the amount of bypassed oil increased. No relationship was observed between the average pore radius and the amount of bypassed oil.

There seems to be little effect of velocity on the amount of bypassed oil. Baker[107] observed essentially no effect in his core floods, while Spence and Watkins[102] found a small decrease in bypassed oil saturation with decreasing displacement velocity.

Effect of Phase Behavior—Fluid Mixing

Residual oil left to miscible flooding also can be caused by precipitation as fluids mix into multiphase regions of the phase diagram. For miscible processes such as the vaporizing-gas drive process or the CO_2 process, solvent and oil are not first-contact miscible but achieve miscibility by solution of solvent into the oil and by vaporization or extraction of hydrocarbon components from the oil into the solvent. As described in Chap. 2, some oil necessarily is left behind the solvent front in these processes, even for simple three-component systems, as a result of transition zone fluids being mixed into the multiphase region by dispersion.

For the condensing-gas drive process, the mechanism by which miscibility is generated and maintained does not leave a residual oil saturation for systems where the phase behavior is characterized by simple two-phase vapor/liquid equilibrium. Precipitation of a residual phase can occur, however, if phase behavior should depart from the simple vapor-liquid type.[17,98]

The magnitude of the residual oil saturation caused by liquid precipitation in CO_2 floods and vaporizing-gas drives depends on the degree of fluid mixing during the displacement and on the size of the multiphase region. These factors control how deeply the profile of overall compositions cuts into the multiphase region. Fluid mixing is caused by micro- and macroscopic transverse and longitudinal dispersion. It also is caused by crossflow of fluids into and out of solvent fingers and by crossflow of oil and solvent between communicating strata. As the factors causing mixing become more severe, the multiphase region is entered more deeply by the overall fluid composition and liquid precipitation increases. For a given reservoir temperature, the size of the multiphase region is affected by pressure. Decreasing pressure increases the size of the multiphase region and increases liquid precipitation.

Blackwell and Koonce's scaled model studies indicate that fluid mixing in a condensing-gas drive also could be caused by capillary forces between the rich gas in forward-projecting fingers and oil in backward-projecting fingers.[109] Although mixing by this mechanism could increase precipitation of asphaltenes in a rich gas/oil system of complex phase behavior, any oil mixed into a gas finger would soon become miscible with the rich gas because of the multiple-contacting mechanism of the process.

A simplified example of the effect of fluid mixing on miscible flood residual oil saturation is given in Figs. 3.38, 3.39, 3.40, and 3.41 for one-dimensional displacement where solvent/oil mixing is caused by longitudinal

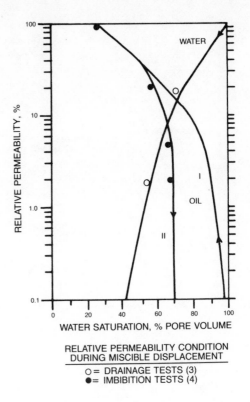

Fig. 3.37—Drainage and imbibition water/oil relative permeability, Berea sandstone, $\phi = 22.2\%$, brine $k = 890$ md (after Ref. 99).

dispersion only. Figs. 3.38 and 3.39 show experimental pseudoternary diagrams at two pressure levels, 1,350 and 2,000 psi, for CO_2 with a west Texas oil.[108] The corresponding pressure-composition diagram for this oil was shown in Chap. 2, Fig. 2.6. At the lower pressure level, three phases (two liquids and a gas) coexist for some overall mixture compositions. At the higher pressure level, only two liquid phases coexist in the multiphase region.

Also shown on these figures are transition zone profiles of the overall composition computed by Gardner, et al.[108] with a numerical simulator for a one-dimensional CO_2 flood. In addition to the two pressures, two values (0.001 and 0.01) for the dimensionless longitudinal dispersion group, K_ℓ/vL, were examined in the calculations. The smaller value was characteristic of CO_2 flood displacements conducted by Gardner et al. at field rates in a 20-ft-long slim-tube laboratory apparatus packed with glass beads.

For all computed cases, the profile of overall compositions cut into the multiphase region and oil was precipitated as a nonflowing liquid phase. For a given level of dispersion, the multiphase region was cut more deeply as pressure decreased and the size of the multiphase region increased, Figs. 3.38 and 3.39. This resulted in lower oil recovery and more oil left as a residual phase as shown in Fig. 3.40. This latter figure also compares computed recovery behavior with experimental behavior observed in the slim tube. Agreement was excellent when the dimensionless longitudinal dispersion group in the calculations was characteristic of

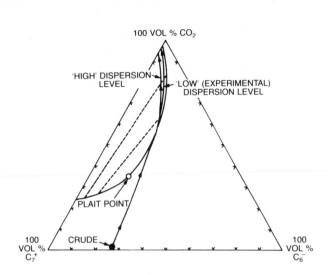

Fig. 3.38—Composition routes calculated with the one-dimensional simulator for displacements of Wasson crude by CO_2 at 2,000 psia and 105°F (after Ref. 108).

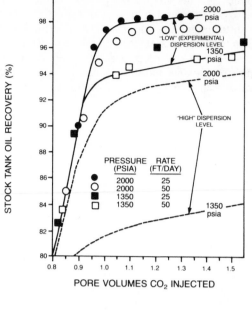

Fig. 3.40—Experimental and calculated CO_2/Wasson crude displacement efficiencies at 105°F. Data points are from slim-tube experiments. Lines are simulation results (after Ref. 108).

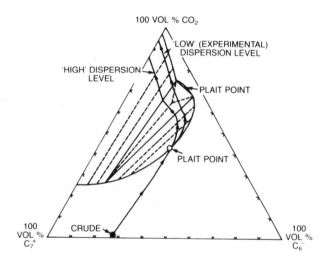

Fig. 3.39—Composition routes calculated with the one-dimensional simulator for displacements of Wasson crude by CO_2 at 1,350 psia and 105°F (after Ref. 108).

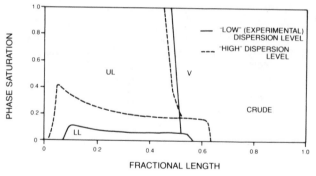

Fig. 3.41—Saturation distributions at 0.5 PV injection calculated with the one-dimensional simulator for displacements of Wasson crude by CO_2 at 1,350 psia and 105°F. Flow is from left to right (after Ref. 108).

the slim tube, thus lending validity to the model calculations.

For a given pressure level, the greater mixing resulting from a higher level of the dimensionless dispersion group caused the multiphase region to be entered more deeply. This again resulted in greater liquid precipitation and decreased recovery. Figs. 3.38, 3.39, and 3.40 illustrate this behavior also.

Fig. 3.41 shows the computed saturation distribution for precipitated liquid at the lower pressure. The results are characteristic of mixing by longitudinal dispersion, with relatively greater mixing and higher saturations near

the inlet and decreasing saturations downstream, leading to a near plateau with distance into the model. A similar profile was projected by Shelton and Yarborough[17] using a simplified theory of solvent/oil mixing and precipitation in a linear displacement. The model calculations of Fig. 3.41 show a small region near the inlet, where partial to total revaporization of precipitated liquid occurred because of the simplified, three-pseudocomponent representation of reservoir fluid by Gardner *et al.*

Limited experimental data, however, are not entirely in agreement with the saturation profile for precipitated oil indicated by Fig. 3.41. Shelton and Schneider[98] published results of condensing-gas drive and CO_2 flood displacements in 16-ft sandstone cores. The same oil was displaced by either gas, and the cores were sectioned after the experiments and were analyzed for residual oil. Relatively constant saturations were found over the core length for both displacements, about 2% PV for the condensing-gas drive and 3% PV for the CO_2 flood.

Gardner and Ypma[47] published the results of experiments and numerical simulator calculations that showed fluid mixing with resulting precipitation of oil in CO_2 and vaporizing-gas drive miscible floods also could be caused by crossflow of oil into viscous fingers. According to the theory proposed by these authors, crossflow of oil into the solvent fingers occurs predominantly in a region near the rear of the finger that already has been swept by the hydrocarbon-enriched compositions at the gas front that are miscible with oil. Oil that crossflows into the rear of the finger mixes with fluid compositions in the trailing edge of the solvent/oil transition zone that do not contain enough extracted hydrocarbons for miscibility, and some oil precipitates. Fig. 3.42 illustrates the effect of this mechanism on displacement efficiency in a two-dimensional, 9-in.-long Berea sandstone core. Oil recoveries are shown in this figure both for first-contact miscible flooding of a laboratory oil by CO_2 and for a displacement of crude oil by CO_2 above the minimum pressure required for dynamic miscibility. Gardner and Ypma postulated that although viscous fingering and crossflow are present in both displacements, a high recovery of the laboratory oil was achieved because the oil that crossflowed into the solvent fingers was first-contact miscible with the CO_2 and was miscibly displaced by the CO_2. This was not the case for the dynamic miscible displacement, and oil that crossflowed into the rear of the viscous fingers in this displacement could not be miscibly displaced because it was behind those transition zone compositions that have been enriched to the point of dynamic miscibility. Gardner and Ypma also simulated viscous fingering with a numerical model and calculated the oil recovery to expect from these experiments. They felt that the results of the calculations, also shown in Fig. 3.42, corroborated the mechanistic view of the displacements just described.

Although the calculations and experiments of Figs. 3.38 through 3.42 give considerable insight into the mechanisms of fluid mixing and liquid precipitation for leaving miscible flood residual oil, the significance of these mechanisms and the magnitude of residual oil saturation left on a reservoir scale are still largely undefined. For a reservoir distance of 1,000 ft, the longitudinal dispersion coefficient would have to be approximately 100-fold higher than its value for microscopic dispersion in laboratory sandstones for the dimensionless longitudinal dispersion group to equal even 0.001. At this dispersion level, the overall compositions were mixed only slightly into the multiphase regions of Figs. 3.38 and 3.39, resulting in precipitation of only 5% or less of the initial oil saturation. However, transverse dispersion in the reservoir can result in significant mixing in viscous fingers or a gravity tongue with large surfaces exposed to oil or in communicating strata

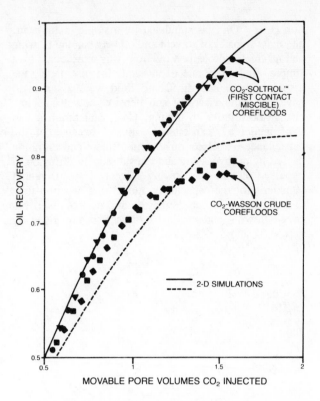

Fig. 3.42—Oil recoveries for CO_2/Soltrol and CO_2/Wasson crude displacements: two-dimensional simulations vs. experiment (after Ref. 47).

through which the solvent is channeling; and, according to the work of Gardner and Ypma, viscous crossflow also may cause significant mixing and liquid dropout. The quantitative effect of these latter mechanisms is not well defined at this time for reservoir-scale systems.

The next subsection describes how some portion of the total residual oil saturation resulting from all the mechanisms described can be recovered by mass transfer. This is followed by a subsection that describes results of attempts to determine the total unrecoverable miscible flood residual oil saturation left in field tests by coring in swept areas behind the solvent front.

Partial Recovery of Miscible Flood Residual Oil

Some of the oil left undisplaced behind the solvent front because of water trapping, dead-end or occluded pore volume, or precipitation may be recovered by mass transfer if it is not completely isolated by water and is in contact with the solvent through oil-filled pores. Molecular diffusion apparently is the predominant cause of this mass transfer, although a velocity dependence of the mass transfer indicates the transfer mechanism also involves mixing other than by diffusion.[107] Evidence for partial recovery of bypassed oil by mass transfer has been found by several investigators.[92,94,98,102,107]

Coats and Smith[110] derived a simple mathematical model for describing the recovery of miscible flood residual oil in a one-dimensional displacement. In this model, a fraction, f, of the hydrocarbon pore volume was assumed to be mobile, while residual oil occupied a fraction $(1-f)$. Oil was simulated by a single component, and mass transfer of solvent and oil between the flowing and residual hydrocarbon fractions was represented by a simple, first-order rate expression. In this model, the displaced fluid and displacing fluid were assumed to have the same viscosity and density, and no volume change occurred with mixing. Coats and Smith called their model a "capacitance" model because of the capacitance-like effect on solvent effluent concentration caused by oil bypassing and the subsequent diffusion of bypassed oil into the flowing solvent. The differential equations for the capacitance model that describe flow and dispersion in one dimension and mass transfer between flowing and residual hydrocarbon are given by Eqs. 3.38 and 3.39.

$$\frac{1}{\gamma}\frac{\partial^2 C}{\partial \xi^2} - \frac{\partial C}{\partial \xi} = f\frac{\partial C}{\partial \tau} + (1-f)\frac{\partial C^*}{\partial \tau} \qquad \ldots\ldots\ldots (3.38)$$

and

$$(1-f)\frac{\partial C^*}{\partial \tau} = \bar{a}(C - C^*), \qquad \ldots\ldots\ldots\ldots (3.39)$$

where

$\bar{a} = K^1 L/v$, dimensionless mass transfer coefficient,

$\gamma = vL/K_\ell$, dimensionless longitudinal dispersion group,

$\xi = x/L$, dimensionless distance,

$\tau = vt/L$, dimensionless time,

C = oil concentration in the mobile hydrocarbon phase,

C^* = oil concentration in the residual hydrocarbon phase, and

K^1 = overall mass transfer coefficient.

A solution to Eqs. 3.38 and 3.39 is given in integral form in the original Coats and Smith article. Brigham[12] pointed out that the solution given by Coats and Smith is for in-situ concentration, and he derived a solution to these equations for flowing concentration, which is typically the concentration measured in laboratory experiments (e.g., in the effluent of corefloods).

Eqs. 3.38 and 3.39 and the solutions to these equations assume first-contact miscibility between solvent and oil—i.e., complete solubility of solvent and oil. For displacements of the dynamic miscibility type, permissible concentrations in the flowing and residual hydrocarbon are limited by the equilibrium phase behavior relationships. Thus, in dynamic miscible displacements, the rate of recovery of residual oil is limited both by the rate of diffusion into the flowing solvent/hydrocarbon phase and by the solubility of the various oil components in this phase.

According to the simplified model of Coats and Smith, the dimensionless mass transfer coefficient $\bar{a}$ governs the efficiency of bypassed oil recovery by mass transfer. As

$\bar{a}$ increases, recovery of residual oil by mass transfer at a given pore volume of fluid injection also increases. A high value for $\bar{a}$ results from a large overall mass transfer coefficient, K^1, and/or a large ratio of displacement distance/velocity, L/v. A large mass transfer coefficient results in a high rate of diffusion, while a large L/v ratio results in a long time available for diffusion to occur.

These remarks are illustrated by Fig. 3.43, which shows results of both experiments and computations for linear displacements conducted under conditions where capacitance affected displacement behavior. In this illustration, the dimensionless mass transfer coefficient was varied by increasing the system length. The experiments were conducted in 8- and 36-ft Berea cores, and a large trapped oil saturation, approximately 72% of the total oil saturation, was established by injecting water and trimethylhexane simultaneously to establish steady-state mobile oil and water saturations.[94] Miscible displacement of the trimethylhexane was accomplished by injecting propane and water simultaneously to maintain a constant, flowing water saturation during the displacement. Other than system length, displacement conditions (such as velocity) were held constant. Fig. 3.43 shows the concentration of propane in the effluent from these displacements. Breakthrough of the 50% solvent concentration in both cores before 1 HCPV of solvent injection shows that oil was trapped by water and was bypassed by the solvent front. The long tailing out of the curves after many HCPV's of solvent injection shows that some of the trapped oil was recovered slowly by mass transfer. The 36-ft core displacement recovered bypassed oil more efficiently than did the 8-ft core displacement. At a given value of pore volumes produced, higher oil recovery was found in the 36-ft core, reflecting the longer time available for diffusion in this experiment.

The best mathematical model representation of the propane effluent concentration data of these experiments was obtained by modifying the Coats and Smith equation to include two trapped oil volumes, each characterized by a separate mass transfer coefficient. Concentration profiles calculated for the best curve-fit values of trapped saturations and corresponding mass transfer coefficients are shown in the figure. The same values for residual volume fractions, mass transfer coefficients, and longitudinal dispersion coefficient were used to characterize both cores, and relatively good agreement was found between the experimental effluent concentration profiles and those calculated by the modified Coats and Smith equation.

A projected concentration profile for a 1,000-ft displacement is also shown in Fig. 3.43. The projection shows that for a 1,000-ft displacement, recovery of the trimethylhexane would be nearly complete if mass transfer in the shorter cores is characterized adequately by the Coats and Smith equation.

Ref. 102 gives another example of improved recovery of bypassed oil with increasing values of the dimensionless mass transfer coefficient. The displacements reported in this work were conducted in cores from two carbonate reservoirs, and the bypassed oil was caused by small-scale heterogeneities. In these cores, higher recovery of the bypassed oil resulted both when displacement velocity was decreased and when displacement

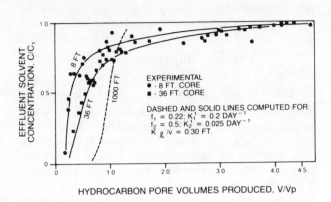

Fig. 3.43—Experimental and computed effluent concentrations for the 8- and 36-ft Berea cores, two stagnant volumes (after Ref. 94).

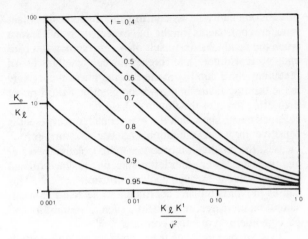

Fig. 3.44—Dependence of the effective dispersion coefficient on the mass-transfer coefficient and flowing fraction (after Ref. 107).

path length was increased, as would be predicted by the capacitance model.

It is interesting to note that the concentration profile in Fig. 3.43 for the 1,000-ft displacement has a shape very similar to the shape that would be expected if longitudinal dispersion were the only mechanism causing solvent/oil mixing, although the length of the mixing zone in this example is substantially larger than would result only from the true longitudinal dispersion coefficient that best characterized these experiments. Baker[107] pointed out that the simple diffusion-convection model (Eqs. 3.5 and 3.10) could be used to calculate mixing zone lengths for displacements with capacitance if an effective dispersion coefficient were used instead of the true longitudinal dispersion coefficient and if the travel distance were sufficiently large. He developed the correlation shown in Fig. 3.44 by calculating mixing zone lengths for various combinations of capacitance model parameters. This correlation relates the ratio of effective dispersion coefficient and true longitudinal dispersion coefficient, K_e/K_ℓ, to the dimensionless group $K_\ell K^1/v^2$ for various values of f. It is valid for $\gamma > 1,000$. Fig. 3.44 shows that capacitance causes the mixing zone length to be greater than would result from longitudinal dispersion alone unless the value of $K_\ell K^1/v^2$ is very large or the quantity of bypassed oil is very small (i.e., if f is large). Length does not enter into the correlation other than through the criterion given above for γ, and K_e/K_ℓ does not diminish with increasing system length.

Yellig and Baker reported observing a small length effect on the effective dispersion coefficient calculated from capacitance model parameters determined from displacements in short laboratory cores.[111] They estimated from mathematical analysis that core length must be greater than about 71 to 146 in. before this effect was negligible. Additional verification of a length effect and its magnitude is warranted because true dispersion in the reservoir could be underestimated by measurements in short laboratory cores because of this effect.

There have been many experimental attempts to determine values for the capacitance model param-

eters.[94,102,107,110-113] These experiments have been conducted with simple laboratory oils rather than with crude oils. Typically, model parameters were determined to give a best fit between effluent concentrations from corefloods and concentrations predicted either by the time domain solution[102,110] of Eqs. 3.38 and 3.39 or the frequency domain solution[107,112,113] of these equations. Yellig and Baker point out that different combinations of model parameters may give equally good fits of the effluent data and that model parameters estimated solely from best-fit techniques may not have physical significance.[111]

Experimental data show the mass transfer coefficient to be velocity dependent, increasing as velocity increases.[102,107] In concept, the mass transfer coefficient should also depend on the effective molecular diffusion coefficient, the area open for diffusion through oil-filled pores, and the length of these pores. Molecular diffusion coefficients increase with decreasing molecular weight of the diffusing species and are sensitive to overall composition in multicomponent mixtures. Thus, in concept there is really no single mass transfer coefficient that characterizes diffusional recovery of a reservoir fluid but a spectrum of such coefficients for the individual components. The lower molecular weight solvents diffuse most rapidly into the residual oil, while the lower molecular weight oil components are recovered most rapidly by diffusion into the flowing solvent.[92,94,98]

In concept, although confirming experimental data are lacking, the mass transfer coefficient probably is affected by rock pore-size distribution and pore geometry. These factors would be expected to affect the accessibility of residual oil by flowing hydrocarbons. Even in a given rock sample, the residual oil very likely is located in sites of varying degrees of accessibility by a diffusing solvent, and a single mass transfer coefficient may not be adequate to characterize this accessibility.

The magnitude of the mass transfer coefficient probably depends on the nature of the residual oil—i.e., whether caused by water trapping, dead-end pore volume, or precipitation. Again, however, there is a lack of experimental data to confirm this hypothesis. Mass transfer coefficients might be expected to be lowest when the residual is the result of water trapping or dead-end pore volume and highest when the result of precipitation. Solubility in the flowing phase most likely is the limiting factor in recovering precipitated oil rather than diffusion into the flowing phase.

Experiments show that the mass transfer coefficient is sensitive to water saturation. As water saturation increases, the residual oil becomes more isolated by water and there are fewer pores left connecting residual oil and flowing solvent through which diffusion occurs. This causes the mass transfer coefficient to decrease as water saturation increases, other factors such as wettability and pore geometry remaining constant.[94,98]

There are no published mass transfer data for reservoir oils left as residual saturations in miscible floods. As indicated by the preceding discussion, recovery of residual oil by diffusion in the reservoir could be considerably more complicated than for the simple hydrocarbon system of Fig. 3.43. With solvent slugs, the contact time between residual oil and solvent is more limited than when solvent is injected continuously. Even with continuous solvent injection and the long times prevailing in a reservoir displacement, the highest molecular weight hydrocarbons may diffuse so slowly (as well as have such low solubility in the flowing phase when the solvent is not first-contact miscible) that they are not recovered to any appreciable extent. Thus, while recovery of a portion of the residual oil by diffusion is a concept that has some experimental verification, the concept has not been extended sufficiently to estimate the magnitude of recovery to expect for various conditions of oil trapping.

Overall Residual Oil Saturations in Field Tests

There have been several field tests where oil saturations left behind the solvent front were determined by coring. Three of the published tests were CO_2 floods and two were condensing-gas drive projects.

Two pressure cores were taken in the Mead-Strawn sandstone behind the CO_2 front at locations approximately 50 and 400 ft from injection wells. Oil saturation averaged 10 and 5% PV stock-tank oil (STO).[114] In this test, pre-water injection was followed by a CO_2 slug driven with water.

A pressure core taken 35 ft from a CO_2 injector in the Willard-Wasson Unit minitest found variable oil saturations.[115] The San Andres carbonate formation in this field is highly stratified, and oil saturations varied from 3 to 30% PV STO among the CO_2-invaded strata. In this test, CO_2 and water were injected alternately.

Residual oil saturations of 2.5 to 4.9% PV STO were determined by pressure coring behind the CO_2 front in the Little Knife pilot test.[116] CO_2 and water were injected alternately into this dolomitized carbonate while the reservoir was undergoing primary depletion.

A pressure core taken behind the enriched-gas front at South Swan Hills field found an average 7.9% PV oil saturation in this carbonate formation.[117]

A core taken 100 ft from an injector in the Seeligson condensing-gas drive flood showed that zones with good permeability were essentially swept clean of oil.[118]

Forty-six samples of core had no oil and 25 samples had saturations no higher than 1% PV. A 52% HCPV slug of rich gas was first driven by residue gas and then by residue gas and water in this sandstone.

3.8 Methods for Improving Displacement Efficiency and Sweepout
Solvent/Water Injection

In 1958, Caudle and Dyes[119] proposed injecting water with solvent to reduce the solvent mobility. Water injection decreases solvent mobility by decreasing the relative permeability of the reservoir rock to solvent. In horizontal or mildly dipping reservoirs, water injection currently is the only method practiced to reduce solvent mobility. In the field, water and solvent are injected in small, alternate slugs rather than simultaneously because (1) solvent and water segregate in the wellbore when injected simultaneously, (2) alternate injection is more convenient operationally than simultaneous injection, and (3) injectivity of either fluid remains higher than would be the case with simultaneous injection. Injectivity remains higher for alternate injection because the saturation and relative permeability of the fluid being injected are higher in the near wellbore region.

The theories of solvent/water injection discussed in the following paragraphs are useful for estimating approximate values for injection ratio and mobility ratio and for teaching principles, but they are oversimplified. An actual displacement can be complicated by many factors such as channeling caused by stratification, crossflow between strata, complete or limited segregation within strata, permeability reduction from precipitation of liquid caused by mixing into a multiphase region, and limited or severe trapping of oil by mobile water, to mention a few. In addition to mobility and sweepout, the effect of alternate injection on injectivity and project life are also factors to consider in selecting an injection ratio. The optimal use of water to improve solvent sweepout in complex reservoir situations is studied best with reservoir simulation models (Sec. 4.2) that account at least in a limited manner for many of the complicating factors.

Nonsegregated Flow. In attempting to analyze solvent/water injection mathematically when gravity segregation is not a factor, two schools of thought have prevailed. One method considers the solvent and water to move through the reservoir as discrete slugs.[120] The other method considers flow as if the solvent and water were injected simultaneously and flowed in a completely dispersed manner. From the discussion in Sec. 3.4 on viscous fingering, it is apparent that the solvent slug will not move through the reservoir in discrete banks. The mobility ratio is unfavorable for a solvent slug displacing a water slug, and the small solvent slugs will finger quickly into the water banks, losing their discrete identity. Flow will not be dispersed completely in a microscopic sense either, but this assumption seems to be the better of the two approaches. In the following paragraphs, this view is used to illustrate principles of mobility ratio improvement by alternate water injection for several idealized displacement situations.

First, consider secondary-recovery miscible flooding in a homogeneous reservoir in which the vertical permeability is so low that no gravity segregation of solvent and water occurs. Eq. 3.40, derived for simultaneous solvent/water injection, gives the condition at which the solvent and water fronts travel at the same velocity.

$$\frac{f_w^* - f_{wi}}{S_w^* - S_{wi}} = \frac{1 - f_w^*}{(1 - S_{orm} + S_{orm} \cdot R_{so} + S_w^* \cdot R_{sw} - S_w^*)},$$

$$\dots \dots \dots \dots \dots \dots \dots \dots \dots (3.40)$$

where

S_w^* = water saturation in the solvent/water region, fraction PV,

S_{wi} = water saturation at initial conditions, fraction PV,

S_{orm} = oil saturation, if any, at reservoir conditions remaining in the solvent/water region because of the factors discussed in Sec. 3.7, fraction PV,

f_w^* = water fractional flow in the solvent/water region,

f_{wi} = water fractional flow at initial conditions,

R_{sw} = solubility of solvent in injected water, RB/RB, and

R_{so} = solubility of solvent in residual oil, RB/RB.

In the derivation of this equation, we assume equilibrium contact of solvent with the miscible flood residual oil.

A method for solving Eq. 3.40 graphically is shown in Fig. 3.45. The solution for f_w^* and S_w^* is the intersection of Line AB with the solvent/water fractional flow curve. Point A is located at $f_w = 1$ and $S_w = (1 - S_{orm} + S_{orm} \cdot R_{so} + S_w^* \cdot R_{sw})$. Point B is located at f_{wi}, S_{wi}. In the example shown in the figure, $S_{orm} = 0$, $f_{wi} = 0$, and $R_{sw} = 0$. When $R_{sw} > 0$, an iterative solution is necessary since S_w^* must be estimated to locate point A. If $S_{orm} > 0$, the solvent/water fractional flow curve used for the graphical construction should be representative of solvent and water flow in the presence of residual oil left to miscible flooding. For this situation, the quantity $1 - S_w$ of the fractional flow graph represents the total nonaqueous saturation—i.e., solvent plus miscible flood residual oil. After solving Eq. 3.40, the solvent/water equal velocity injection ratio, in reservoir barrels per reservoir barrel, is given by $(1 - f_w^*)/f_w^*$.

The condition of equal velocity flow was chosen for the following reasons. When too little water is injected, the solvent front travels faster than the water front and a bank of solvent forms ahead of the solvent/water region. When this happens, the mobility ratio at the solvent/oil bank front is just the normal viscosity ratio between solvent and oil, and solvent fingers into the oil. Even so, the lower solvent mobility in the solvent/water region still causes the overall volumetric sweepout of solvent to be greater than would have been the case without water injection.[75,84,*] If water is overinjected, the water front

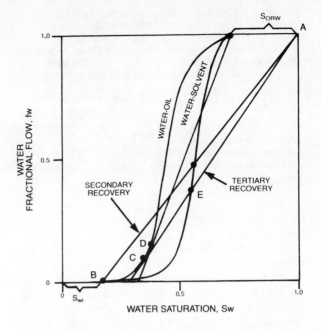

Fig. 3.45—Estimation of equal velocity solvent/water ratio.

travels faster than the trailing edge of the oil bank. A higher water saturation is established at the solvent/oil front than would have been the case with solvent overinjection or equal velocity injection, and this mobile water has the potential for trapping some of the oil at the trailing edge of the oil bank. Oil trapping by water may or may not be significant in the particular reservoir rock of interest at the saturation anticipated from limited overinjection of water, and this aspect of solvent/water injection must be evaluated.

Table 3.3 compares calculated mobility ratios for secondary recovery miscible flooding when solvent and water are injected at the equal velocity conditions. Examples are shown for oils with viscosities of 1.2 and 5 cp displaced by a 0.06-cp viscosity solvent. Two mobility ratios are calculated for the equal velocity condition—one assuming the total mobility of the solvent/water region governs solvent fingering and sweepout, the other assuming solvent mobility alone governs fingering and sweepout. Relative permeabilities for these illustrations were a composite of actual measurements on cores from a carbonate reservoir. The oil/water and solvent/water fractional flow curves calculated from these relative permeabilities are shown in Fig. 3.45.

The example shown in Table 3.3 is for a situation where both the miscible flood residual oil saturation and solubility of solvent in water are negligibly small. If solvent only were injected, the mobility ratio would be 20 for the 1.2-cp oil and 83 for the 5-cp oil. A substantial reduction in mobility ratio is calculated for solvent/water injection. The calculated injection ratio for equal velocity flow is 1.08 RB solvent/RB water, and the calculated mobility ratio at this condition for displacement of the 1.2-cp oil is only 0.27 or 0.53, depending on whether solvent mobility only or total mobility of the solvent/water region is chosen for the mobility ratio calculation. For the 5-cp oil, the calculated mobility

*This statement also is supported by unpublished experiments with first-contact miscible fluids in a five-spot model where a high-mobility solvent slug was displaced with a low-mobility miscible drive fluid — also by unpublished numerical simulator calculations for a highly stratified reservoir.

**TABLE 3.3—MODERATION OF MOBILITY RATIO IN
SECONDARY RECOVERY MISCIBLE FLOODING BY
ALTERNATE SOLVENT/WATER INJECTION**
[equal-velocity injection ratio (Eq. 3.40) = 1.08 RB
solvent/RB water; $S_{orm} = 0$; $R_{sw} = 0$; initial water
saturation at the irreducible value]

	Calculated Mobility Ratio of Displacement	
	$\mu_o = 1.2$ cp	$\mu_o = 5$ cp
With solvent mobility only	0.27	1.1
With total mobility of solvent/water region	0.53	2.2

ratio is 1.1 or 2.2, depending on the basis for calculating displacing fluid mobility.

Miscible flood residual oil and solvent solubility in water may cause the calculated injection ratio to decrease or increase. For example, if $S_{orm} = 0.1$ and $R_{so} = 0.5$ RB/RB, the calculated equal velocity injection ratio decreases to 0.98. If $S_{orm} = 0$ but solvent solubility in water is 0.07 RB/RB, the calculated injection ratio increases to 1.2.

In stratified reservoirs, solvent enters the higher-permeability layers at a greater rate than would be predicted from the kh ratio of the layers. Water is distributed between the layers in a manner more closely approximating the kh distribution. The result is a solvent/water ratio in the higher permeability layers that is higher than the injected ratio and a solvent/water ratio in the lower-permeability layers that is less than the injected ratio.[121] At the overall ratio calculated from Eq. 3.40, solvent will advance faster than water in the more permeable strata, whereas water will advance faster than solvent in the less permeable strata. In this situation it may be desirable to inject at a lower overall solvent/water injection ratio than calculated by Eq. 3.40 to prevent a high-mobility solvent bank from forming in the more permeable strata that contribute the bulk of miscible displacement oil recovery.

Solvent/water injection also can be of benefit in tertiary recovery miscible flooding even though injection of more water into rock where the oil saturation already has been driven to its residual value by water intuitively may seem counterproductive. Consider first the injection of solvent without any water. Solvent initially displaces water and reduces the water saturation. As the water saturation is reduced, the solvent displaces that portion of the residual oil that it can contact. The resulting mixture of displaced oil and solvent as well as some solvent-free oil driven ahead of the solvent/oil mixture forms the "tertiary oil bank." At this point, further displacement of water and reduction of water saturation is accomplished by the tertiary oil bank.

Fig. 3.46 shows a profile of fluid saturations for an idealized linear displacement of waterflood residual oil by solvent/water injection when the solvent and water are injected at the equal velocity ratio for tertiary recovery flooding. Gravity segregation and viscous fingering are assumed not to occur in this idealized displacement, and there is assumed to be no mixing between solvent and oil. Water saturation and water fractional flow at the leading edge of the tertiary oil bank are determined by the point of tangency to the water/oil frac-

tional flow curve of a line passing through the point $f_w = 1$, $S_w = 1 - S_{orw}$. The point of tangency is shown as Point D in Fig. 3.45. Water saturation and water fractional flow at the trailing edge of the tertiary oil bank are determined by the point of tangency to the water/oil fractional flow curve of a line passing through Point A located at $f_w = 1$ and $S_w = (1 - S_{orm} + S_{orm} \cdot R_{so} + S_w^* \cdot R_{sw})$. This latter point of tangency is shown as Point C in Fig. 3.45. Point E in this figure is the point on the water/solvent fractional flow curve that defines the equal velocity condition for simultaneous solvent/water injection, and the solvent/water equal-velocity injection ratio, in reservoir barrels per reservoir barrel, is given by $(1 - f_w^*)/f_w^*$. For this example, $S_{orm} = 0$ and $R_{sw} = 0$. As discussed for secondary recovery flooding by solvent/water injection, when $S_{orm} > 0$, the solvent/water fractional flow curve used for the graphical construction should be representative of solvent and water flow in the presence of residual oil left to miscible flooding.

If the solvent is overinjected, a zone of high solvent concentration will form between the solvent/water region and the rear of the tertiary oil bank. If water is overinjected, water will invade the rear of the tertiary oil bank, causing the water saturation to be higher than determined from Point C in Fig. 3.45.

In practice, displacement of the residual oil by solvent and formation of a tertiary oil bank is more complicated than this simplified discussion. Solvent will penetrate a significant fraction of the oil bank width because of viscous fingering and because of the effects of macroscopic heterogeneities and water saturation on solvent dispersion. This reduces the effective viscosity of the oil bank, which in turn reduces the effectiveness of the oil bank in reducing water saturation. The result in practice is to cause the water saturation at the trailing edge of the oil bank to be higher than calculated by the method of Fig. 3.45 using pure oil viscosity to estimate the water/oil fractional flow curve. The injection ratio calculated using pure oil viscosity will tend to err in the direction of overinjecting water and forcing a higher water saturation to be maintained in the tertiary oil bank than otherwise would have been the case with solvent injection only. The higher water saturation increases the potential for oil to be trapped by water in the oil bank because solvent cannot "see" the oil and displace it.

There is no simple method available for estimating the effective viscosity of the tertiary oil bank and for calculating the appropriate water/oil fractional flow curve that defines saturations at the trailing edge of the tertiary oil bank. If oil trapping by mobile water in the reservoir rock of interest is believed severe, judgment as well as more sophisticated analysis methods (Sec. 4.2) should be used to decide how much the solvent/water ratio calculated with Eq. 3.40 should be increased to achieve a proper balance between miscible sweepout and unit displacement efficiency.

Table 3.4 compares calculated mobility ratios for tertiary recovery miscible flooding when solvent only is injected and when solvent and water are injected at the equal-velocity ratio. In this illustration, the saturation at the trailing edge of the oil bank was calculated two ways: (1) effective oil bank viscosity that of pure oil and (2) effective oil bank viscosity equivalent to the viscosity of a

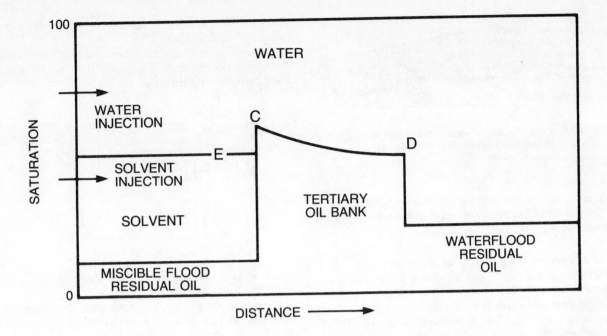

Fig. 3.46—Idealized saturation profile for the linear displacement of waterflood residual oil by simultaneous solvent/water injection at the equal-velocity ratio.

50/50 oil/solvent mixture. The total mobility of the solvent/water region was used in calculating mobility ratios for Table 3.4.

For conditions of negligible solvent solubility in water and negligible miscible flood residual oil saturation, the calculation shows that an injection ratio of 1.7 RB solvent/RB water would be required for equal velocity flooding if the solvent did not mix materially with the displaced tertiary oil and decrease the effective viscosity of oil in the tertiary oil bank from 1.2 cp. A ratio of 2.6 RB/RB would be necessary if the effective viscosity of oil in the tertiary oil bank were that of a 50/50 solvent/oil mixture. In this example, the unfavorable mobility ratio between solvent and tertiary oil bank is reduced by a factor of three to eight by alternate water injection, whereas

the mobility ratio between solvent and the waterflooded region ahead of the tertiary oil bank is reduced by a factor of three to seven by alternate water injection.

Because of the density difference between miscible solvents of interest and water, the condition of nonsegregated flow assumed in Eq. 3.40 is never found entirely in practice. Segregation is minimized in some reservoirs where the ratio of effective vertical to horizontal permeabilities is exceptionally low, less than about 0.01, because of numerous and scattered shale laminae, impermeable dense streaks, randomly dispersed permeability heterogeneities that make flow highly tortuous, or other lithological or petrographic factors. An example of low effective vertical permeability is found in certain carbonate reservoirs in the Permian Basin

**TABLE 3.4—MODERATION OF MOBILITY RATIO IN TERTIARY
RECOVERY MISCIBLE FLOODING BY ALTERNATE
SOLVENT/WATER INJECTION**
($S_{orm} = 0$; $R_{sw} = 0$)

	Minor Solvent/Oil Mixing ($\mu_{ob} = 1.2$ cp)	50:50 Solvent/Oil Mixing ($\mu_{ob} = 0.3$ cp)
Solvent Injection Only		
solvent mobility:oil bank mobility	18	4.3
solvent mobility:waterflooded region mobility	12	12
Solvent/Water Injection		
solvent/water injection ratio	1.7	2.6
solvent/water region mobility:oil bank mobility	2.2	1.3
solvent/water region mobility:waterflooded region mobility	1.5	3.5

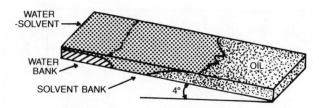

Fig. 3.47—Typical profile of water/solvent mixture displacing low-viscosity oil downdip (after Ref. 84).

where dense streaks provide barriers to vertical flow, although some segregation may still occur between dense streaks.

Fully Segregated Flow.

Laboratory experiments[84] and reservoir simulations[75] both show that improved oil recovery may still result from alternate solvent/water injection even when gravity causes the solvent and water to segregate. Fig. 3.47 is a pictorial representation of model experiments run by Blackwell *et al.*[84] to study solvent/water segregation in secondary-recovery miscible displacement. By keeping a small bank of solvent ahead of the advancing water front, most of the oil was displaced miscibly. The advancing water immiscibly displaced solvent as well as some oil at the bottom of the model that had not been displaced completely by solvent before water-front arrival. These authors proposed the following equation, which is modified here to include the effects of miscible flood residual oil and solvent solubility in water, for calculating the *minimum* solvent/water injection ratio for conditions where complete segregation of injected water and solvent occurs.

$$\frac{q_s}{q_w} = (1 - S_{w,rs} - S_{orm} + S_{orm} \cdot R_{so} + S_{wr} \cdot R_{sw}) \div$$

$$\left[(S_{w,rs} - S_{wr}) - \left(\frac{k_{w,rs}}{k_{s,wr}}\right)\left(\frac{\mu_s}{\mu_w}\right) \right.$$

$$\left. \cdot (1 - S_{wr} - S_{orm} + S_{orm} \cdot R_{so} + S_{wr} \cdot R_{sw}) \right]. \quad (3.41)$$

This equation was derived for secondary-recovery miscible flooding in homogeneous porous media when the water saturation at the start of the solvent injection is at its irreducible value. The relative permeabilities $k_{w,rs}$ and $k_{s,wr}$ are evaluated at $S_{w,rs} = (1 - S_{srw} - S_{orm})$ and $S_s = (1 - S_{orm} - S_{wr})$, where S_{srw} is the solvent saturation trapped by water. At the injection ratio calculated by this equation, solvent is supplied to the solvent/water front at the same rate that solvent is trapped by the advancing water front. In practice, an injection ratio greater than the calculated value is desired for an excess quantity of solvent to precede the water front and to maintain an adequate bank of solvent ahead of it. For example, Blackwell *et al.*[84] found that an injection ratio about 25% greater than the calculated value gave best oil recovery results in their laboratory model with a 0.3-cp oil, whereas best results were achieved with a 4-cp oil

when the injection ratio was two to three times the calculated value.

Blackwell *et al.*[84] also recommended that the effective mobility ratio for fully segregated flow be estimated from the following equation.

$$M = \left[\frac{1 + \dfrac{q_w}{q_s}}{\left(\dfrac{k_{w,rs}}{k_{s,wr}}\right)\left(\dfrac{\mu_s}{\mu_w}\right) + \dfrac{q_w}{q_s}} \right]$$

$$\cdot \left(\frac{k_{w,rs}}{k_{o,wr}}\right)\left(\frac{\mu_o}{\mu_w}\right), \quad \ldots\ldots\ldots\ldots\ldots (3.42)$$

where $k_{o,wr}$ is the relative permeability to oil at the irreducible water saturation.

Numerical simulator calculations show that when solvent and water segregate in tertiary-recovery miscible flooding, the overriding solvent displaces oil but pushes a significant fraction of the tertiary oil bank down toward the middle of the cross section, as shown in Fig. 3.31.[75,76] No analytical expression is available for estimating the optimal solvent/water ratio in this situation. If oil trapping by mobile water is not significant, simulations show that improved recovery performance may still result from alternate water injection.[75,81]

Viscosity Grading

In this method, a transition zone of displaced-fluid/displacing-fluid compositions is injected ahead of the pure displacing fluid. The idea is to cause a gradual change in viscosity from displaced to displacing fluid rather than an abrupt change and either to prevent the formation of viscous fingers or to moderate their growth. The method is costly for oilfield flooding with miscible solvents since a substantial amount of oil must be injected at the leading edge of the transition zone. As a result, it has seen no application in miscible flood field projects. It is more applicable in polymer flooding and in micellar/polymer flooding for preventing drive water from fingering through the polymer bank and has seen some field application with these processes.

Perrine[52] and Chuoke[44] both derived equations from perturbation analysis for calculating the transition-zone concentration gradients for continuous viscosity grading that would prevent fingering in linear displacements. Laboratory experiments subsequently showed that transition-zone lengths calculated with these equations were highly conservative and that adequate control of fingering could be accomplished with much shorter viscosity-graded zones.[33,122,123] For a given application, actual viscosity gradients required to suppress fingering should be determined experimentally in laboratory models.

Continuous viscosity grading is awkward and inconvenient to accomplish in field operations. Some use has been made of the method, however, in laboratory experiments designed to measure unit displacement efficiency.[124]

A more convenient method for injecting a viscosity-graded transition zone is to inject discrete slugs of graded concentration and viscosity. In this method, the viscosity ratio for any particular slug displacing the preceding slug is unfavorable, and finger zones develop at the front and back of each intermediate slug. Claridge[125] recommends sizing the slugs so that when the slug reaches the producing well, the forward-projecting fingers originating at the trailing edge of a slug just rouch but do not overlap the backward-projecting fingers originating at the leading edge of the slug. He also recommends calculating the length of each fingered zone by the Koval method discussed in Sec. 3.4.

Gravity Stabilization

This method was discussed in Sec. 3.6, which the reader should consult for details. In some reservoirs with dip, gravity segregation can be used to advantage to prevent viscous fingering or gravity tonguing. To accomplish this in the strictest sense, the displacement rate must be below the stable rate given by Eq. 3.31. In practice, the critical rate, Eq. 3.32, is often an adequate criterion.

Foams

There has been considerable laboratory research to decrease the mobility of injection gases by foaming them with water and surfactants.[91,126-141] Mobility of a gas in the presence of a foaming agent can be adjusted to some extent through foaming agent concentration and the gas/liquid ratio.[127] Primary requirements of a foam for reservoir application are long-term chemical stability of the surfactant, acceptable levels of surfactant adsorption on the reservoir rock, and long-term foam stability at the desired level of effective mobility. Results of laboratory corefloods have been encouraging and there has been one published report of an injection test,[142] but otherwise the effectiveness and economic attractiveness of the method have not been proved by field testing.

References

1. Craig, F.F.: *The Reservoir Engineering Aspects of Waterflooding,* Monograph Series, SPE, Dallas (1971) **3.**
2. Craig, F.F. Jr., Geffen, T.M., and Morse, R.A.: "Oil Recovery Performance of Pattern Gas or Water Injection Operations from Model Tests," *Trans.,* AIME (1955) **204,** 7-15.
3. Collins, R.E.: *Flow of Fluids Through Porous Media,* Reinhold Publishing Co., New York City (1961) 201.
4. Perkins, T.K. and Johnston, O.C.: "A Review of Diffusion and Dispersion in Porous Media," *Soc. Pet. Eng. J.* (March 1963) 70-84; *Trans.,* AIME, **228;** *Miscible Processes,* Reprint Series, SPE, Dallas (1965) **8,** 77-92.
5. Crank, J.: *The Mathematics of Diffusion,* second edition, Oxford University Press, New York City (1975).
6. Brigham, W.E., Reed, P.W., and Dew, J.N.: "Experiments on Mixing During Miscible Displacement in Porous Media," *Soc. Pet. Eng. J.* (March 1961) 1-8; *Trans.,* AIME, **222.**
7. Crane, F.E. and Gardner, G.H.F.: "Measurements of Transverse Dispersion in Granular Media," *J. Chem. Eng. Data* (1961) **6,** 283.
8. van der Poel, C.: "Effect of Lateral Diffusivity on Miscible Displacement in Horizontal Reservoirs," *Soc. Pet. Eng. J.* (Dec. 1962) 317-26; *Trans.,* AIME, **225.**
9. Klinkenberg, L.J.: "Analog Between Diffusion and Electrical Conductivity in Porous Rocks," *Bull.,* GSA (1951) **62,** 559.
10. Scott, D.S. and Dullien, F.A.L.: "Diffusion of Ideal Gases in Capillaries and Porous Solids," *AIChE J.* (1962) **8,** 113.
11. Aris, R. and Amundson, N.R.: "Some Remarks on Longitudinal Mixing or Diffusion in Fixed Beds," *AIChE J.* (1957) **3,** 280.
12. Brigham, W.E.: "Mixing Equations in Short Laboratory Cores," *Soc. Pet. Eng. J.* (Feb. 1974) 91-99; *Trans.,* AIME, **257.**
13. Blackwell, R.J.: "Laboratory Studies of Microscopic Dispersion Phenomena," *Soc. Pet. Eng. J.* (March 1962) 1-8; *Trans.,* AIME, **225;** *Miscible Processes,* Reprint Series, SPE, Dallas (1965) **8,** 69-76.
14. Hassinger, R.C. and von Rosenberg, D.V.: "A Mathematical and Experimental Examination of Transverse Dispersion Coefficients," *Soc. Pet. Eng. J.* (June 1968) 195-204; *Trans.,* AIME, **243.**
15. Pozzi, A.L. and Blackwell, R.J.: "Design of Laboratory Models for Study of Miscible Displacement," *Soc. Pet. Eng. J.* (March 1963) 28-40; *Trans.,* AIME, **228.**
16. Warren, J.E. and Skiba, F.F.: "Macroscopic Dispersion," *Soc. Pet. Eng. J.* (Sept. 1964) 215-30; *Trans.,* AIME, **231.**
17. Shelton, J.L. and Yarborough, L.: "Multiple Phase Behavior in Porous Media During CO_2 or Rich Gas Flooding," *J. Pet. Tech.* (Sept. 1977) 1171-78.
18. Deans, H.A. and Shallenberger, L.K.: "Single Well Chemical Tracer Method to Measure Connate Water Saturation," paper SPE 4755 presented at the 1974 SPE Improved Oil Recovery Symposium, Tulsa, April 22-24.
19. Sheely, C.Q.: "Description of Field Tests to Determine Residual Oil Saturation by Single Well Tracer Method," paper SPE 5840 presented at the 1976 SPE Improved Oil Recovery Symposium, Tulsa, March 22-24.
20. Deans, H.A. and Majoros, S.: *The Single-Well Chemical Tracer Method for Measuring Residual Oil Saturation,* Final Report for U.S. DOE, Contract No. DE-AS19-79BC20006 performed at Rice U. (Oct. 1980).
21. Tomich, J.F. *et al.*: "Single Well Tracer Method to Measure Residual Oil Saturation," *J. Pet. Tech.* (Feb. 1973) 211-18; *Trans.,* AIME, **225.**
22. *Tables of the Error Function and Its Derivative,* Natl. Bureau of Standards, Applied Mathematics Series No. 41 (Oct. 1954); for sale by the Superintendent of Documents, U.S. Government Printing Office, Washington, DC.
23. Brigham, W.E.: "Mixing Equations in Various Geometries," paper SPE 4585 presented at the 1973 SPE Annual Meeting, Las Vegas, Sept. 30-Oct. 3.
24. Crane, F.E., Kendall, H.A., and Gardner, G.H.F.: "Some Experiments of the Flow of Miscible Fluids of Unequal Density Through Porous Media," *Soc. Pet. Eng. J.* (Dec. 1963) 277-80; *Trans.,* AIME, **228.**
25. Blackwell, R.J., Rayne, J.R., and Terry, W.M.: "Factors Influencing the Efficiency of Miscible Displacement," *Trans.,* AIME (1959) **216,** 1; *Miscible Processes,* Reprint Series, SPE, Dallas (1965) **8,** 197-204.
26. Craig, F.E. Jr. *et al.*: "A Laboratory Study of Gravity Segregation in Frontal Drives," *Trans.,* AIME (1957) **210,** 275-82.
27. Elkins, L.F. and Skov, A.M.: "Some Field Observations of Heterogeneity of Reservoir Rocks and Its Effects on Oil Displacement Efficiency," paper SPE 282 presented at the 1962 SPE Production Research Symposium, Tulsa, April 12-13.
28. Habermann, B.: "The Efficiencies of Miscible Displacement as a Function of Mobility Ratio," *Trans.,* AIME (1960) **219,** 264; *Miscible Processes,* Reprint Series, SPE, Dallas (1965) **8,** 205-14.
29. Mahaffey, J.L., Rutherford, W.M., and Matthews, C.W.: "Sweep Efficiency by Miscible Displacement in a Five-Spot," *Soc. Pet. Eng. J.* (March 1966) 73-80; *Trans.,* AIME, **237.**
30. Benham, A.L. and Olson, R.W.: "A Model Study of Viscous Fingering," *Soc. Pet. Eng. J.* (June 1963) 138-44; *Trans.,* AIME, **228.**
31. Greenkorn, R.A., Johnson, C.R., and Haring, R.E.: "Miscible Displacement in a Controlled Natural System," *J. Pet. Tech.* (Nov. 1965) 329-35; *Trans.,* AIME, **234.**
32. Offeringa, J. and van der Poel, C.: "Displacement of Oil from Porous Media by Miscible Fluids," *Trans.,* AIME (1954) **201,** 310; *Miscible Processes,* Reprint Series, SPE, Dallas (1965) **8,** 227-233.
33. Kyle, C.R. and Perrine, R.L.: "Experimental Studies of Miscible Displacement Instability," *Soc. Pet. Eng. J.* (Sept. 1965) 189-95; *Trans.,* AIME, **234.**
34. Bilhartz, H.L. Jr. and Charlson, G.S.: "Field Polymer Stability

Studies," paper SPE 5551 presented at the 1975 SPE Annual Technical Conference and Exhibition, Dallas, Sept. 28–Oct. 1.

35. Slobod, R.L. and Thomas, R.A.: "Effect of Transverse Diffusion on Fingering in Miscible Displacement," *Soc. Pet. Eng. J.* (March 1963) 9–15.

36. Perkins, T.K. and Johnston, O.C.: "A Study of Immiscible Fingering in Linear Models," *Soc. Pet. Eng. J.* (March 1969) 39–48.

37. Perkins, T.K., Johnston, O.C., and Hoffman, R.N.: "Mechanics of Viscous Fingering in Miscible Systems," *Soc. Pet. Eng. J.* (Dec. 1965) 301–17; *Trans.*, AIME, **234**.

38. Dougherty, E.L.: "Mathematical Model of an Unstable Miscible Displacement," *Soc. Pet. Eng. J.* (June 1963) 155–65; *Miscible Processes,* Reprint Series, SPE, Dallas (1965) **8**, 151–60.

39. Hurst, W.: "Determination of Performance Curves in Five-Spot Waterflood," *Pet. Eng.* (1953) **25**, 840–46.

40. Perrine, R.L.: "The Development of Stability Theory for Miscible Liquid-Liquid Displacement," *Soc. Pet. Eng. J.* (March 1961) 17–25.

41. Perrine, R.L.: "A Unified Theory for Stable and Unstable Miscible Displacement," *Soc. Pet. Eng. J.* (Sept. 1963) 205–13; *Trans.*, AIME, **228**.

42. Peaceman, D.W. and Rachford, H.H. Jr.: "Numerical Calculation of Multidimensional Miscible Displacement," *Soc. Pet. Eng. J.* (Dec. 1962) 327–39; *Trans.*, AIME, **225**.

43. Koval, E.J.: "A Method for Predicting the Performance of Unstable Miscible Displacement in Heterogeneous Media," *Soc. Pet. Eng. J.* (June 1963) 145–54; *Trans.*, AIME, **228**.

44. Chuoke, R.L., van Meurs, P., and van der Poel, C.: "The Instability of Slow, Immiscible Viscous Liquid-Liquid Displacements in Permeable Media," *Trans.*, AIME (1959) **216**, 188–94.

45. Outmans, H.D.: "Nonlinear Theory for Frontal Stability and Viscous Fingering in Porous Media," *Soc. Pet. Eng. J.* (June 1962) 165–76.

46. Chandrasekhar, S.: *Hydrodynamic and Hydromagnetic Stability,* Oxford Clarendon Press, London (1961).

47. Gardner, J.W. and Ypma, J.G.J.: "An Investigation of Phase Behavior-Macroscopic Bypassing Interaction in CO_2 Flooding," paper SPE 10686 presented at the 1982 SPE Enhanced Oil Recovery Symposium, Tulsa, April 4–7.

48. Bentsen, R.G. and Nielson, R.F.: "A Study of Plane Radial Miscible Displacement in a Consolidated Porous Medium," *Soc. Pet. Eng. J.* (March 1965) 1–5; *Trans.*, AIME, **234**.

49. Dyes, A.B., Caudle, B.H., and Erikson, R.A.: "Oil Production After Breakthrough—As Influenced by Mobility Ratio," *Trans.*, AIME (1954) **201**, 81–86.

50. Hall, H.N. and Geffen, T.M.: "A Laboratory Study of Solvent Flooding," *Trans.*, AIME (1957) **210**, 48–57; *Miscible Processes,* Reprint Series, SPE, Dallas (1965) **8**, 133–42.

51. Lacey, J.W., Draper, A.L., and Binder, G.G. Jr.: "Miscible Fluid Displacement in Porous Media," *Trans.*, AIME (1958) **213**, 76–79.

52. Perrine, R.L.: "Stability Theory and Its Use to Optimize Solvent Recovery of Oil," *Soc. Pet. Eng. J.* (March 1961) 9–16; *Trans.*, AIME, **222**.

53. Buckley, S.E. and Leverett, M.C.: "Mechanisms of Fluid Displacement in Sands," *Trans.*, AIME (1942) **146**, 107–116.

54. Handy, L.L.: "An Evaluation of Diffusion Effects in Miscible Displacement," *Trans.*, AIME (1959) **216**, 61–63.

55. Claridge, E.L.: "Prediction of Recovery in Unstable Miscible Flooding," *Soc. Pet. Eng. J.* (April 1972) 143–55.

56. Todd, M.R. and Longstaff, W.J.: "The Development, Testing, and Application of a Numerical Simulator for Predicting Miscible Flood Performance," *J. Pet. Tech.* (July 1972) 874–82.

57. Dykstra, H. and Parsons, R.L.: "The Prediction of Oil Recovery by Waterflooding," *Secondary Recovery of Oil in the United States,* second edition, API, Dallas (1950) 160.

58. Claridge, E.L.: "Design of Graded Viscosity Banks for Enhanced Recovery Processes," PhD dissertation, U. of Houston (July 1979).

59. Muskat, M. and Wyckoff, R.D.: "A Theoretical Analysis of Waterflooding Networks," *Trans.*, AIME (1934) **107**, 62–76.

60. Botset, H.G.: "The Electrolytic Model and Its Application to the Study of Recovery Problems," *Trans.*, AIME (1946) **165**, 15–25.

61. Ramey, H.J. Jr. and Nabor, G.W.: "A Blotter-Type Electrolytic Model Determination of Areal Sweeps in Oil Recovery by In-Situ Combustion," *Trans.*, AIME (1954) **201**, 119–23.

62. Burton, M.B. Jr. and Crawford, P.B.: "Application of the Gelatin Model for Studying Mobility Ratio Effects," *Trans.*, AIME (1956) **207**, 33–37.

63. Lee, B.D.: "Potentiometric Model Studies of Fluid Flow in Petroleum Reservoirs," *Trans.*, AIME (1948) **174**, 41–66.

64. Paulsell, B.L.: "Areal Sweep Performance of Five-Spot Pilot Floods," MS thesis, Pennsylvania State U., University Park (Jan. 1958).

65. Nobles, M.A. and Janzen, H.B.: "Application of a Resistance Network for Studying Mobility Ratio Effects," *Trans.*, AIME (1958) **213**, 356–58.

66. Slobod, R.L. and Caudle, B.H.: "X-Ray Shadowgraph Studies of Areal Sweepout Efficiencies," *Trans.*, AIME (1952) **195**, 265–70.

67. Fay, C.H. and Prats, M.: "The Application of Numerical Methods to Cycling and Flooding Problems," *Proc.*, Third World Pet. Cong. (1951) **2**, 555–63.

68. Aronofsky, J.: "Mobility Ratio—Its Influence on Flood Patterns During Water Encroachment," *Trans.*, AIME (1952) **195**, 15–24.

69. Morel-Seytoux, H.J.: "Analytical-Numerical Method in Waterflooding Predictions," *Soc. Pet. Eng. J.* (Sept. 1965) 247–58.

70. Jacquard, P.: "Calculs Numériques de Déplacements de Fronts," *Proc.*, Seventh World Pet. Cong. (1963) paper 7.

71. Morel-Seytoux, H.J.: "Unit Mobility Ratio Displacement Calculations for Pattern Floods in Homogeneous Medium," *Soc. Pet. Eng. J.* (Sept. 1966) 217–27.

72. Lacey, J.W., Faris, J.E., and Brinkman, F.H.: "Effect of Bank Size on Oil Recovery of the High Pressure Gas-Driven LPG-Bank Process," *J. Pet. Tech.* (Aug. 1961) 806–12; *Trans.*, AIME, **222**; *Miscible Process,* Reprint Series, SPE, Dallas (1965) **8**, 215–26.

73. Kimbler, O.K., Caudle, B.H., and Cooper, H.E. Jr.: "Areal Sweepout Behavior in a Nine-Spot Injection Pattern," *J. Pet. Tech.* (Feb. 1964) 199–202; *Trans.*, AIME, **231**.

74. Claridge, E.L.: "A Trapping Hele-Shaw Model for Miscible-Immiscible Flooding Studies," *Soc. Pet. Eng. J.* (Oct. 1973) 255–61.

75. Warner, H.R. Jr.: "An Evaluation of Miscible CO_2 Flooding in Waterflooded Sandstone Reservoirs," *J. Pet. Tech.* (Oct. 1977) 1339–47.

76. Miller, M.C. Jr.: "Gravity Effects in Miscible Displacement," paper SPE 1531 presented at the 1966 SPE Annual Meeting, Dallas, Oct. 2–5.

77. Spivak, A.: "Gravity Segregation in Two-Phase Displacement Processes," *Soc. Pet. Eng. J.* (Dec. 1974) 619–27.

78. Dietz, D.N.: "A Theoretical Approach to the Problem of Encroaching and Bypassing Edge Water," *Proc.*, Acad. Sci. Amst. B (1953) **56**, 83.

79. Hawthorne, R.G.: "Two-Phase Flow in Two-Dimensional Systems—Effects of Rate, Viscosity, and Density on Fluid Displacement in Porous Media," *Trans.*, AIME (1960) **219**, 81–87.

80. Gardner, G.H.F., Downie, J., and Kendall, H.A.: "Gravity Segregation of Miscible Fluids in Linear Models," *Soc. Pet. Eng. J.* (June 1962) 95–104.

81. Youngren, G.K. and Charlson, G.S.: "History Match Analysis of the Little Creek CO_2 Pilot Test," *J. Pet. Tech.* (Nov. 1980) 2042–52.

82. Dumoré, J.M.: "Stability Considerations in Downward Miscible Displacement," *Soc. Pet. Eng. J.* (Dec. 1964) 356–62; *Trans.*, AIME, **231**.

83. Hill, S.: "Génie Chemique," *Chem. Eng. Sci.* (1952) **1**, No. 6, 246.

84. Blackwell, R.J. *et al.*: "Recovery of Oil by Displacement with Water-Solvent Mixtures," *Trans.*, AIME (1960) **21**, 293–300; *Miscible Processes,* Reprint Series, SPE, Dallas (1965) **8**, 103–110.

85. Koonce, K.T. and Blackwell, R.J.: "Idealized Behavior of Solvent Banks in Stratified Reservoirs," *Soc. Pet. Eng. J.* (Dec. 1965) 318–28; *Trans.*, AIME, **234**.

86. Gardner, A.O., Peaceman, D.W., and Pozzi, A.L. Jr.:

"Numerical Calculation of Multidimensional Miscible Displacement by the Method of Characteristics," *Soc. Pet. Eng. J.* (March 1964) 26–36; *Trans.*, AIME, **231**; *Miscible Processes*, Reprint Series, SPE, Dallas (1965) **8**, 161–72.

87. Simmons, J. *et al.*: "Swept Areas After Breakthrough in Vertically Fractured Five-Spot Patterns," *Trans.*, AIME (1959) **216**, 73–77.

88. Hartsock, J.H. and Slobod, R.L.: "The Effect of Mobility Ratio and Vertical Fractures on the Sweep Efficiency of a Five-Spot," *Prod. Monthly* (Sept. 1961) **26**, No. 9, 2–7.

89. Hutchinson, C.A. Jr.: "Reservoir Inhomogeneity Assessment and Control," *Pet. Eng.* (Sept. 1959) **31**, B19–26.

90. Warren, J.E. and Cosgrove, J.J.: "Prediction of Waterflood Performance in a Stratified System," *Soc. Pet. Eng. J.* (June 1964) 149–57; *Trans.*, AIME, **231**.

91. Marsden, S.S. Jr. *et al.*: "Use of Foam in Petroleum Operations," *Proc.*, Seventh World Pet. Cong., Mexico City, April 2–7, 1967.

92. Raimondi, P., Torcaso, M.A., and Henderson, J.H.: "The Effect of Interstitial Water on the Mixing of Hydrocarbons During a Miscible Displacement Process," Mineral Industries Experimental Station Circular No. 61, Pennsylvania State U., University Park, Oct. 23–25, 1961.

93. Thomas, G.H., Countryman, G.R., and Fatt, I.: "Miscible Displacement in a Multiphase System," *Soc. Pet. Eng. J.* (Sept. 1963) 189–96; *Trans.*, AIME, **228**.

94. Stalkup, F.I.: "Displacement of Oil by Solvent at High Water Saturation," *Soc. Pet. Eng. J.* (Dec. 1970) 337–48.

95. Fitzgerald, J.G. and Nielsen, R.F.: "Oil Recovery from a Consolidated Core by Light Hydrocarbons in the Presence of High Water Saturations," Mineral Industries Experimental Station Circular No. 66, Pennsylvania State U., University Park (1964) 171.

96. Tiffin, D.L. and Yellig, W.F.: "Effects of Mobile Water on Multiple Contact Miscible Gas Displacements," paper SPE 10687 presented at the 1982 SPE/DOE Enhanced Oil Recovery Symposium, Tulsa, April 4–7.

97. Raimondi, P. and Torcaso, M.A.: "Distribution of the Oil Phase Obtained Upon Imbibition of Water," *Soc. Pet. Eng. J.* (March 1964) 49–55; *Trans.*, AIME, **231**.

98. Shelton, J.L. and Schneider, F.N.: "The Effects of Water Injection on Miscible Flooding Methods Using Hydrocarbons and Carbon Dioxide," *Soc. Pet. Eng. J.* (June 1975) 217–22.

99. Shelton, J.L. and Schneider, F.N.: "The Effects of Water Injection on Miscible Flooding Methods Using Hydrocarbons and Carbon Dioxide," paper SPE 4580 presented at the 1973 SPE Annual Meeting, Las Vegas, Sept. 30–Oct. 3.

100. Griffith, J.D. and Horne, A.L.: "South Swan Hills Solvent Flood," *Proc.*, Ninth World Pet. Cong. (1975) **4**, 269–78.

101. Christian, L.D. *et al.*: "Planning a Tertiary Oil Recovery Project for Jay-Little Escambia Creek Fields Unit," *J. Pet. Tech.* (Aug. 1981) 1535–44.

102. Spence, A.P. and Watkins, R.W.: "The Effect of Microscopic Core Heterogeneity on Miscible Flood Residual Oil Saturation," paper SPE 9229 presented at the 1980 SPE Annual Technical Conference and Exhibition, Dallas, Sept. 21–24.

103. DeLaney, R.P. and Fish, R.M.: "Judy Creek CO_2 Flood Performance Predictions," paper 80-31-23 presented at the 1980 Annual Technical Meeting of the Petroleum Society of CIM, Calgary, May 25–28.

104. Mungan, N.: "Certain Wettability Effects in Laboratory Waterfloods," *J. Pet. Tech.* (Feb. 1966) 247–52; *Trans.*, AIME, **237**.

105. Holm, L.W. and Josendal, V.A.: "Mechanisms of Oil Displacement by Carbon Dioxide," *J. Pet. Tech.* (Dec. 1974) 1427–35; *Trans.*, AIME, **257**.

106. "Wizard Lake D-3A Pool-Engineering Study of Primary Performance and Miscible Flooding," Texaco Exploration Co. Application to Alberta Oil and Gas Conservation Board, Calgary (Dec. 1968).

107. Baker, L.E.: "Effects of Dispersion and Dead-End Pore Volume in Miscible Flooding," *Soc. Pet. Eng. J.* (June 1977) 219–27.

108. Gardner, J.W., Orr, F.M., and Patel, P.D.: "The Effect of Phase Behavior on CO_2 Flood Displacement Efficiency," *J. Pet. Tech.* (Nov. 1981) 2067–81.

109. Blackwell, R.J. and Koonce, K.T.: "Laboratory Model Studies: Utility and Limitations in Predicting Reservoir Performance," *Proc.*, Seventh World Petroleum Congress (1967) 169–79.

110. Coats, K.H. and Smith, B.D.: "Dead-End Pore Volume and Dispersion in Porous Media," *Soc. Pet. Eng. J.* (March 1964) 73–84; *Trans.*, AIME, **231**.

111. Yellig, W.F. and Baker, L.E.: "Factors Affecting Miscible Flooding Dispersion Coefficients," *J. Cdn. Pet. Tech.* (Oct.–Dec. 1981) 69–75.

112. Goddard, R.R.: "Fluid Dispersion and Distribution in Porous Media Using the Frequency Response Method With a Radioactive Tracer," *Soc. Pet. Eng. J.* (June 1966) 143–52; *Trans.*, AIME, **237**.

113. Goss, M.J.: "Determination of Dispersion and Diffusion of Miscible Liquids in Porous Media by a Frequency Response Method," paper SPE 3525 presented at the 1971 SPE Annual Technical Conference and Exhibition, New Orleans, Oct. 3–6.

114. Holm, L.W. and O'Brien, L.J.: "Carbon Dioxide Test at the Mead-Strawn Field," *J. Pet. Tech.* (April 1971) 431–42.

115. Bilhartz, H.L. and Charlson, G.S.: "Coring for In-Situ Saturations in the Willard Unit CO_2 Flood Mini-Test," paper SPE 7050 presented at the 1978 SPE Symposium on Improved Methods for Oil Recovery, Tulsa, April 16–18.

116. Desch, J.B. *et al.*: "Enhanced Oil Recovery by CO_2 Miscible Displacement in the Little Knife Field, Billings County, North Dakota," paper SPE 10696 presented at the 1982 SPE/DOE Enhanced Oil Recovery Symposium, Tulsa, April 4–7.

117. Griffith, J.D. and Cyca, L.G.: "Performance of South Swan Hills Miscible Flood," *J. Pet. Tech.* (July 1981) 1319–26.

118. Walker, J.W. and Turner, J.L.: "Performance of Seeligson Zone 20B-07 Enriched-Gas Drive Project," *J. Pet. Tech.* (April 1968) 369–73.

119. Caudle, B.H. and Dyes, A.B.: "Improving Miscible Displacement by Gas-Water Injection," *Trans.*, AIME (1958) **213**, 281–84; *Miscible Processes*, Reprint Series, SPE, Dallas (1965) **8**, 111–14.

120. Fitch, R.A. and Griffith, J.D.: "Experimental and Calculated Performance of Miscible Floods in Stratified Reservoirs," *J. Pet. Tech.* (Nov. 1964) 1289–98.

121. McLeod, J.G.F.: "Behavior of Enriched Gas Injected Alternately with Water in Horizontal Miscible Floods," *J. Cdn. Pet. Tech.* (April–June 1980) 51–57.

122. Slobod, R.I.: "Use of Graded Viscosity Zone to Reduce Fingering in Miscible Phase Displacement," *Prod. Monthly* (Aug. 1960) 12.

123. Mungan, N.: "Programmed Mobility Control in Polymer Floods," RR-1, Petroleum Recovery Inst., Calgary (March 1968).

124. Watkins, R.W.: "A Technique for the Laboratory Measurement of Carbon Dioxide Unit Displacement Efficiency in Reservoir Rock," paper SPE 7474 presented at the 1978 SPE Annual Technical Conference and Exhibition, Houston, Oct. 1–4.

125. Claridge, E.L.: "A Method for the Design of Graded Viscosity Banks," *Soc. Pet. Eng. J.* (Oct. 1978) 315–24.

126. Bernard, G.G. and Holm, L.W.: "Use of Surfactant to Reduce CO_2 Mobility in Oil Displacement," *Soc. Pet. Eng. J.* (Aug. 1980) 281–92.

127. Holm, L.W.: "The Mechanism of Gas and Liquid Flow Through Porous Media in the Presence of Foam," *Soc. Pet. Eng. J.* (Dec. 1968) 359–69.

128. Bernard, G.G.: U.S. Patent No. 3,330,351 (1967).

129. Bond, D.C. and Holbrook, O.C.: U.S. Patent No. 2,866,507 (1958).

130. Craig, F.F. Jr. and Lummus, J.L.: U.S. Patent No. 3,185,634 (1965).

131. Jacobs, W.L. and Bernard, G.G.: U.S. Patent No. 3,330,346 (1967).

132. Fried, A.N.: "The Foam Drive Process for Increasing the Recovery of Oil," RI 5866, USBM (1961).

133. Bernard, G.G.: "Effect of Foam on Recovery of Oil by Gas Drive," *Prod. Monthly* (1963) **27**, No. 1, 18–21.

134. Bennett, G.S.: "A Study of the Foam-Drive Process for Removal of Brine from a Consolidated Sandstone Core," MS thesis, Pennsylvania State U., University Park (1963).

135. Kolb, G.E.: "Several Parameters Affecting the Foam-Drive

Process for the Removal of Water from Consolidated Porous Media," MS thesis, Pennsylvania State U., University Park (1964).

136. Deming, J.R.: "Fundamental Properties of Foams and Their Effects on the Efficiency of the Foam Drive Process," MS thesis, Pennsylvania State U., University Park (1964).

137. Bernard, G.G. and Holm, L.W.: "Effect of Foam on Permeability of Porous Media to Gas," *Soc. Pet. Eng. J.* (Sept. 1964) 267–74.

138. Bernard, G.G., Holm, L.W., and Jacobs, W.L.: "Effect of Foam on Trapped Gas Saturation and on Permeability of Porous Media to Water," *Soc. Pet. Eng. J.* (Dec. 1965) 295–300.

139. Marsden, S.S. and Khan, S.A.: "The Flow of Foam Through Short Porous Media and Apparent Viscosity Measurements," *Soc. Pet. Eng. J.* (March 1966) 17–25.

140. Bond, D.G. and Bernard, G.G.: "Rheology of Foams in Porous Media," paper presented at the SPE-AIChE Joint Symposium, AIChE 58th Annual Meeting, Dallas, Feb. 7–10, 1966.

141. Raza, S.H. and Marsden, S.S.: "The Streaming Potential and the Rheology of Foam," *Soc. Pet. Eng. J.* (Dec. 1967) 359–68.

142. Holm, L.W.: "Foam Injection Test in the Siggins Field, Illinois," *J. Pet. Tech.* (Dec. 1970) 1499–1506.

143. Raimondi, P., Gardner, G.H.F., and Petrick, C.B.: "Effect of Pore Structure and Molecular Diffusion on the Mixing of Miscible Liquids Flowing in Porous Media," paper 43 presented at AIChE-SPE Joint Symposium, San Francisco, Dec. 6–9, 1959.

144. Orlob, G.T. and Radhakrishna, G.N.: "The Effects of Entrapped Gases on the Hydraulic Characteristics of Porous Media," *Trans.*, AGU (1958) 39, 648.

145. Singer, E. and Wilhelm, R.H.: "Heat Transfer in Packed Beds; Analytical Solution and Design Method; Fluid Flow, Solids Flow, and Chemical Reaction," *Chem. Eng. Prog.* (1950) 46, 343.

146. Fahien, R.W. and Smith, J.M.: "Mass Transfer in Packed Beds," *AIChE J.* (1955) 1, 28.

147. Latinen, G.A.: "Mechanism of Fluid Phase Mixing of Fixed and Fluidized Beds of Uniformly Sized Spherical Particles," PhD dissertation, Princeton U., Princeton, NJ (1951).

148. Claridge, E.L.: "CO_2 Flooding Strategy in a Communicating Layered Reservoir," *J. Pet. Tech.* (Dec. 1982) 2746–62.

Chapter 4
Design Considerations and Predictive Methods

This chapter is concerned with the reservoir engineering aspects of miscible flood design and performance prediction. Important factors that need to be considered in a design and performance prediction are reviewed in the first section. Subsequent sections are devoted to discussion of different methods of predicting oil production and overall miscible flood performance. As yet there is no generalized engineering method or model that adequately accounts for all factors that usually need to be considered in a design and performance prediction. Each predictive method tends to emphasize one or more aspects of the displacement while neglecting other aspects for the sake of tractability. For this reason, a good understanding of what the various methods do and don't account for is necessary as is a good understanding of basic miscible flood principles. The engineer must exercise good judgment in selecting a method appropriate for the job at hand and in evaluating the degree and direction of uncertainty in the resulting projections.

4.1 Some Important Considerations in Miscible Flood Design and Performance Prediction

The purpose of this section is to set the stage for the discussion of predictive methods that follows by reviewing some of the important factors that usually need to be evaluated in the design of a miscible flood and in prediction of miscible flood performance. Procedures or guidelines for evaluating these factors are not discussed in this section, and the reader is referred to the methods discussed in other sections of the chapter as well as to the discussion in Chaps. 2 and 3. It is hoped that from the discussion in this section the reader can see how the different models discussed in the remainder of the chapter measure up to the predictive capability ideally desired. It may not always be possible to evaluate quantitatively all the factors judged to be important for a given project because of limitations of the available predictive methods or because of uncertainty or even unavailability of the data needed for predictions. Nevertheless, an awareness of these limitations is important for the reservoir engineer to judge the quality of and potential degree of uncertainty in his design or prediction.

A miscible flood design/performance prediction should take into account the extent of miscible sweepout, and immiscible sweepout as well if miscible displacement does not occur throughout the total swept volume. In addition to pore volumes of solvent and drive fluid injected, miscible sweepout is affected by pressure distribution; size of the solvent slug; type of drive fluid; mobilities of the solvent, drive fluid, and reservoir fluids; and reservoir heterogeneity. A miscible flood design/performance prediction also should take into account the unit displacement efficiency of both miscible and immiscible displacement and should evaluate how much of the displaced oil actually will be captured by producing wells. The rate of oil recovery is important. Producing rate is affected by fluid mobilities, reservoir properties, and well injectivities and productivities. Adequate reservoir description is fundamental to any design or prediction. Additional aspects of these basic considerations are discussed in the following subsections.

Reservoir Description

The importance of starting with an adequate reservoir description cannot be overemphasized. The accuracy of any predictive technique is limited directly by the accuracy with which the reservoir can be described. Because of the high cost of miscible injection fluids compared with water and the effect of this cost on project economics, a higher degree of accuracy is required for miscible flood performance projections than for waterflooding. If an unrealistically low oil recovery is predicted, a potential project may not look attractive, whereas waterflood economics may have withstood such an error. Also, failure to achieve in practice an unrealistically high recovery prediction can be tolerated to a greater degree in waterflooding than in miscible flooding before unfavorable economics result. As a result, reservoir studies that were adequate for waterflooding may not be adequate for miscible flooding.

Availability of data for reservoir description can vary substantially from one reservoir to another. Ideally, log, core, well tests, and production data should be used to synthesize the best possible description of (1) spatial distribution of macroscopic permeability variations within the pay, (2) continuity of pay between wells, (3) distribution of fractures and faults, (4) locations of shales, (5) fluid saturations and their distribution, (6) well injectivities and productivities, and (7) rock properties such as relative permeabilities. Recent publications emphasize the importance of utilizing geological information, such as knowledge of depositional environment, in helping to construct a realistic reservoir description.[1-13] In some instances, seismic techniques can be utilized to obtain an improved reservoir characterization.[14-19] Special additional data may be required, such as additional cores, special logs, tracer tests, and well tests. Matching producing history with reservoir simulators also can be invaluable both for deriving a reservoir description and for ensuring that the description is consistent with past performance.

Miscible Sweepout

Reservoir pressure must be above the MMP at the displacing solvent front for miscible displacement to occur. Pressure distribution and the reservoir volume above the MMP can be determined from well tests or estimated with reservoir simulators. If miscible sweepout is limited too severely by reservoir pressure, some repressuring could be required.

Although the correlations described in Chaps. 6, 7, and 8 and calculations with compositional simulators are useful for screening purposes, dynamic miscibility conditions for design work and field implementation should be determined experimentally unless reservoir pressure is sufficiently high to exceed safely the miscibility pressure estimated from these correlations when the maximum likely error in the estimate is accounted for. The procedure outlined in Chap. 2 is recommended. If a gas-driven slug process is to be practiced, miscibility conditions must be determined at the trailing edge of the slug also.

Volumetric sweepout in horizontal miscible floods can be affected by gravity tonguing, viscous fingering, and channeling caused by fractures and permeability stratification. Gravity tonguing or viscous fingering may occur simultaneously with channeling caused by fractures and permeability stratification. In very thin beds, transverse dispersion may be important and can retard the growth of a gravity tongue. It is important to account for all these phenomena in miscible flood predictions. A realistic reservoir description is especially important for modeling these phenomena.

Realistic relative permeability data must be used in evaluating volumetric sweepout. The effects of multiphase flow in the solvent/oil transition zone,[20,21] miscible flood residual oil, and asphaltene precipitation

on phase permeabilities should be evaluated.

Realistic relative permeability data also are important for evaluating the proper injection ratio and the potential benefit of alternate solvent/water injection for reducing solvent mobility and improving volumetric sweepout.[22] Equally important when evaluating the solvent/water process is a realistic assessment of the influence of gravity segregation and the potential for oil trapping.

The importance of injection well condition and pattern geometry on miscible sweepout should not be underemphasized. The target interval should be isolated sufficiently that expensive injection fluids are not lost to other horizons at the wellbore. Completion and operating techniques should ensure good vertical coverage of injection fluids. The influence of irregular pattern geometry should be evaluated to ensure that producers nearest to solvent injection wells will not produce injection fluids excessively and interfere with solvent sweepout toward other producers.

The size of the solvent slug, the choice of fluid to drive the slug, and the method of operation all affect the extent of miscible sweepout achieved by the slug. Solvent slugs may be gas- or water-driven. Water may be injected alternately with the solvent. The displacement may be essentially horizontal or may be of the gravity-stable type in a steeply dipping reservoir. Each of these situations requires different design and predictive considerations.

Unequal distribution of the slug into strata of differing permeabilities is a problem encountered in any slug process if the solvent is injected into the full pay interval with no zonal control. At the end of solvent injection the lower permeability strata will have smaller slug sizes (as a percent of pore volume) than the higher permeability strata. As a result, slugs in the lower permeability strata will be dissipated in a shorter travel distance than slugs in higher permeability strata. The choice of overall slug size will depend on the strategy and criteria for optimizing project performance and economics. If the slug in the most permeable stratum is sized to be dissipated by the time it reaches the producing well, miscible sweepout in the lower permeability strata could be poor, depending on the permeability distribution. Design for high miscible sweepout in the lower permeability strata could result in substantial oversizing and production of the slugs in the higher permeability strata.

The minimum solvent concentration that can be tolerated before losing miscibility and the rate at which the solvent becomes diluted in flowing through the reservoir are two important considerations that need to be addressed in predictions involving gas-driven slugs. In gravity-stable displacements, solvent dilution results from mixing at both the leading and trailing edges of the slug. In displacements where viscous fingering, gravity tonguing, or channeling because of permeability stratification are important, mixing also occurs along the sides of the solvent protrusions. Mixing can result from molecular diffusion and from microscopic and macroscopic convective dispersion. A high water saturation, caused either by failure of a tertiary oil bank to displace water saturation back to the irreducible value or by alternate injection of water and solvent slugs to lower the solvent mobility, can increase the effective microscopic dispersion coefficient. Diffusion of bypassed oil into the slug and diffusion of solvent into by-passed oil also may contribute to solvent dilution.

The minimum permissible dilution depends on the phase behavior of the oil/solvent/drive-gas system, which can be determined experimentally or calculated satisfactorily in some cases. Methods are relatively well developed for estimating slug dilution in a gravity-stable flood (see Secs. 3.2, 3.6, and 4.6). Methods for calculating slug dilution in a finger-dominated displacement are not well developed.

With water-driven slugs, displacement at the trailing edge is immiscible. The slug is dissipated by being left as a nonflowing residual phase behind the drive-water front. If the slug size in a given stratum is too small, drive water may overtake the leading edge of the slug before the desired miscible sweepout is achieved.

With CO_2 slugs, the solubility of CO_2 in reservoir and injected water must also be taken into account in calculating slug dissipation.

Unit Displacement Efficiency

Miscible flood unit displacement efficiency may be less than 100%, and a realistic accounting for unit displacement efficiency should be made in a miscible flood prediction. The amount of residual oil saturation left in rock swept by a miscible injection fluid depends on the type of miscible process employed and on other factors such as rock properties, phase behavior, and the level of dispersion.

Oil in that reservoir volume below the MMP will be displaced immiscibly. Mechanisms such as oil swelling and viscosity reduction caused by solvent dissolving in the oil as well as the mechanisms of extraction and vaporization of hydrocarbons by the solvent affect the unit displacement efficiency of the immiscible portion of the displacement and should be taken into account. Failure to do so could cause an overly pessimistic estimate of oil recovery during any period of immiscible displacement, especially if it occurs only a few hundred pounds per square inch or so below the MMP.

Crossflow and Capture Efficiency

All the oil that is displaced in a miscible flood may not be produced.[23-25] In secondary recovery floods, some oil banked up in high-permeability regions will resaturate low-permeability regions that were depleted by solution gas drive and that may not be contacted effectively by the solvent. In tertiary recovery floods, some of the tertiary oil bank may crossflow into regions that were swept by water but may not be swept by the solvent because of limited slug size, unfavorable mobility ratio, etc. These sources of potential loss of displaced oil should be evaluated. The ability of producers to capture and to produce the displaced oil must be given careful attention in designs and predictions to maximize recovery of displaced oil and to account for losses into stagnant flow areas where the oil may be difficult to produce.

Injectivity

Injection well injectivity affects project life and can have a significant effect on project economics both through the discounting of revenues and through the drilling of sufficient injectors to achieve a reasonably short project life. It should be evaluated both for continuous solvent injection and for alternate injection of water and solvent slugs.[24] Both an oil saturation left behind the solvent front and asphaltene precipitation act to reduce injectivity, even under continuous solvent injection. A realistic representation of relative permeability is needed. Lower-than-expected injectivities, for example, have been found in at least two field tests where CO_2 was injected continuously.[26,*] Reduced injectivity is to be expected from alternate water/solvent injection, although simple two-phase relative permeability curves obtained from conventional relative permeability measurements may not be adequate for describing this method of operation.[22] Water used to drive CO_2 slugs will dissolve residual CO_2 around the wellbore gradually, which causes injectivity to increase, other factors being equal.

Field tests and/or special relative permeability measurements and laboratory testing may be required to evaluate injectivity.

Gravity-Stable Displacements

In addition to proper sizing of the solvent slug, important design considerations in gravity-stable flooding are (1) evaluation of the stable rate, (2) initial placement of the slug, and (3) coning of gas, solvent, and water.

Conditions required for gravity-stable miscible displacement were discussed in Sec. 3.6. The stable rate and critical rate for solvent displacing oil and for gas displacing solvent can be calculated with Eqs. 3.31 and 3.32. When the flood is in a previously waterflooded reservoir, the critical velocities for oil immiscibly displacing water (leading edge of tertiary oil bank) and for solvent immiscibly displacing water should be calculated by the equation given by Chuoke.[27] If the solvent/oil density difference is not great enough to give an acceptably large stable rate, it may be possible to increase the density difference by adding methane to the solvent.[28,29] In one project, butane had to be added to preserve CO_2 miscibility as methane was added to decrease CO_2 density.[28]

In gravity-stable flooding, the solvent slug is injected above the oil zone to be displaced. Drive gas then is injected above the slug to force it downward through the reservoir. If a gas cap already overlies the oil column, the solvent slug must be injected at the gas/oil contact in such a manner that it is allowed to spread entirely across the gas/oil contact before downward displacement begins. Careful attention must be given to location and completion of injection wells, injection rates, operation of producers, and time requirements to accomplish the placement of solvent.

Adequate reservoir description is crucial for evaluating the gravity-stable rate, which varies directly with rock permeability. If average permeability varies with depth because of changing lithology or because of a change in factors that affect permeability development within a lithology, the gravity-stable rate also will vary. Gravity-stable displacement is required at both the leading and trailing edges of the solvent slug. The need for an adequate reservoir description when designing a gravity-stable displacement was illustrated vividly by the performance of the Golden Spike flood, where barriers to ver-

*Results of one test are unpublished.

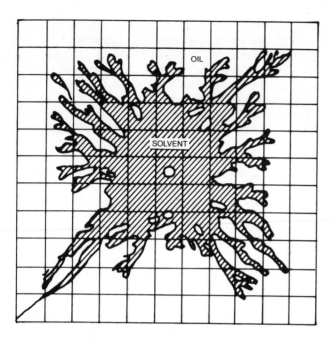

Fig. 4.1—Swept zone at solvent breakthrough for a five-spot pattern flood as observed in a Hele-Shaw (parallel plate) model, $M \cong 15$ (after Ref. 50).

tical flow, unsuspected at the project start but distributed throughout the reef, held up the solvent slug and destroyed its integrity.[30] As a result, the solvent slug was diluted rapidly, and miscible displacement was lost.

Careful attention also must be given to producing well rates during the downward displacement to avoid coning the solvent slug and drive gas into producers and thereby producing-off and dissipating the slug. Critical coning rate depends on vertical permeability in the vicinity of the producer and on height of the oil column above the completion interval.[31-38] It can vary from producer to producer as well as with life of the project. If bottom-water is present, water coning also must be considered. Ultimately, the oil column will become so thin that coning is inevitable, leading to some loss of oil.

4.2 Numerical Finite-Difference Simulators

Finite-difference mathematical simulators are probably the most commonly used tools today for projecting reservoir performance of miscible floods. In these simulators, finite differences are used to approximate derivatives in those differential equations chosen to represent the fluid flow and other mechanisms being modeled. The resulting finite-difference equations are solved by various numerical methods using high-speed computers.[39,40] In practice, the reservoir is represented by discrete grid blocks, as shown in Fig. 4.1 for two-dimensional areal modeling, and mass transport of each constituent being tracked is computed across each block face for a succession of small time increments. Pressure distribution also is calculated for each time increment.

Simulation of complex and realistic flooding conditions and operation of the model wells in a manner that simulates realistic operation of a miscible flood are the chief advantages of finite-difference simulation. Finite-difference simulators can calculate multiphase fluid flow

in two and three dimensions. They can treat large areas of a field with multiple wells, and wells can be managed in the simulator much as they would be in practice — i.e., added or abandoned with time, completed in different intervals, injection or producing rates and pressures varied with time. These simulators can incorporate the spatial distribution of rock heterogeneity into a calculation (to the extent that it is known), and they can simulate its influence and that of injection and producing well locations on sweepout, crossflow, and capture efficiency. Finite-difference simulators can calculate gravity segregation and gravity tonguing, and they can simulate miscible flooding starting at various stages of primary depletion or waterflooding.

However, a general finite-difference simulator that simultaneously models all of the important aspects of miscible displacement discussed in Chaps. 2 and 3 has not been feasible yet because of computing time, storage, and cost considerations. Instead, finite-difference modeling techniques to date have been limited to describing only selected features of process behavior. These techniques have resulted in simulators that generally fall into one of two categories: modified black-oil simulators and compositional simulators.[41] Utility of each kind of simulator is discussed in the following subsections, but a discussion of the finite-difference approximations and solution methods, and of the many other techniques that are important for constructing a workable finite-difference simulator, is not given. Several reference texts are available for this.[39,40]

Modified Black-Oil Simulators

From the discussion in Sec. 3.2 concerning fluid mixing during miscible displacement caused by diffusion and convective dispersion, the diffusion/convection equation would seem to be a plausible starting point for developing a miscible flood simulator. Peaceman and Rachford[42] took this approach and approximated the diffusion/convection equation with finite differences. In their finite-difference simulator, solvent was first-contact miscible with a single-component "oil." They demonstrated that with a sufficiently large number of grid blocks their simulator accurately calculated the analytical solution for one-dimensional miscible displacement and also adequately calculated the behavior of two-dimensional laboratory displacements that were dominated by viscous fingering. Viscous fingers in the simulator calculations were initiated by small random variations in grid-block permeabilities. However, there has been limited use of this modeling approach, partly because standard finite-difference approximations introduce a numerical dispersion that usually is many times greater than the true physical dispersion of solvent and oil unless a very large number of grid blocks are used in the simulation. Reducing numerical dispersion to manageable levels by increasing the number of grid blocks generally has been impractical in the past for field problems because of computer storage and computing time requirements.

Garder *et al.*[43] proposed a method of moving points incorporated into a finite-difference simulation for accurately calculating mixing caused by physical dispersion in a miscible displacement. This technique is based on the method of characteristics and introduces no numerical dispersion, but it has been so difficult to apply for complex flow geometries and for solvent slug calculations that it too has seen only limited application in miscible flood prediction.

Other techniques have been proposed for dealing with numerical dispersion in finite-difference solutions of the diffusion-convection equation, but they have proved so far to be of limited usefulness for reservoir-size problems.[44,45,53]

Interestingly, simulators based on solutions of immiscible flow equations containing no physical dispersion term have been used more extensively in the past for reservoir calculations of miscible displacement than simulators that are solutions of the diffusion/convection equation. For many years, black-oil simulators have been used to calculate reservoir performance under solution gas drive, immiscible gas flooding, and waterflooding. In these simulators, reservoir oil is represented by two components, stock-tank oil (or black oil) and gas, and differential equations are solved describing immiscible flow of reservoir oil, water, and gas.[47,48] Lantz[49] noted that a rigorous analogy exists between the equations for two-phase flow of gas and oil in black-oil simulators and the diffusion/convection equation for first-contact miscible displacement providing mixing during miscible displacement results primarily from molecular diffusion, and providing the expressions for gas and oil relative permeabilities and capillary pressure in the black oil model take the form

$$k_{rg} = \mu_g S_g / \mu(S_g) ,$$

$$k_{ro} = \mu_o (1 - S_g) / \mu(S_g) ,$$

and

$$P_c = \frac{D_o \phi}{k} \int \frac{\mu(S_g) dS_g}{S_g (1 - S_g)} + \Delta \rho \int dh(x, y). \quad \ldots \ldots (4.1)$$

In Eq. 4.1, S_g, the gas saturation in the immiscible flow equations, also represents the volume fraction of solvent in a miscible flood simulation, and $\mu(S_g)$ is the viscosity of the solvent/oil mixture containing volume fraction S_g of solvent. By modifying the black-oil simulator relative permeability and capillary pressure functions in this manner, the piston-like feature of first-contact miscible displacement is simulated through the relative permeability curves, and mixing by diffusion is simulated through the capillary pressure function. However, as in solution of the diffusion/convection equation, numerical dispersion again typically causes much larger mixing zones than would be expected from the physical diffusion being simulated through the capillary pressure function; and because of the exaggerated mixing caused by numerical dispersion, modified black-oil simulators typically do not calculate the effect of physical dispersion on flow behavior accurately, as, for example, the dilution of solvent slugs by mixing or solvent/oil mixing in a gravity tongue.

Large grid blocks cause a different problem in the simulation of viscous finger-dominated displacements. Accurate calculation of sweepout and oil recovery requires that the fine structure of the unstable front be reproduced in detail, which necessitates use of so many grid blocks that it has been impractical in the past for all but the simplest reservoir systems. The problem is illustrated in Fig. 4.1, which shows both the solvent-invaded region observed in a Hele-Shaw five-spot model and a network of simulation grid blocks superimposed on the five-spot pattern. The 11×11 grid system obviously is too coarse to resolve the fine structure of the viscous fingers even though it is still much more detailed than those usually employed in multiwell simulations. It is apparent that many of the grid blocks represent a Hele-Shaw model volume that contains solvent fingers of relatively high solvent concentration, oil fingers of relatively high oil concentration, and the mixed zones between these fingers. In reality, these solvent and oil fingers in the model volume represented by a grid-block flow at different relative velocities. However, a basic assumption of both the diffusion/convection simulator and the modified black-oil simulator is that the solvent and oil within a grid block are homogeneously mixed, and, as a consequence, these components are computed to flow at the same velocity. Thus, even when permeability perturbations are introduced among the grid blocks in these simulators to initiate fingering, frontal instability will be substantially retarded in coarsely gridded simulations, causing optimistic sweepout to be computed.

Todd and Longstaff proposed to improve the simulation of sweepout and oil recovery in viscous finger-dominated displacements with modified black-oil simulators by assuming partial mixing when calculating solvent and oil component viscosities.[50] They recommended modifying gas and oil relative permeabilities for immiscible flow simulators according to the following expressions.

$$k_{ro} = \frac{S_o}{S_n} \cdot k_{rn}$$

and

$$k_{rg} = \frac{S_g}{S_n} \cdot k_{rn} , \quad \ldots \ldots \ldots \ldots \ldots \ldots (4.2)$$

where $S_n = S_o + S_g$, and $k_{rn} = k_{rn}(S_w)$ is the imbibition relative permeability of the nonwetting phase. They also recommended calculating effective viscosities for the oil and gas equations by the following formulas.

$$\mu_{ge} = \mu_g^{1-\omega} \cdot \mu_m^{\omega}$$

and

$$\mu_{oe} = \mu_o^{1-\omega} \cdot \mu_m^{\omega} , \quad \ldots \ldots \ldots \ldots \ldots \ldots (4.3)$$

where μ_m is the viscosity of the mixture resulting from homogeneous mixing of the total oil and solvent volumes contained within a grid block and ω is a mixing equation parameter. A value of $\omega = 1$ corresponds to complete mixing within a grid block, whereas a value of $\omega = 0$ corresponds to the assumption of negligible mixing. When ω has a value less than one, the effective viscosity of the solvent component in a given grid block will be less than the effective viscosity of the oil component. Solvent then will flow out of the grid block at a higher velocity than the oil component, thereby simulating the behavior of viscous fingering.

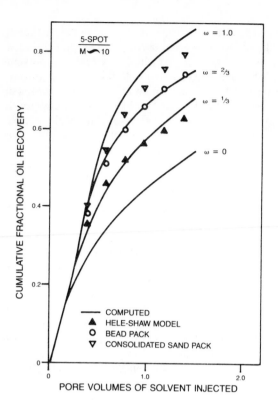

Fig. 4.2—Miscible displacement performance in a confined five-spot pattern for $M \cong 10$ (after Ref. 50).

Expressions for the effect of partial mixing on densities of the solvent and oil fingers also were suggested by Todd and Longstaff[50] and by Aziz and Settari.[39]

The mixing parameter method is analogous to the Koval method discussed in Sec. 3.4 for simulating viscous fingering in that both methods simulate viscous fingering through a solvent/oil fractional flow curve.* The mixing parameter method does not calculate the detailed structure of the viscous fingers. Instead, the technique is a means for approximating the *influence* of viscous fingering on sweepout and oil recovery in coarsely gridded simulations provided that a value for the mixing parameter can be estimated to characterize the particular problem of interest adequately.

Adequate estimation of ω, however, is a significant limitation of the method currently. There is only a limited amount of published material to aid in this estimation. Warner[23] concluded from a theoretical analysis of linear and radial flow data at unfavorable mobility ratios that a value of 0.8 might be an upper bound. Todd and Longstaff[50] estimated ω values that characterized unstable two-component displacements in various laboratory five-spot models by comparing computed and experimental sweepout behavior. Fig. 4.2 shows their comparison for displacements at a mobility ratio of 10. The degree of reservoir mixing modeled by the homogeneous laboratory experiments of this figure ranges from mixing caused by molecular diffusion (Hele-Shaw model) to mixing great enough to be caused by a large degree of macroscopic dispersion (small, con-

solidated sand model). Mixing parameter values ranging from one-third to greater than two-thirds were required for the Todd-Longstaff black oil simulator calculations to span the range of experimental sweepout data.

Todd and Longstaff[50] also reported that the value of ω required to characterize displacements in a given five-spot model was relatively insensitive to the nominal mobility ratio of the displacement as long as a sufficient number of grid blocks were used to obtain a convergent solution with the mixing parameter finite-difference simulator. A 5×5 grid was reported adequate for a quarter five-spot simulation with mobility ratios between 40 and 85, but a 7×7 grid was required for a mobility ratio of 10.

The mixing parameter estimates discussed in the preceding paragraphs utilized data from first-contact miscible displacements with no mobile water or gas present. Conceivably, the type of displacement (first-contact vs. dynamic) and initial saturation distribution also could influence the mixing parameter, as might pattern geometry. However, investigations are lacking to confirm or disprove this speculation.

Several authors[51,52] have attempted to estimate a value for the mixing parameter by history-matching field pilot tests. The history match gives a value of the mixing parameter that may be used for making predictions of full-scale performance, providing that (1) the mixing parameter was used to account for viscous fingering only in the pilot test and not effects of other influences on oil recovery and (2) process variables and reservoir heterogeneity do not change appreciably from test conditions.

When other data are lacking, the limited work to date suggests a value in the range of 0.5 to 0.7 as a first approximation.

Other features that are sometimes included in mixing parameter black-oil simulators are (1) residual oil saturation to miscible flooding, (2) miscibility or complete immiscibility as a function of pressure, and (3) water blocking of oil.[51,52]

Finite-difference simulators, black oil as well as compositional, also may suffer from a numerical phenomenon called the "grid orientation effect," which causes convergence of the solution to two different answers, depending on whether the grid is oriented parallel or diagonal to injector/producer pairs.[39,46,53,54] The grid orientation effect tends to become more pronounced as the mobility ratio becomes more adverse. The severity of the effect is illustrated in Fig. 4.3 for simulations of first-contact miscible displacement in a five-spot pattern. Comparisons with variational model calculations, Sec. 4.4, are shown on this figure also.

The grid orientation effect appears to be caused by inadequate estimation of the transmissibility between grid points when the five-point finite-difference formula is used for the finite-difference approximation of spatial derivatives.[39] With this finite-difference scheme, which has been used extensively in reservoir simulators to date, fluids cannot flow into or out of grid blocks diagonally. Several transmissibility weighting schemes to reduce the grid orientation effect have had limited success.[53,55]

The grid orientation effect can be reduced by either variational or nine-point finite-difference approxima-

---

*There are important differences, however. The Koval method utilizes an equation with fixed constants to characterize the effective mobility ratio, whereas in mixing parameter models, ω is a parameter that can be adjusted to characterize the degree of mixing occurring in the particular system of interest.

tions.[39,54,56,57] Thus far, variational methods have been too cumbersome except for relatively simple problems (Sec. 4.4). The nine-point finite-difference scheme is somewhat more complex to program than the five-point scheme, and Coats[58] reports that it causes a moderate to large increase in computing time, depending on the technique used to solve the difference equations. The nine-point scheme also does not extend easily to three dimensions. When it is necessary to simulate unstable displacements with the conventional five-point approximation, the use of a parallel grid orientation gives more realistic results.[54]

Another limitation of the black-oil simulator for miscible flood performance predictions is that important near-miscible and immiscible effects resulting from interphase mass transfer are not represented at all. Solvent and oil are assumed to be first-contact miscible and the multicontact mechanism is neglected for the dynamic miscibility processes. Although this may be an acceptable approximation in many instances,[59] a more serious limitation is the inability to calculate displacement behavior satisfactorily in regions where reservoir pressure may fall below that required for miscibility but still may be sufficiently high that immiscible displacement efficiency is improved by mechanisms such as oil swelling, viscosity reduction, and extraction.

Calculation of miscible flood performance with black-oil simulators requires large high-speed computers. Computer storage and computing time requirements are greater (usually significantly greater) than for the simpler but less flexible predictive models described in Secs. 4.3 and 4.6. However, black-oil simulators are less complicated and easier to use than the compositional simulators described in the remainder of this section. Computer storage and computing time requirements are not nearly as great as for the compositional simulators.

Compositional Simulators

In compositional simulators, the oil and gas phases are represented by multicomponent mixtures rather than by the single or binary component representation of black-oil simulators. The flux of each component is computed from one grid block to another as the component's volume, mass, or molar fraction times the flux of the phase. Ordinarily, the component flow equations do not include a physical dispersion term, although dispersion recently was included in a modified compositional simulator designed for chemical flooding.[60] Equilibrium compositions of the phases coexisting in each grid block are determined from flash calculations using K-value correlations[61-70] or from solving equations of state at the thermodynamic equilibrium condition of equal fugacity of a given component in each phase present.[71-76] Phase properties are computed from correlations or equations of state. Thus, in principle, the effect of phase behavior on development of miscibility and on the displacement efficiency of both miscible and immiscible displacement can be taken into account with compositional simulators.

Compositional simulation in principle offers some significant advantages over black-oil simulation. This is particularly true for situations involving immiscible or near-miscible displacement behavior when compositional-dependent mechanisms such as vaporization, conden-

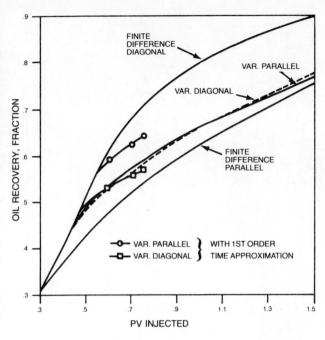

Fig. 4.3—Oil recovery in a five-spot with M = 41 calculated using different grid orientations (after Ref. 54).

sation, and oil swelling are important or when displacement behavior is affected significantly by a compositional dependence of viscosities and densities. Effect of composition on IFT and the effect of IFT on residual oil saturation, while not included routinely in most compositional simulators, can be handled in a relatively straightforward manner.[76] Historically, compositional simulators have seen greatest use in problems involving primary depletion or immiscible gas injection in volatile oil reservoirs or pressure depletion or gas cycling in condensate reservoirs.

Compositional simulators also offer some significant potential advantages for miscible displacement calculations. In principle, they can account for the effects of phase behavior, multicontact miscibility, and compositional-dependent phase properties, such as viscosity and density, on miscible sweepout and unit displacement efficiency. They also can account for compositional effects that improve immiscible displacement efficiency in regions of the reservoir that may be close to but still below miscibility pressure. However, because of a number of limitations that have not yet been resolved, application of fully compositional simulations to miscible displacement has been limited and generally of a research nature,[69-71,74-77] although there have been a few miscible flood field studies reported.[78,79]

Calculation of phase behavior generally is more demanding for miscible displacement simulations than for simulations of composition-dependent immiscible displacement in volatile oil or gas condensate reservoirs. This is because those processes that attain dynamic miscibility, as well as some first-contact miscible displacements, require calculation of phase compositions and properties in the plait-point region, whereas immiscible displacement phase behavior usually is removed from the plait point. Consistent, stable convergence of gas and oil phase compositions, densities,

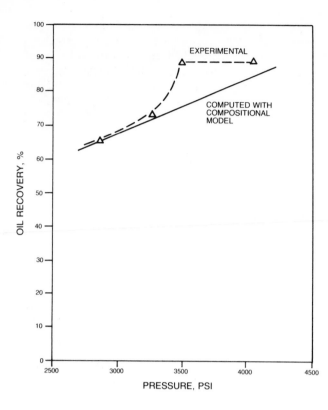

Fig. 4.4—Comparison of experimental and computed ultimate recovery for vaporizing-gas drive displacements in a long sandpacked tube.

and viscosities as the plait point is approached can be difficult to achieve in compositional simulation of dynamic miscible processes. Moreover, K-value and physical property correlations as well as equations of state invariably are less accurate in the plait-point region, and experimental data in this region against which to calibrate the predictive equations typically are lacking.

Errors made in the calculation of plait-point region phase behavior can result in an erroneous calculation of the pressure or gas composition at which miscibility is attained, which in turn results in an erroneous calculation of the effect of pressure and/or gas composition on displacement efficiency. In addition to being tested against static-cell PVT data, compositional simulator calculations also should be tested against experimental slim-tube data to validate the simulator's ability to calculate miscibility conditions. For example, Fig. 4.4 compares a series of one-dimensional compositional simulator calculations with results of experimental vaporizing-gas drive displacements in a long, sandpacked tube. The compositional simulator erroneously calculated a displacement efficiency characteristic of miscible displacement to occur first at a pressure almost 800 psi higher than observed in the sandpack displacements, even though the phase behavior package of the simulator was calibrated with experimental PVT data.* However, when phase behavior can be predicted with sufficient accuracy, it is possible to simulate slim-tube miscible displacement behavior adequately, although a large number of cells may be required.[76,77,80]

Calculation of reservoir oil/solvent phase behavior, in the plait-point region as well as in regions removed from the plait point, has improved steadily since compositional simulators began to be considered seriously for miscible displacement calculations, and it continues to improve. Recent developments in calculating complex phase behavior with equations of state show considerable promise even with CO_2 as the solvent.[72,73,81-85] The state of the art continues to be such, however, that calibration of the phase behavior packages against experimental PVT data within or close to the phase behavior region of interest and testing the simulator against slim-tube results are advisable.

Other limitations of current compositional simulators for miscible displacement application include (1) the large number of components that may be required for accurate calculation of complex phase behavior, (2) inaccuracy caused by numerical dispersion, and (3) inaccuracy caused by failure to model viscous fingering. As a result of these limitations, computing time and storage requirements for acceptable accuracy may become excessive, depending on the nature of the problem being solved.

Fussel,[81] for example, using a modified version of the Redlich-Kwong equation of state, found that tens of components were required to predict accurately the complex phase behavior observed behind some CO_2/oil displacement fronts. Watkins[59] reported that a nine-component compositional simulator similar to the one described by Nolen[65] required approximately 20 times more computing time than a modified black-oil simulator* to predict miscible displacement behavior in a vertical cross section represented by 6 vertical cells and 10 horizontal cells. Young and Stephenson,[86] however, described a CO_2 flood problem in which the phase behavior and physical properties could be represented satisfactorily by a technique involving three pseudocomponents. For this problem, they concluded that black-oil and compositional simulator computing times were similar.

Numerical dispersion affects compositional simulator calculations in a number of ways. Laboratory experiments such as slim-tube displacements show that dynamic miscibility is achieved over a distance that is small compared with reservoir dimensions, perhaps on the order of several feet or less (Sec. 2.8). Calculating the attainment of dynamic miscibility over this short a distance is typically unfeasible with compositional simulators. Several authors[75,87] have reported that so many grid blocks were required to achieve dynamic miscibility within a few feet of the injection well that the simulation was unreasonable for field simulations. Since in field simulations there are typically a relatively small number of grid blocks between wells, a significant distance between wells may be traveled by the solvent before dynamic miscibility is achieved. Watkins[59] reported an example in which reservoir pressure was "well above" the dynamic miscibility pressure and in which miscibility was developed after the solvent had traveled through only one or two grid blocks. However, even in this example, each grid block represented a

---

*Unpublished data, ARCO Oil & Gas Co.

*Sequential, semi-implicit solution method.

travel distance of 264 ft.

Coats[71] reported one-dimensional condensing-gas drive calculations that showed the influence of numerical dispersion on the rate of advance of the miscible front and on oil recovery. Results did not stabilize with 40 grid blocks between wells or perhaps even with as many as 80 grid blocks, and Coats concluded that "further effort is required to analyze and reduce this numerical dispersion." Other authors report as many as 25 to 300 grid blocks being required for simulation of dynamic miscibility in one-dimensional displacements.[69,75,76]

Numerical dispersion also should be expected to affect the computed mixing of transition-zone fluids into the multiphase region, the computed dilution of solvent slugs by mixing with oil and drive gas, and the slug size computed as being required to achieve a given miscible sweepout.[54] Attempts to reduce numerical dispersion in compositional simulators have been reported by several authors,[69,70,87] but numerical dispersion remains one of the chief accuracy problems in modeling miscible displacement.

The problem of modeling viscous fingering already has been discussed for black-oil simulators. Complete mixing of fluids within a grid block typically has been assumed for compositional simulators to date. Thus, the simulators suffer from an inability to represent viscous fingering and its effect on sweepout and oil recovery except with very fine and usually impractical grids.

Todd[41] proposed a number of simplifications and modifications to compositional simulators to take advantage of the best features of both compositional and black-oil simulators for miscible displacement problems. His proposals include the following.

1. Neglect the multicontact mechanism and calculate all displacements as first-contact miscible when local conditions of pressure and composition exceed miscibility requirements as determined from laboratory experiments.

2. Represent reservoir fluids by a minimum number of components determined from experimental phase behavior data, and make compositional calculations to account for oil swelling, light and intermediate component extraction, CO_2 solubility, and viscosity and density modification when local conditions do not meet miscibility criteria.

3. Represent residual oil to miscible flooding by an immobile phase, represent its effect on mobile phase transport through a resistance, and represent solvent solubility in the residual phase through K-values, all experimentally determined.

4. Represent water blocking by an experimentally derived equilibrium function relating trapped oil to water saturation.

5. Represent viscous fingering by the mixing parameter method.

Chase and Todd[88] describe the formulation of such a model, and Todd et al.[24] and Claridge[25] describe reservoir studies with this model.

Such an approach requires substantial experimental data. Todd[41] proposed estimating the mixing parameter by running a simplified high-resolution simulator built both for speed and a large number of grid blocks. His idea was to calculate viscous fingering behavior explicitly with the high-resolution simulator and then to estimate the mixing parameter by history-matching results of the high-resolution simulator with the mixing parameter field simulator. There would be only two phases in the high-resolution simulator, wetting and nonwetting, and the nonwetting phase would contain only two components, oil and solvent. The high-resolution simulator would represent only a fraction of a pattern element and would be gridded finely enough to initiate viscous fingers and calculate finger growth. The validity of the high-resolution approach for estimating the mixing parameter has not yet been demonstrated.

Compositional simulators or modified compositional simulators undoubtedly will be used increasingly for reservoir problems as improvements continue.

Field Studies With Finite-Difference Simulators

Guidelines for selecting a compositional instead of a modified black-oil simulator for miscible displacement calculations are not well defined nor generally accepted. Watkins[59] compared predictions made with both types of simulators for two vertical cross-sectional test problems. Both problems were CO_2 miscible displacements after waterflooding. Model properties and test conditions were such that miscible flood performance was dominated by gravity tonguing. In the first problem, reservoir pressure was "well above" the miscibility pressure and oil rate predictions of the two simulators were comparable. In the second problem, reservoir pressure was close to the miscibility pressure and free hydrocarbon gas was present. Moderately higher oil rates were predicted by the black-oil model for about half the producing life. Peak oil rates differed by about 25%. However, there was no way to tell which simulator prediction was most nearly correct.

Common practice when making field studies with finite-difference simulators is to calculate performance for one or more representative patterns or field elements and to scale up the results for these calculations to reflect larger-scale flood behavior. This approach not only makes a study more tractable with regard to computer storage and computing time but also allows more grid blocks between wells, which helps reduce the effect of numerical dispersion and improve the accuracy of the calculation. Occasionally the entire field may be modeled, if it is small enough for the simulation to be tractable. Until recently, most black-oil simulation studies of immiscible displacement used up to about 3,000 grid blocks.[58] The trend is toward larger simulations, and two studies using more than 10,000 grid blocks have been reported.[58,89] The number of grid blocks feasible for a black-oil simulation of immiscible displacement also should be feasible, at least approximately, for black-oil simulation of miscible displacement. Because of the increased storage requirement, compositional model studies have been limited to substantially fewer grid blocks.

Because of cost and complexity, three-dimensional simulations have been used sparingly in field studies. Usually, the bulk of field study calculations have been made in two-dimensional areal and/or vertical cross-sectional modes to study performance sensitivity to process and reservoir variables. Two-dimensional calculations also have been used extensively for final performance estimates. In some studies, final projections have

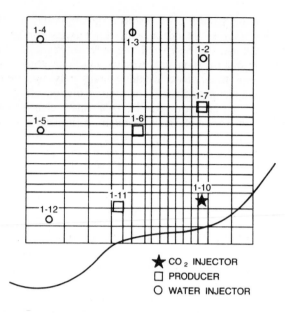

Fig. 4.5—16 × 16 areal grid (after Ref. 52).

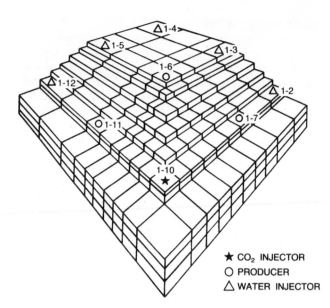

Fig. 4.6—Three-dimensional 10 × 10 grid for variable thickness case illustrating validity of pseudorelative permeability technique for estimating three-dimensional behavior (after Ref. 52).

been made with a three-dimensional model after sensitivity studies were performed with two-dimensional models.

The remainder of this section discusses some recent, selected examples of miscible flood field studies to illustrate the types of problems receiving finite-difference simulator application and to illustrate current practice in using these simulators.

Example 1

CO₂ Flood Study With a Mixing-Parameter Type Black-Oil Simulator; Horizontal Flooding in a Relatively Homogeneous Sandstone Reservoir. Youngren and Charlson[52] used a mixing-parameter type simulator to history match performance of the Little Creek CO_2-flood pilot test and to project performance for various slug sizes and CO_2/water injection ratios. In this tertiary recovery test (Fig. 4.5), three producers surrounded a single injector to the north and the west, and the reservoir shale-out boundary confined fluids to the east and the south. Backup water injectors helped confine fluids to the test area. CO_2 was injected continuously, without alternate water injection, into the 30-ft average thickness sandstone.

These authors used two 2-dimensional cross-sectional models with 7×10 and 7×5 gridding (X-Z) to develop pseudofunctions[90] to account for vertical flow phenomena in a subsequent areal model. The cross-sectional calculations showed that tertiary oil recovery would be dominated by gravity tonguing and crossflow of displaced oil to below the gravity tongue. A 16×16 grid areal model (Fig. 4.5) with pseudofunctions was used to represent the pilot test area, including all wells, and to calculate pilot test performance. A value of 0.7 for the mixing parameter was estimated by simulating laboratory nine-spot sweepout data with the model. This value of ω was used in subsequent calculations of pilot

test performance with the areal model. Performance computed with the areal model using pseudofunctions was compared with performance computed with a 10×10×10 grid three-dimensional model (Fig. 4.6) to test the validity of the two-dimensional areal calculations. Validity was established by comparisons such as the one shown in Fig. 4.7 for total oil production.

Figs. 4.8a and 4.8b show the final history match of pilot test oil and gas production made with the areal model. The most important history match variable was the east/west-to-north/south permeability ratio. These comparisons show the mixing-parameter black-oil simulator did a good job of representing miscible flood behavior in this pilot test.

Example 2

Vaporizing-Gas Drive Study With a Modified Black-Oil Simulator; Horizontal Flooding in a Stratified Carbonate Reservoir. Christian et al.[91] described performance estimates for a proposed vaporizing-gas drive flood with nitrogen in the Jay-Little Escambia Creek Fields Unit. The reservoir is a stratified limestone-dolomite of about 350-ft average thickness. The field was being waterflooded on a 3:1 staggered line-drive pattern of 640 acres when the simulation study was made. Fig. 4.9 shows the location of wells and size of the unit.

These authors used a black-oil simulator modified according to Lantz[49] and including the provision to specify a minimum miscible-flood residual oil saturation. They felt that it was not necessary to use a compositional simulator because the average reservoir pressure was approximately twice the MMP. Most of the simulation work was done with three 2-dimensional vertical cross sections of variable widths.

The cross sections were constructed to model a one-fourth symmetry element of the 3:1 line drive (Figs.

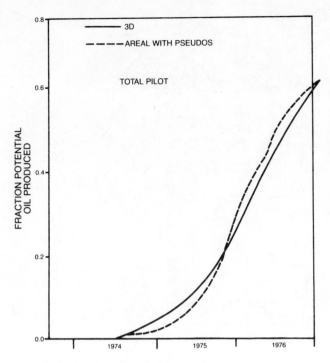

Fig. 4.7—Comparison of oil production, three-dimensional and areal with pseudo-functions (after Ref. 52).

4.10 and 4.11) and to represent the reservoir in different areas of the field (Fig. 4.9). The largest of the cross sections had a 21×30 (X-Z) gridding (Fig. 4.11).

Christian *et al.*[91] calculated miscible flood performance with these models for a 20% HCPV nitrogen slug injected alternately with water at various water/gas injection ratios. The nitrogen slug was driven ultimately with water. Waterflooding history was simulated before beginning miscible flood calculations. The simulations showed that miscible displacement behavior was influenced in a complex way by stratification, gravity segregation, and oil crossflow between strata. Calculated incremental recovery varied from 5.2 to 11.2% OOIP for the three models. Two 3-dimensional models with eight vertical layers (X-Y gridding not given) were constructed to evaluate the impact of areal sweepout on the cross-sectional calculations.

The cross-sectional results were scaled up and combined to estimate full-scale performance by a method described by these authors.

Example 3

CO$_2$ Flood Study With a Mixing Parameter-Type Black-Oil Simulator; Horizontal Flooding in a Stratified Carbonate Reservoir. Another example of miscible flood performance simulation for a stratified reservoir was given by Bilhartz *et al.*[51] for CO$_2$ flooding the San Andres carbonate formation in the Willard Unit of Wasson field. This reservoir had porous and permeable zones interspersed with dense zones, and the dense zones were considered effective barriers to vertical flow.

These authors used a mixing-parameter type simulator modified in such a way that (1) a residual oil saturation could be specified for the miscible flood and (2) the

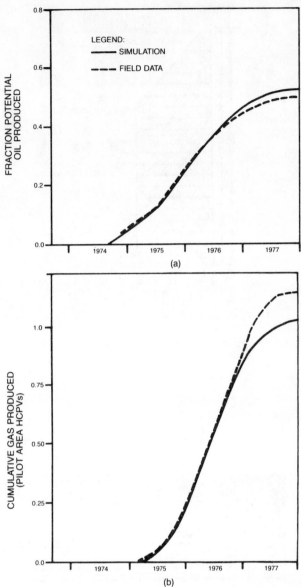

Fig. 4.8—Comparison of simulated and observed production, total pilot area (after Ref. 52).

calculation would revert to conventional immiscible gas drive for those regions of the model that were below miscibility pressure. They used a vertical cross-sectional model, gridded with 16 vertical and 20 horizontal cells, both to history match previous waterflooding performance and to predict CO$_2$-flood performance for selected areas of the field. Layer permeabilities were the primary history-match variables. Thus, since a cross-sectional model was used to simulate waterflood history rather than a three-dimensional model, the adjusted stratification derived from history matching reflected the influence of areal as well as vertical sweepout on waterflood performance.

The CO$_2$-flood predictions were made for a 20% HCPV CO$_2$ slug alternately injected with water slugs and ultimately driven with water. A value for the mixing parameter was derived by history matching the saturation

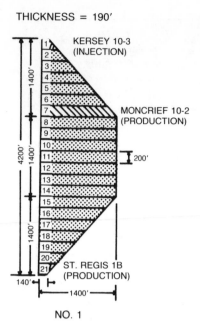

Fig. 4.9—Location of field two-dimensional cross-sectional models (after Ref. 91).

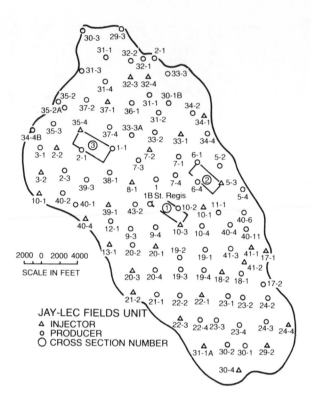

Fig. 4.11—Profiles at nitrogen gas breakthrough in St. Regis 1-B, Cross Section No. 1 (after Ref. 91).

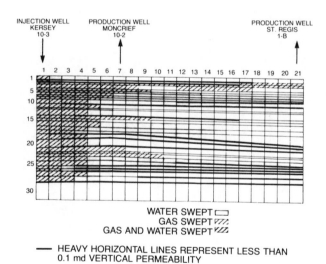

Fig. 4.10—Areal layout of cross-sectional models (after Ref. 91).

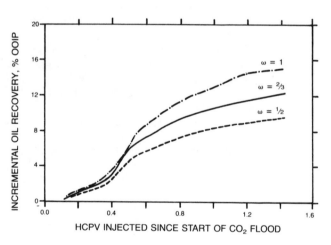

Fig. 4.12—Effect of mixing-parameter value on incremental oil recovery (after Ref. 51).

changes observed in the logging observation well of a three-well pilot test. A cross-sectional model also was used to represent the pilot test area for this history match.

Fig. 4.12 shows how the mixing parameter affected computed incremental recovery for large-scale flooding. Over a range of mixing parameters from one-half to unity, computed incremental recovery varied from about 10 to 15% OOIP. The pilot test history match indicated a mixing parameter value of about two-thirds best represented CO_2 flooding in this reservoir. The simulator calculations showed that channeling caused by permeability stratification would be a major factor in CO_2 flooding this reservoir and that oil-bank crossflow also would affect oil recovery.

Example 4

Condensing-Gas Drive Study With a Black-Oil Simulator Using Modified Relative Permeability Curves; Downward Displacement in a Reef. Gillund and Patel[92] used a three-phase black-oil simulator to calculate behavior of a vertical condensing-gas drive flood in the small, 285-ft-thick Nisku B pool. In this simulation, a solvent slug was driven with lean gas, and the three phases of the simulator represented oil, solvent, and gas. These authors simulated miscible displacement by modifying the solvent and oil relative permeability curves in such a way that the simulator calculated an efficient displacement of oil from a grid block by solvent.

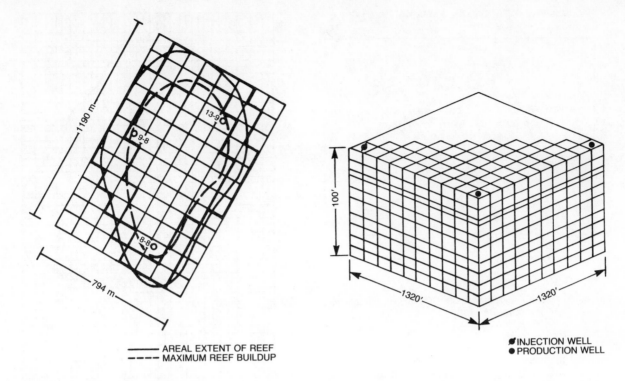

Fig. 4.13—West Pembina two-dimensional Bigoray Nisku B (after Ref. 92).

Fig. 4.14—Three-dimensional grid configuration (after Ref. 78).

— AREAL EXTENT OF REEF
---- MAXIMUM REEF BUILDUP

🗡 INJECTION WELL
● PRODUCTION WELL

Simulator equations were formulated with a nine-point finite difference approximation to overcome the grid orientation problem.[56]

The entire B pool was modeled with a three-dimensional, $8 \times 12 \times 10$ grid. Fig. 4.13 shows the areal gridding of this model and locations of the three wells. Tendency for the solvent to cone into the producers was evaluated with a separate coning simulator. The solvent slug was sized independently of the numerical simulation with analytical equations proposed by Brigham.[93] Presumably this was done because the black-oil simulator could not calculate a realistic dilution of the solvent slug by physical dispersion. An overall recovery factor of 72% was estimated by this study for an 18% HCPV slug.

Example 5

CO_2 Study With a Compositional Simulator; Horizontal Flooding in a Stratified Carbonate Reservoir. Spivak *et al.*[78] described a compositional simulator study to evaluate past performance and to project future performance of the giant SACROC Unit CO_2 flood. This study was made for the Phase 1 project area where CO_2 and water were injected alternately into areas on the flanks of the field that had not been waterflooded. The reservoir is a highly stratified carbonate reef. There had been numerous CO_2 breakthroughs at the time of the study (1974), and these authors felt it was necessary to use a compositional simulator to study performance because "in order to track the movement of CO_2, it is necessary to distinguish between hydrocarbons and CO_2 in the gas phase."

A first step was to adjust K-value, density, and viscosity correlations to calculate laboratory PVT data obtained for CO_2 and SACROC reservoir fluid. Ten components were used in subsequent reservoir simulations. The authors used a combination of one-, two-, and three-dimensional simulations to make the study. The purpose of one-dimensional simulations was to determine the effect of pressure on displacement efficiency. A series of two-dimensional calculations was made on a 20×10 grid cross-sectional model to investigate the effects of pressure level, CO_2/water injection ratio, size of the alternate CO_2 and water slugs, total CO_2 slug size, and reservoir heterogeneity on performance. A three-dimensional model (Fig. 4.14) was used for final performance predictions. This model represented one-eighth of a nine-spot pattern and was intended to represent an average of the multiple Phase 1 patterns. Layer permeabilities for this model were based on measured CO_2 and water injection profiles as well as on results of calculations for the two-dimensional model. A waterflood case and projections for various total CO_2 slug sizes were computed for the three-dimensional model.

Fig. 4.15 compares actual Phase 1 CO_2 production with compositional simulator projections for the three-dimensional model. Agreement is acceptable. Incremental recovery over waterflood of about 10% OOIP was predicted for a 16% HCPV slug. The basic curves generated by the compositional simulator were used to predict detailed performance for the SACROC Unit through an elaborate scale-up procedure.[94] In a revised forecast in 1977, the original type curves computed with the compositional simulator were adjusted to reflect producing history to that date, which resulted in an in-

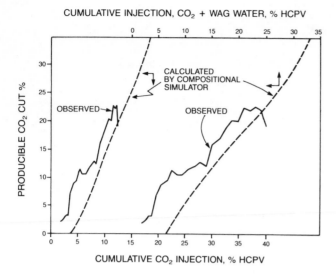

Fig. 4.15—Phase I area producible CO_2 cut vs. cumulative HCPVI (after Ref. 78).

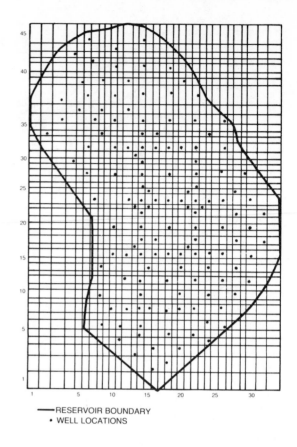

——RESERVOIR BOUNDARY
• WELL LOCATIONS

Fig. 4.16—Block 31 compositional model study on 45 × 32 areal grid (after Ref. 95).

cremental recovery prediction of 7% OOIP.[94] This latter prediction also reflected the injection of slug sizes in the 12-to-15% HCPV range.

Example 6

Vaporizing-Gas Drive Study With a Compositional Simulator; Horizontal Flooding in a Moderately Stratified Reservoir. Warner *et al.*[79,95] used a compositional simulator to history match and to predict future performance for the University Block 31 vaporizing-gas drive project. In this project, lean hydrocarbon gas is injected above miscibility pressure into one area of the field while flue gas is injected above miscibility pressure into the remainder of the field. There are four distinct pay zones, largely isolated from one another, in the 170-ft-thick Middle Devonian reservoir. The reservoir has never been waterflooded.

A compositional simulator was selected because of the objectives of the study: (1) determine effects of varying composition of the produced fluids on the gasoline plant production and plant fuel requirements, (2) predict behavior of continued current operation, (3) predict partial blowdown behavior with flue gas plant shut down, and (4) predict total blowdown behavior. The authors chose to simulate behavior of the entire field with a 45 × 32 grid two-dimensional areal model. There were enough grid blocks in this model for each of the more than one hundred wells to be represented by a separate grid block and still have several grid blocks between wells (Fig. 4.16). The authors felt the reservoir description did not warrant multilayers, even though the Middle Devonian reservoir consisted of four zones. They felt that the effects of multiple layers could be accounted for in the areal simulation by adjustments of the gas/oil relative permeability curves.

The first step in the study was to adjust the K-value and physical property correlations to match data from PVT experiments with methane, flue gas, and reservoir oil. Calculations showed that 15 components were required to match adequately both the experimental PVT

data and compositions of residual material left in gassed-out Block 31 cores. Simulation of experimental displacements in sandpacked tubes showed the simulator calculated reasonable ultimate recoveries at various pressures; however, vaporizing-gas drive miscibility was never developed fully in the two-dimensional areal calculations. Grid-block pore volumes, gas/oil relative permeability curves, and grid-block interface transmissibilities were the primary history-match variables. Five sets of relative permeability curves were used, each applied to a different region of the field. More than 90% of the GOR history matches for individual wells were judged excellent or acceptable. Fig. 4.17 shows the overall GOR history match for the field.

Ultimate recovery for continued current operation was projected to be about 60% OOIP.

Other Examples of Field Studies With Finite-Difference Simulators

Some other examples of miscible flood field studies with modified black-oil simulators can be found in Refs. 26 and 96 through 99. Some other examples of compositional simulator field studies are given in Refs. 100 and 101. Examples of reservoir studies with the modified, mixing-parameter-type compositional simulator are given in Refs. 24 and 25. In these latter two studies, the simulations attempted to account for CO_2 solubility in water, waterblocking of oil from contact by the invading solvent, asphaltene drop-out with a resulting decrease in phase permeabilities, residual oil saturation to CO_2

flooding, and enhanced immiscible displacement. In addition, the effect of viscous fingering on sweepout was simulated by the mixing parameter method.

4.3 Streamtube Models

Higgins and Leighton[102,103] originally developed the streamtube technique for making waterflood calculations, and Faulkner[104] described its application to miscible flood problems. The method involves (1) calculating streamlines describing the areal flow of fluids between wells, (2) generating streamtubes, or channels for flow, for a single layer from these streamlines, (3) calculating flow performance in each streamtube and summing to get layer performance, and (4) summing layer performances. In the method described by Faulkner, the streamtubes are divided into discrete cells with each cell having a different value, if desired, of porosity, permeability, and thickness. Flow between adjacent cells in a streamtube is described by Darcy's law for linear flow and can be calculated by numerical methods including finite-difference simulation, although simpler material balance methods for tracking fronts, approximating flow resistances, and calculating flow rate also could be used.

In streamtube models, the reservoir is represented by multiple layers that communicate at wellbores but otherwise are not in communication between wells. There is no flow of fluids by gravity segregation or crossflow between layers. Each layer can have a different permeability, and other properties such as thickness and porosity also may vary. Although this representation of a reservoir may be inappropriate in particular cases, it is not arbitrary since most oil reservoirs are essentially successive layers of sediment that may or may not be in communication.

A streamtube model calculates one-dimensional displacement behavior for each streamtube in a layer. It calculates fluid injectivity, pressure distribution, solvent sweepout, and oil, gas, and water production for a layer by summing performances of the individual streamtubes of that layer. If the streamtubes are based on a streamline model that assumes a mobility ratio of unity, the miscible sweepout calculated for an unfavorable mobility ratio will be optimistic.[105] Gravitational forces are not considered, so the method does not account for any gravity segregation or gravity override that might occur within a layer. It also does not account for the effect of viscous fingering on sweepout for those conditions where viscous fingering is an important mechanism. The method does not allow for any crossflow or transverse mixing between streamtubes, although fluid mixing along the direction of the streamtube can be incorporated into the one-dimensional streamtube calculation.

The streamtube method is not feasible for hand calculations and requires that the model equations be solved by computer. Even so, a major advantage of the method is its simplicity compared with large finite-difference simulators and the relatively small computer storage and computing time requirements compared with finite-difference simulators.

In summary, because of these restrictions, the streamtube method is best suited for problems involving water-driven solvent slugs in stratified reservoirs that have only minor communication between strata and in which both

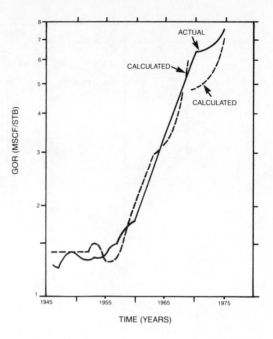

Fig. 4.17—Block 31 field GOR vs. time (after Ref. 95).

gravity segregation within a stratum is minor and the mobility ratio of the displacement is near unity (as might occur in some instances when alternate solvent/water injection is practiced). The more the intended application departs from these conditions, the less suited the streamtube model becomes for the calculations.

Faulkner[104] described the application of a streamtube model for predicting behavior of a CO_2-flood pilot test. This test was to be run in a stratified west Texas reservoir; and when Faulkner's calculations were made, the test pattern had been waterflooded, although CO_2 injection had not yet begun. Fig. 4.18 compares actual pilot test waterflood performance with performance calculated by the model after adjustment of reservoir description data. Agreement is generally good for oil rate through about 800 days of predicted performance and for water production through about 525 days. After this latter time, the author felt that the divergence between actual and calculated water production was caused by water leaving the test pattern after fillup, which the streamtube model, limited to fixed streamlines, could not simulate.

Fig. 4.18 also shows CO_2 flood performance predicted with this model for a 20% HCPV CO_2 slug injected alternately with water at a ratio of about 1.0 RB/RB and driven with a 30% HCPV slug of hydrocarbon gas injected alternately with water at a 0.8 RB/RB ratio. In these calculations, an effort was made to account for the effects of dispersion and fingering on mixing zone growth at the leading and trailing edges of the slug through a correlation[106] developed from miscible displacement experiments in long, small-diameter laboratory cores.* The predicted incremental recovery for this mode of operation was 33% OOIP.

Drennon *et al.*[105] also discuss the calculation of

---

*Considering the discussion of Secs. 2.8 and 3.4 on viscous fingering in small-diameter cores compared with reservoir-scale systems, the applicability of such a correlation seems questionable when viscous fingering is a factor affecting mixing zone growth.

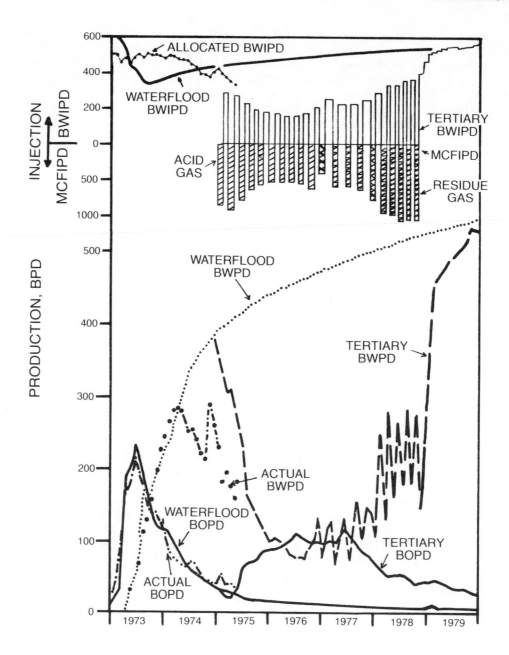

Fig. 4.18—Historical waterflood match and tertiary prediction for Well 2-79 (after Ref. 104).

CO_2-flood performance in stratified reservoirs with a streamtube model. These authors describe a method for grouping areas of similar waterflood performance that can be represented by average patterns in CO_2-flood predictions of full-scale performance. They evaluated 11 properties by this method and predicted incremental CO_2-flood recoveries in the range of 15 to 25% OOIP for CO_2 slug sizes that were calculated to be economically optimum.

4.4 Finite-Element Models

Several authors have applied finite-element methods to miscible displacement problems.[54,107-111] These methods divide the flow space into a grid network that is coarse compared with finite-difference techniques. Solutions to the flow equations are approximated with basis functions that are defined at points on the grid and that approximate the solution in a continuous manner between grid points.

The advantage of finite-element methods compared with finite-difference methods is that significantly fewer elements are required to solve problems to the same level of accuracy when the mobility ratio is unfavorable (e.g., calculation of miscible sweepout) and when reservoir-scale dispersion is important (e.g., dilution of solvent slugs, viscosity moderation in a gravity tongue or viscous fingers). For some problems, accurate calculation with a finite-difference simulator may not be feasible. Settari *et al.*[54] gave an example of slug dilution in a homogeneous five-spot pattern, computed with a 10×10 gridded finite-element model, that they estimated would have required 20,000 to 40,000 finite-difference grid blocks for comparable accuracy.* These authors also reported finite-element model calculations for continuous solvent injection at a mobility ratio of 40 into a heterogeneous five-spot. The finite-element model computed a displacement front that had the approximate character expected for viscous fingering and computed a sweepout that compared favorably with experimental data.**

Another significant advantage of some finite-element methods is the near elimination of the grid orientation effect characteristic of five-point finite-difference approximations at unfavorable mobility ratios.[54,109,110] This is illustrated in Fig. 4.3.

Some major disadvantages of finite-element methods are complexity of programming and the large amount of computing time required for generation of the basis-function coefficients. Computing time can be prohibitive for large problems, and, as a result, finite-element models to date have not been widely applied to miscible displacement calculations and have been limited to one- or two-dimensional pattern element problems of first-contact miscible, single-phase flow of two or three components.

4.5 Scaled Physical Models

Physical modeling was one of the earliest methods for evaluating performance of miscible displacement processes. In theory, flow behavior in a laboratory model can be made to duplicate that in a reservoir prototype if certain dimensionless groups characterizing the displacement have the same value in the model as in the prototype. The appropriate dimensionless groups for miscible displacement have been derived by a combination of inspectional analysis and dimensional analysis.[112,113] For convenience, the basic scaling criteria are divided into similar groups or classifications as follows.

1. Some groups describe *geometry* of the well pattern and large-scale heterogeneities — i.e., the length-to-height or thickness ratio, L/h; the length-to-width ratio, L/W; and the angle of dip, α.

2. One group relates *viscous and gravitational forces* (the viscous/gravity ratio); $u\mu_o L/kg\Delta\rho h$.

3. Some groups describe *boundary and initial conditions*.

4. Some groups relate *fluid properties*; for example, the dimensionless viscosity, μ_m/μ_s (ratio of solvent/oil mixture viscosity to solvent viscosity), dimensionless density $(\rho_m - \rho_s)/(\rho_o - \rho_s)$, and dimensionless diffusion coefficient $(D_m - D_s)/(D_o - D_s)$ must be congruent functions of concentration in model and prototype.

5. Some groups *scale effects of mixing or microscopic dispersion* (longitudinal and transverse mixing groups); (K_l/vL), $(K_t L/vh^2)$, and $(K_t L/vW^2)$.

Pozzi and Blackwell[113] point out that when mixing is dominated by convective dispersion rather than molecular diffusion, it cannot be assumed that the dispersion coefficient relationships for fluids having unequal viscosities and densities are identical to those discussed in Sec. 3.2 for equal viscosity and equal density fluids. These authors empirically determined a correlation for the effective transverse dispersion coefficient in packs of unconsolidated sand and glass beads for displacements with fluids of unequal densities and viscosities, and they discuss details of designing laboratory models when convective dispersion of fluids of unequal densities and viscosities is the dominant mixing mechanism. Ref. 113 should be consulted for details.

In practice, it generally is not possible to scale all important dimensionless groups simultaneously. For example, miscible displacement studies usually require a model that is impractically large as well as long experiment times to match simultaneously the scaling groups for longitudinal and transverse mixing, geometry, and viscous/gravity ratio in both model and prototype.

For some problems, however, it may be permissible to scale only those groups believed to have a significant effect on performance. Thus, transverse mixing may be left unscaled in a gravity-stable flood, while longitudinal mixing may be left unscaled in a horizontal displacement dominated by gravity or viscous fingering. According to Pozzi and Blackwell,[113] the group describing geometry in cross-sectional models, L/h, need not be scaled as long as flow is within flow Regions I and II as defined in Sec. 3.3, and the viscous/gravity ratio need not be scaled for displacements in Flow Region II.

Other difficulties are associated with modeling fluid and phase behavior properties, modeling tertiary recovery, and in model construction. It usually is difficult to find laboratory fluids with viscosities, densities, diffusion coefficients, etc. such that pertinent scaling groups vary with concentration in a similar manner in both model and prototype. The same is true for modeling phase behavior in slug processes and thereby modeling the resulting loss of miscibility caused by mixing and solvent dilution.

Other scaling difficulties are associated with modeling both immiscible displacement and miscible displacement in the same model. Such problems arise in modeling tertiary recovery and recovery behavior after breakdown of a miscible slug. Relative permeability and capillary pressure/saturation relationships of unconsolidated sand or glass-bead packs differ appreciably from those needed to simulate immiscible displacement processes in consolidated porous media. On the other hand, naturally occurring consolidated porous media usually have nonuniform and often unknown permeability distributions and have limited flexibility for modeling macroscopic permeability heterogeneity. Artificially consolidated media may introduce problems of rock wettability.

Hele-Shaw models have been useful for scaling the mixing caused by molecular diffusion.[115] These models are constructed of parallel plates of glass with a small fluid-filled gap between the plates. Scaling is accomplished by varying the dimensions of the model and size of the gap. Claridge devised a trapping Hele-Shaw model for studying tertiary recovery.[127] Notches in the

---

*This comparison probably gives a worst case for finite-difference methods. Methods for reducing numerical dispersion, such as two-point upstream weighting, would result in a smaller number of grid blocks being required for comparable accuracy with a finite-difference simulator. However, the number of grid blocks required for comparable accuracy still may be excessive for most problems.

**Heterogeneity was introduced by randomly varying permeability no more than ± 20% from the average value. However, Settari *et al.*[54] found that their model also computed fingers for a problem with homogeneous permeability if the dispersion coefficient was sufficiently small, apparently because numerical errors introduced enough perturbation for fingers to be initiated. A similar observation also was made by Russell.[110] Young[109] found no such behavior for a homogeneous problem as long as the numerical approximation was sufficiently accurate; thus the dominant factors causing fingering in the Settari *et al.* heterogeneous model are not clearly established.

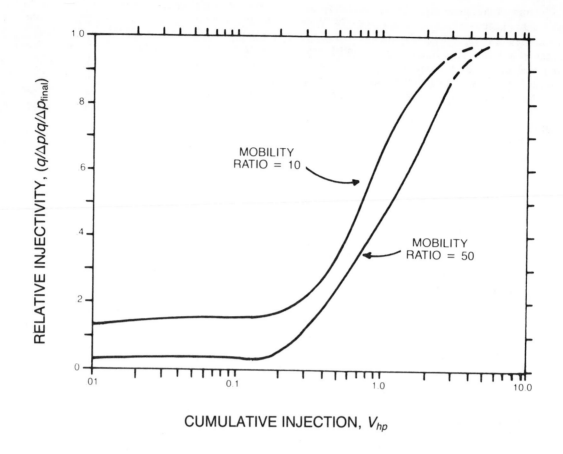

Fig. 4.19—Relative injectivity vs. volume injected, five-spot (after Ref. 121).

upper plate allowed the oil phase to be trapped when the model was waterflooded and thus simulate a waterflood residual oil saturation.

Laboratory-scaled models can be difficult, expensive, and time consuming to construct, depending on the nature of the problem being studied. Since it is impractical to satisfy all scaling criteria simultaneously and since construction techniques further restrict the scope of problems that can be physically modeled, the situation being modeled must be analyzed carefully to determine which scaling criteria may and may not be relaxed without destroying the usefulness of results.

Physical models were used in the 1950's and early 1960's to study miscible sweepout for relatively simple reservoir descriptions. To date they have had greatest application for studying secondary recovery with continuous injection of first-contact miscible fluids. They are not used widely today because of their limitations and because finite-difference simulators are usually much more powerful tools for predicting reservoir performance. However, physical models may be useful tools for special situations where finite-difference simulation is judged to be too limited (e.g., some problems involving viscous fingering or mixing by dispersion) and for providing experimental data that can be used to verify a finite-difference simulator for a particular application.

4.6 Multilayer Calculations Involving Physical Model Data for Single-Layer Performance

Craig[116] discussed a number of layered techniques for making calculations of waterflood performance in his monograph.[117-120] The waterflooding methods were adapted and modified for miscible flooding by several authors.[104,121-123] Miscible flood calculations require information on performance of the miscible process under consideration in a single layer for the pattern element in question. Preferably, the single-layer data are obtained from the type of scaled-physical-model experiments described in Sec. 4.5. Overall performance for a multilayered reservoir is calculated by making the following estimates after each small increment of total injection: (1) total injection rate and fraction of the total injection stream that enters each layer, (2) sweepout and producing cuts for each layer, and (3) overall recovery and producing well cuts. This latter information is obtained by summing the layer performances.

To calculate production vs. time, data are required relating injectivity as well as sweepout to the volume of fluid injected. Several publications discuss approximate methods for estimating single-layer injectivity vs. cumulative injection for nonsegregated flow and continuous solvent injection.[118,121,123,124] Fig. 4.19, for example, shows experimentally measured relative-injectivity data published by Fitch and Griffith[121] for a five-spot pattern at mobility ratios of 10 and 50. Single-layer injectivity vs. cumulative injection can be estimated from these data by employing the steady-state flow capacity equations of Muskat[125] or Prats *et al.*[118] to calculate endpoint injectivity. Thus, for a five-spot

$$\frac{q}{\Delta p} = \frac{0.003541\,kh}{\mu\,[\ln(L/r_w)-0.619]}, \quad \ldots\ldots\ldots\ldots (4.4)$$

where

L = distance between wells, ft, and

r_w = injection well radius, ft,

and Δp, k, h, μ, and q are in units of pounds per square inch, millidarcies, feet, centipoise, and reservoir barrels per day, respectively. For a given cumulative injection, the injection rate into each layer i is estimated from Fig. 4.19 and Eq. 4.4, and total injection rate is

$$q = \sum_i q_i.$$

Dimensionless injectivities for line-drive and nine-spot patterns also are given in Ref. 121.

An advantage of this method compared with streamtube models is that viscous fingering, gravity effects, and mixing phenomena within a single layer can be accounted for to the extent that these phenomena can be modeled physically. Major disadvantages include the expense and time required to obtain the scaled-model, single-layer data as well as the limitations discussed in Sec. 4.5 inherent with physical modeling. The method, of course, does not account for any communication between layers except at the wellbore, and in this respect has limitations similar to the limitations already discussed for the multilayered streamtube model.

The calculations for the multilayer method described above are extremely laborious and usually are impractical for hand calculation, requiring that the method be programmed for computer solution.

4.7 Single-Layer Desk Top Methods

In some instances, valuable insight about a potential miscible flood application can be obtained by judicious use of available sweepout and unit displacement data and by application of basic principles to make rough estimates of oil recovery vs. fluid injection. From these rough estimates it may be possible to identify those factors that will affect oil recovery most significantly, which in turn may aid the planning of more sophisticated evaluation methods, such as finite-difference simulation. Estimates of this nature also may be useful in some instances for quick, screening-type evaluations to determine whether a prospect merits a more detailed study. However, good judgment and caution must be exercised when using the type of simplified calculations illustrated in this section to evaluate potential process behavior even for screening purposes; for the simple calculations to have meaning, the application being evaluated must not violate appreciably the simplifying assumptions that are made in the calculation procedure.

There is no fixed procedure for making such estimates; sound application of principles and available data are the

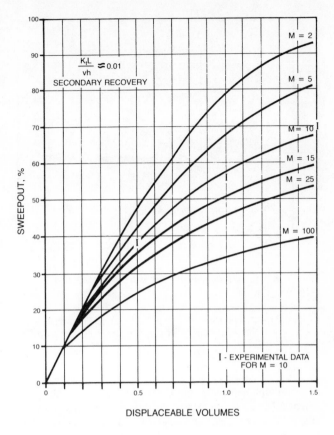

Fig. 4.20—Volumetric sweepout in a normal five-spot pattern for Region II flow.

chief requirements. The following subsections illustrate several desk top estimates.

Continuous Solvent Injection, Secondary Recovery

Sec. 3.3 gives guidelines for determining when a displacement will be dominated by viscous fingering and when it will be dominated by gravity tonguing. If volumetric sweepout is dominated by viscous fingering, the sweepout curves of Fig. 3.20 can be used in a straightforward manner. In this figure, volumetric sweepout is given vs. displaceable volumes of solvent injected for different mobility ratios. Displaceable volumes of solvent injected can be calculated by the following equation.

$$D_{vs} = \frac{Q_s}{V_p(1-S_{orm}-S_{wr})} = \frac{Q_s}{V_{Ds}}, \quad \ldots\ldots\ldots (4.5)$$

where

D_{vs} = displaceable volumes of solvent injected at reservoir conditions,

Q_s = volume of solvent injected, RB,

V_{Ds} = displaceable volume of pattern, RB,

V_p = pattern pore volume, RB,

S_{orm} = miscible flood residual oil saturation, at reservoir conditions, fraction, and

S_{wr} = irreducible water saturation, fraction.

In Eq. 4.5, Q_s is the volume of injected solvent that goes to occupy the pore space filled with solvent at saturation $1 - S_{orm} - S_{wr}$. If the solvent is appreciably soluble in reservoir brine or miscible flood residual oil, the actual

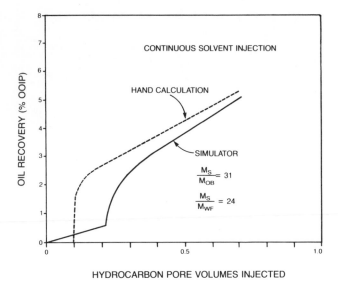

Fig. 4.21—Comparison of desk top and finite-difference simulator calculations.

volume of solvent injection will be greater than Q_s by the amount necessary to satisfy the solubility requirement. If the solvent is soluble in water, a correction to the total injected volume should be made to account for this loss when estimating displaceable volumes injected. Another correction should be made to account for solvent solubility in the miscible flood residual oil, if this information is available. Oil recovery is estimated by

$$N_p = V_p E_v (\bar{S}_o - \bar{S}_{orm}), \quad \ldots\ldots\ldots\ldots\ldots (4.6)$$

where

$\quad N_p$ = cumulative recovery from the start of solvent injection, STB,

$\bar{S}_o, \bar{S}_{orm}$ = average oil saturation at the start of injection and miscible flood residual oil saturation, both at stock-tank conditions, fraction, and

$\quad E_v$ = volumetric sweepout, fraction.

When sweepout is dominated by gravity tonguing, sweepout curves must be estimated because there is very little experimental sweepout data available. Fig. 4.20 shows volumetric sweepout estimated for a five-spot pattern when sweepout is dominated by a single gravity tongue in the Region II flow regime and when the effect of transverse dispersion on sweepout is negligible. The curves of this figure were estimated by the following procedure.

1. Construct curves of vertical sweepout vs. displaceable volumes injected for mobility ratios of 1.85, 16.3, and 69 from the data of Pozzi and Blackwell.[113]

2. Calculate the effective mobility ratios necessary to make the Dietz equation, Eq. 3.29, best predict these curves.

3. Construct a plot of effective mobility ratio (Dietz equation) vs. nominal mobility ratio and calculate vertical sweepout curves for other nominal mobility ratios

using Eq. 3.29. (Computed curves departed from the Pozzi and Blackwell data by 0.04 at D_v of 1.5, and computed curves were adjusted downward by this amount at $D_v = 1.5$ and by lesser amounts for $0.5 < D_v < 1.5$.)

4. Calculate volumetric sweepout from $E_v = E_I E_A$. Areal sweepout data for $M = 1$ were used for this latter calculation, a decision prompted by observations of areal sweepout in a five-spot laboratory model for a few unpublished volumetric sweepout experiments for Region II flow.*

Fig. 4.20 compares the sweepout curve for $M = 10$ computed by the procedure described above with the previously unpublished experimental volumetric sweepout data.

Fig. 4.20 can be used with Eqs. 4.5 and 4.6 to estimate oil recovery vs. solvent volume injected.

Continuous Solvent Injection, Tertiary Recovery

There are no published experimental sweepout curves for this situation. Mobilities of the waterflooded region, tertiary oil bank, and solvent region all influence the sweepout of solvent and recovery of displaced oil. Oil displaced by the solvent invades and resaturates waterflooded rock while it also is being swept into the producing well by solvent.

Rough estimates of oil recovery vs. solvent injected can be made with single-front sweepout curves such as those of Figs. 3.20 or 4.20 by making the following assumptions.

1. Waterflood residual oil is banked up into a tertiary oil bank that displaces water with a sharp front.

2. Solvent displaces the tertiary oil bank with a sharp front.

3. The mobility ratio governing sweepout of the solvent front is the ratio of solvent-invaded region mobility to waterflood region mobility.

4. The sweepout curve for solvent also describes sweepout of the tertiary oil bank.

5. For a given displaceable volumes of solvent injected, sweepout of the leading edge of the oil bank is estimated by calculating a pseudodisplaceable volume injected as follows.**

$$D_{vob} = D_{vs} + \Delta D_{vob}, \quad \ldots\ldots\ldots\ldots\ldots (4.7a)$$

$$\Delta D_{vob} = D_{vs} \frac{(S_{orw} - S_{orm})}{(S_{ob} - S_{orw})}, \quad \ldots\ldots\ldots\ldots (4.7b)$$

and

$$D_{vs} = \frac{Q_s}{V_p(1 - S_{orm} - S_{wt})}, \quad \ldots\ldots\ldots\ldots (4.7c)$$

---

*White, G.L.: unpublished data, ARCO Oil & Gas Co. (1979).

**To estimate sweepout of the leading edge of the tertiary oil bank, the oil displaced by solvent is assumed to correspond to an equivalent volume of injected fluid. For example, the volume of oil displaced by solvent is $D_{vs} V_p(S_{orw} - S_{orm})$. If oil actually were injected into the pattern, the total displaceable volume for oil would be $V_p(S_{ob} - S_{orw})$. To estimate sweepout of the oil bank front, the solvent-displaced oil is assumed to correspond to an equivalent displaceable volume of fluid injected—i.e.,

$$\Delta D_{vob} = \frac{D_{vs}(S_{orw} - S_{orm})}{(S_{ob} - S_{orm})}$$

and the total pseudodisplaceable volume of fluid injected is $D_{vs} + \Delta D_{vob}$.

where

D_{vob} = pseudodisplaceable volumes for estimating oil bank sweepout,

S_{ob} = average oil saturation in the oil bank, fraction,

S_{wt} = water saturation at the trailing edge of the oil bank, fraction, and

S_{orw} = waterflood residual oil saturation, fraction.

Oil bank saturations can be estimated by the Welge tangent method (see Fig. 3.45).

6. Oil recovery is calculated from

$$N_p = V_p E_s (\bar{S}_{orw} - \bar{S}_{orm}) - V_p (E_{ob} - E_s)$$
$$\cdot (S_{ob} - S_{orw}) \frac{1}{B_o} , \quad \ldots\ldots\ldots\ldots\ldots (4.8)$$

where

E_s = volumetric sweepout of solvent, fraction,

E_{ob} = volumetric sweepout of leading edge of the oil bank, fraction, and

B_o = oil formation volume factor at beginning of miscible flood, RB/STB.

In actuality, the mobility ratio affecting solvent sweepout changes during the displacement. The mobility of the waterflooded region should exert a larger influence early in the displacement, while the oil bank mobility becomes more important as the oil bank grows. The effective mobility of the tertiary oil bank may be influenced appreciably by solvent/oil mixing as described in Sec. 3.8. Mobility ratio between the tertiary oil bank and waterflooded region usually will be more favorable than the average mobility ratio estimated by this procedure. As a result, too-early oil bank breakthrough and too-optimistic initial production of tertiary oil are calculated.

There are few data against which the accuracy of this procedure can be tested. Fig. 4.21 compares the method with the result of an unpublished vertical cross-sectional calculation made with a finite-difference miscible flood simulator for Region II gravity-tongue-dominated flow. Example calculations are given in Appendix A. In this example, first production of incremental oil was estimated by the simple method to occur nearly 0.1 HCPV of solvent injection earlier than calculated by the simulator. After about 0.3 HCPV of injection, oil recovery calculated by the two methods after a given hydrocarbon pore volume of injection was in fair agreement. Excessive numerical dispersion in the simulator may have moderated the effective solvent/oil viscosity ratio unduly and caused solvent sweepout in the numerical model to be too large.

Fig. 4.22 compares actual oil recovery from the Little Creek CO_2-flood pilot test[52,126] with oil recovery estimated by this simple procedure. Details of the calculations are given in Appendix B. In this example, also, the desk top method gave an optimistic prediction — i.e., predicted recovery occurred sooner than in the pilot test and was higher than in the pilot test for a given hydrocarbon pore volume of solvent injected up to 1 HCPV.

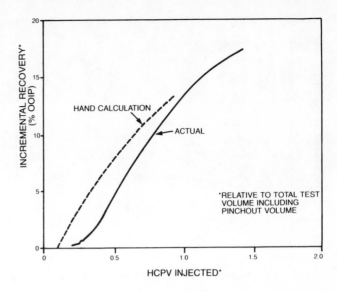

Fig. 4.22—Incremental oil production in the Little Creek CO_2-flood pilot test.

Displacement With Solvent Slugs

Three factors complicate making performance estimates for solvent slugs compared with continuous solvent injection: (1) estimating sweepout of the drive fluid, (2) estimating the influence of drive fluid on sweepout of other fronts, and (3) estimating slug dilution and loss of miscibility caused by dispersion.

Solvent Slugs Driven With Water. When the solvent slug is driven by water, loss of miscibility from mixing is not a factor. In this mode of operation, slug destruction occurs from immiscible displacement of slug by water, which leaves solvent trapped as a residual saturation. For this situation, a suggested method for making recovery estimates is as follows.

1. Use the procedure outlined for continuous solvent injection until all of the slug is injected.

2. Assume that drive-water breakthrough occurs when all remaining solvent has been left as a residual saturation.[127,128] During this period, the same sweepout curve is used for solvent and oil bank fronts that was used during solvent injection. The displaceable volumes for calculating sweepout are as follows.

Sweepout of drive-water front.

$$D_{vw} = \frac{Q_w}{V_p (1 - S_{orm} - S_{srw} - S_{wt})} = E_w , \quad \ldots\ldots (4.9)$$

where

D_{vw} = displaceable volumes of water injected,

S_{srw} = residual solvent saturation to drive water, fraction, and

Q_w = volume of water injected, res bbl.

Sweepout of solvent front.

$$D_{vs} = D'_{vs} + \Delta D_{vs} , \quad \ldots\ldots\ldots\ldots\ldots\ldots (4.10)$$

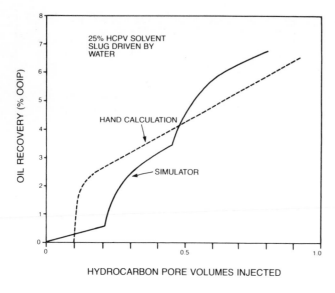

Fig. 4.23—Comparison of desk top and finite-difference simulator calculations.

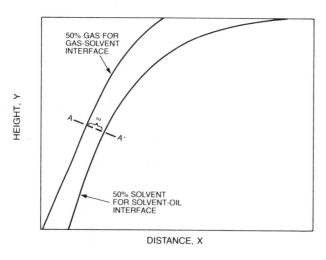

Fig. 4.24—Estimated positions of fronts for leading and trailing edges of slug.

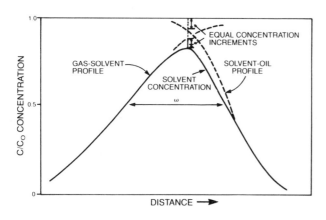

Fig. 4.25—Estimated solvent concentration along line AA′ in Fig. 4.24.

where

$$\Delta D_{vs} = D_{vw} \frac{(S_s - S_{srw})}{(1 - S_{orm} - S_{wt})}$$

$$= D_{vw} \frac{(1 - S_{orm} - S_{wt} - S_{srw})}{(1 - S_{orm} - S_{wt})} , \quad \dots\dots\dots (4.11)$$

D'_{vs} is the total solvent slug size expressed as displaceable volumes of solvent,

$$\frac{Q_s}{V_p(1 - S_{orm} - S_{wt})} , \quad \dots\dots\dots\dots\dots (4.12)$$

and S_s is solvent saturation, fraction.

Sweepout of leading edge of tertiary oil bank.

$$D_{vob} = D_{vs} + \Delta D_{vob} . \quad \dots\dots\dots\dots\dots (4.13)$$

where ΔD_{vob} is calculated from Eq. 4.7b.

3. When al the slug is left as a residual phase, continued water injection sweeps out the remaining oil bank. The mobility ratio for this displacement is more favorable than for solvent sweeping the oil bank, so an appropriate sweepout curve for this new mobility ratio must be "guesstimated." During this phase of the displacement, the pseudodisplaceable volume for estimating oil bank sweepout is

$$D_{vob} = D'_{vob} + \Delta D_{vw} , \quad \dots\dots\dots\dots\dots (4.14)$$

and, for estimating sweepout of the water front, is

$$D_{vw} = D'_{vw} + \Delta D_{vw} , \quad \dots\dots\dots\dots\dots (4.15)$$

D'_{vob} = pseudodisplaceable volumes for estimating oil bank sweepout when last of solvent slug is left as a residual phase,

ΔD_{vw} = displaceable volumes of water injected after all of solvent slug is left as a residual phase, and

D'_{vw} = displaceable volumes of water injected when last of solvent slug is left as a residual saturation.

An example calculation is given in Appendix C. The calculation is for tertiary recovery by a 25% HCPV solvent slug driven by water in a vertical cross section where the displacement is dominated by gravity tonguing. Fig. 4.23 compares oil production computed for this problem by the desk top method with oil production computed for the same problem by a finite-difference miscible flood simulator.

Solvent Slugs Driven With Miscible Gas. When the drive fluid initially is miscible with the solvent slug, loss of miscibility by dilution of the solvent slug into the multiphase region must be taken into consideration to estimate the extent of miscible sweepout. Unfortunately, no analytical methods for doing this are available when flow is characterized by viscous fingering. Helpful experimental data are similarly lacking except for the data of Lacey et al.[129] These authors conducted experiments in a high-pressure, five-spot model where propane slugs displaced a methane-saturated refined oil, and the propane slugs in turn were driven by methane. Although the phase behavior wasn't reported, Lacey *et al.* stated that it was similar to that of some reservoir oil/LPG/natural gas systems. In these experiments the viscosity ratio was 10 at the leading edge of the slug and 8.5 at the trailing edge. Recovery also was determined experimentally for immiscible displacement with the drive gas and for continuous injection of a miscible solvent at the mobility ratio of drive gas to oil. Recovery histories for the Lacey *et al.* experiments are shown in Fig. 3.23. For each slug experiment an approximate value for miscible sweepout can be calculated, which is shown in Fig. D1 of Appendix D.

Fig. D1 can be incorporated with the method discussed above for water-driven slugs to estimate recovery history of a gas-driven slug process. An example calculation is given in Appendix D. Obviously, the method is *very approximate*. Microscopic transverse dispersion in the Lacey *et al.* model probably was large relative to that expected in a homogeneous reservoir. While this would dilute the slug more rapidly in the model, it also would moderate the viscous fingering to some extent. Whether the model data are optimistic or pessimistic is difficult to judge. Presumably, the phase behavior for these experiments is typical of an LPG flood. Other factors being equal, the relationship of Fig. D1 should be optimistic for rich-gas or CO_2 slugs that can't be diluted as much as an LPG slug before miscibility is lost. The Fig. D1 relationship should be pessimistic for displacements at less adverse mobility ratios.

When flow is dominated by a single gravity tongue, rough estimates of miscible sweepout at the point of first dilution into the multiphase region can be made by application of the Dietz[130] or Hawthorne[131] theories for flow in the vertical cross section along with Eqs. 3.10 or 3.38 for calculating mixing zones at the leading and trailing edges of the slug. For example, according to Dietz, the distance traveled by a point in linear flow on the solvent/oil interface at a fractional height y above the base of the cross section is

$$x_{so} = \frac{Q}{A\phi(1-S_w-S_{orm})} \cdot \frac{\bar{M}}{[\bar{M}(1-y)+y]^2} , \quad \ldots (4.16)$$

where A is the cross-sectional area open to flow and Q is the total volume of solvent and drive gas injected. This equation was derived for displacements with sharp fronts but may be used to estimate the position of the 50% solvent concentration front after different amounts of total fluid injection (see Fig. 4.24). For greatest accuracy, the effective mobility ratio, $\bar{M}$, should be adjusted to make Eq. 3.29 predict experimental data for vertical sweepout, such as the data of Pozzi and Blackwell.[113] A similar procedure can be followed for estimating the position of the 50% drive gas concentration front.

Once the positions of the solvent/oil and gas/solvent 50% concentration fronts have been estimated, mixed zones can be superimposed on these fronts. For flooding at reservoir rates, the mixing caused by microscopic convective dispersion is usually small compared with the mixing caused by molecular diffusion. If mixing of bypassed oil into the solvent slug is negligible also,

growth of the mixed zones at the leading and trailing edges of the slug may be estimated by

$$C = \frac{1}{2}\text{erfc}\left(\frac{\delta}{2\sqrt{Dt}}\right) , \quad \ldots (4.17)$$

where δ is the distance of dimensionless concentration, C, *normal* to the 50% concentration front. Combining the mixed zones for the leading and trailing edges of the slug as shown in Fig. 4.25 allows estimation of the peak slug concentration at various fractional heights.

The method is adapted easily for radial flow. Pearson *et al.*[132] discuss a similar but more complicated method for estimating slug dilution in radial flow when the displacement is dominated by a single gravity tongue. In this method an attempt is made to account for the effects of mixing and resulting viscosity and density changes on movement of the solvent/oil and gas/solvent fronts.

Gravity-Stable Flooding

When velocity in a dipping stratum is below the critical rate defined by Eq. 3.32 at both the leading and trailing edges of the slug, the solvent/oil and gas/solvent interfaces assume angles of tilt relative to the horizontal that can be calculated by Eq. 3.30. For a completely stable displacement, the velocity must be below the stable rate defined by Eq. 3.31. When velocity is between the stable and critical rates at either front, limited fingering occurs in a portion of that transition zone, although overall stability between displaced and displacing fluid is preserved as long as displacement velocity is below the critical rate — i.e., no bulk fingering of displacing fluid occurs. The transition zone fingering, however, should cause more rapid mixing of fluids than would result from dispersion only.

For displacements below the stable rate, dilution of the solvent slug may be estimated in a manner similar to that described for unstable gravity tonguing. Since this method assumes mixing results from dispersion only, it is optimistic for velocities between the stable and critical rate to whatever extent transition zone fingering causes additional mixing.

References

1. Wadman, D.E., Mrosovsky, I., and Lamprecht, D.E.: "Reservoir Description Through Joint Geologic/Engineering Analysis," *J. Pet. Tech.* (July 1979) 933–40.

2. Richardson, J.G., Harris, D.G., and Rossen, R.S.: "Synergy in Reservoir Studies," SPE 6700 presented at the 1977 SPE Annual Technical Conference and Exhibition, Denver, Oct. 8–12.

3. Craig, F.F. Jr. *et al.*: "Optimized Recovery Through Cooperative Geology and Reservoir Engineering," *J. Pet. Tech.* (July 1977) 755–60.

4. Harris, D.G. and Hewitt, C.H.: "Synergism in Reservoir Management—The Geologic Perspective," *J. Pet. Tech.* (July 1977) 761–70.

5. Sneider, R.M., Tinker, C.N., and Mackel, L.D.: "Deltaic Environment Reservoir Types and Their Characteristics," *J. Pet. Tech.* (Nov. 1978) 1538–46.

6. LeBlanc, R.J.: "Distribution and Continuity of Sandstone Reservoirs," *J. Pet. Tech.* (July 1977) 776–804.

7. Jardine, D. *et al.*: "Distribution and Continuity of Carbonate Reservoirs," *J. Pet. Tech.* (July 1977) 873–85.

8. Tillman, R.W.: "Geology of El Dorado Field, a Deltaic Resevoir: An Application in Enhanced Oil Recovery," paper SPE 6704 presented at the 1977 SPE Annual Technical Conference and Exhibition, Denver, Oct. 8–12.

9. Sneider, R.M. *et al.*: "Predicting Reservoir Rock Geometry and Continuity in Pennsylvanian Reservoirs, Elk City Field, Oklahoma," *J. Pet. Tech.* (July 1977) 851–66.

10. Weber, K.J.: "Influence on Fluid Flow of Common Sedimentary Structures in Sand Bodies," paper SPE 9247 presented at the 1980 SPE Annual Technical Conference and Exhibition, Dallas, Sept. 21–24.

11. Lapre, J.F.: "Reservoir Potential as a Function of the Geologic Setting of Carbonate Rocks," paper SPE 9246 presented at the 1980 SPE Annual Technical Conference and Exhibition, Dallas, Sept. 21–24.

12. Richardson, J.G. *et al.*: "The Effect of Small, Discontinuous Shales on Oil Recovery," *J. Pet. Tech.* (Nov. 1978) 1531–37.

13. Dickey, P.A.: *Petroleum Development Geology*, The Petroleum Publishing Co., Tulsa (1979).

14. Whipple, A.P., Galloway, W.E., and Yancey, M.S.: "Seismic Stratigraphic Model of a Depositional Platform Margin Eastern Anadarko Basin, Oklahoma," paper SPE 7555 presented at the 1978 SPE Annual Technical Conference and Exhibition, Houston, Oct. 1–4.

15. Farr, J.B.: "High Resolution Seismic as a Reservoir Analysis Technique," paper SPE 7440 presented at the 1978 SPE Annual Technical Conference and Exhibition, Houston, Oct. 1–4.

16. Ausburn, B.E., Nath, A.K., and Wittick, T.R.: "Modern Seismic Methods—An Aid to the Petroleum Engineer," *J. Pet. Tech.* (Nov. 1978) 1519–30.

17. Albright, J.N. and Pearson, C.F.: "Location of Hydraulic Fractures Using Microseismic Techniques," paper SPE 9509 presented at the 1980 SPE Annual Technical Conference and Exhibition, Dallas, Sept. 21–24.

18. Bain, J.S., Nordberg, M.O., and Hamilton, T.M.: "3D Seismic Applications in the Interpretation of Dunlin Field, U.K. North Sea," *J. Pet. Tech.* (March 1981) 407–12.

19. Balch, A.H. *et al.*: "The Use of Vertical Seismic Profiles and Surface Seismic Profiles to Investigate the Distribution of Aquifers in the Madison Group and Red River Formation, Powder River Basin, Wyoming-Montana," paper SPE 9312 presented at the 1980 SPE Annual Technical Conference and Exhibition, Dallas, Sept. 21–24.

20. Giraud, A. *et al.*: "A Laboratory Investigation Confirms the Relative Inefficiency of True Miscible Drives and Outlines New Concepts for Maximizing Oil Recovery by Gas Injection," paper SPE 3486 presented at the 1971 SPE Annual Meeting, New Orleans, Oct. 3–6.

21. Henry, R.L. and Metcalfe, R.S.: "Multiple Phase Generation During CO_2 Flooding," paper SPE 8812 presented at the 1980 SPE/DOE Enhanced Oil Recovery Symposium, Tulsa, April 20–23.

22. Schneider, F.N. and Owens, W.W.: "Relative Permeability Studies of Gas-Water Flow Following Solvent Injection in Carbonate Rocks," *Trans.*, AIME (1976) **261**, 23–30.

23. Warner, H.R.: "An Evaluation of Miscible CO_2 Flooding in Waterflooded Sandstone Reservoirs," *J. Pet. Tech.* (Oct. 1977) 1339–47.

24. Todd, M.R., Cobb, W.M., and McCarter, E.D.: "CO_2 Flood Performance Evaluation for the Cornell Unit, Wasson San Andres Field," *J. Pet. Tech.* (Oct. 1982) 2271–82.

25. Claridge, E.L.: "CO_2 Flooding Strategy in a Communicating Layered Reservoir," *J. Pet. Tech.* (Dec. 1982) 2746–56.

26. Pontius, S.B. and Tham, M.J.: "North Cross (Devonian) Unit CO_2 Flood—Review of Flood Performance and Numerical Simulation Model," *J. Pet. Tech.* (Dec. 1978) 1706–16.

27. Chuoke, R.L., Van Meurs, P., and Van der Poel, C.: "The Instability of Slow, Immiscible, Viscous Liquid-Liquid Displacements in Permeability Media," *Trans.*, AIME (1959) **216**, 188–94.

28. Cardenas, R.L. *et al.*: "Laboratory Design of a Gravity Stable, Miscible CO_2 Process," paper SPE 10270 presented at the 1981 SPE Annual Technical Conference and Exhibition, San Antonio, Oct. 5–7.

29. Perry, G.E. *et al.*: "Weeks Island 'S' Sand Reservoir B Gravity Stable Miscible CO_2 Displacement," *Proc.*, Fourth Annual DOE Symposium on Enhanced Oil and Gas Recovery and Improved Drilling Methods, Tulsa (1978) **13**.

30. Reitzel, G.A. and Callon, G.O.: "Pool Description and Performance Analysis Leads to Understanding Golden Spike's Miscible Flood," *J. Cdn. Pet. Tech.* (April-June 1977) 39–48.

31. Sobocinski, D.P. and Cornelius, A.J.: "A Correlation for Predicting Water Coning Time," *Trans.*, AIME (1965) **234**, 594–600.

32. Bournazel, C. and Jeanson, B.: "Fast Water-Coning Evaluation Method," paper SPE 3628 presented at the 1971 SPE Annual Meeting, New Orleans, Oct. 3–6.

33. Telkov, A.D.: "Solution of Static Problems in Coning with Account of Changes in Thickness of the Oil-Saturated Portion of a Formation," *Neft. Khoz.* (1971) **49**, No. 1, 34–37.

34. Cottin, R.H.: "Optimization of the Completion and Exploitation of a Well Under Two Phase Coning," *Rev. L'Inst. Franc. Petrol.* (1971) **26**, 171–97.

35. Chappelear, J.E. and Hirasaki, G.J.: "A Model of Oil-Water Coning for 2-D Areal Reservoir Simulation," *Trans.*, AIME (1976) **261**, 65–72.

36. Meyer, H.I. and Gardner, A.D.: "Mechanics of Two Immiscible Fluids in Porous Media," *J. Applied Physics* (1954) **25**, No. 11, 1400–05.

37. Muskat, M. and Wyckoff, R.D.: "An Approximate Theory of Water Coning in Oil Production," *Trans.*, AIME (1935) **114**, 144–63.

38. Collona, J., Iffly, R., and Millet, J.L.: "Water Coning in Underground Gas Reservoirs," *Rev. L'Inst. Franc. Pet.* (1969) **24**, No. 1, 121–44.

39. Aziz, K. and Settari, A.: *Petroleum Reservoir Simulation*, Applied Science Publishers Ltd., London (1979).

40. Thomas, G.W.: *Principles of Hydrocarbon Reservoir Simulation*, U. of Norway, Trondheim, Tapier (1977).

41. Todd, M.R.: "Modeling Requirements for Numerical Simulation of CO_2 Recovery Processes," paper SPE 7988 presented at the 1979 SPE California Regional Meeting, Ventura, April 18–20.

42. Peaceman, D.W. and Rachford, H.H. Jr.: "Numerical Calculation of Multidimensional Miscible Displacement," *Soc. Pet. Eng. J.* (Dec. 1962) 327–39; *Trans.*, AIME, **225**.

43. Garder, A.O. Jr. *et al.*: "Numerical Calculation of Multidimensional Miscible Displacement by the Method of Characteristics," *Soc. Pet. Eng. J.* (March 1964) 26–36; *Trans.*, AIME, **231**; *Miscible Processes*, Reprint Series, SPE, Dallas (1965) **8**, 161–71.

44. Chaudhari, N.M.: "An Improved Numerical Technique for Solving Multidimensional Miscible Displacement Equations," *Soc. Pet. Eng. J.* (Sept. 1971) 277–84.

45. Larson, R.G.: "Controlling Numerical Dispersion by Variably Timed Flux Updating in One Dimension," *Soc. Pet. Eng. J.* (June 1982) 399–408.

46. Coats, K.H., George W.D., and Marcum, B.E.: "Three Dimensional Simulation of Steamflooding," *Soc. Pet. Eng. J.* (Dec. 1974) 573–92; *Trans.*, AIME, **257**.

47. Coats, K.H. *et al.*: "Simulation of Three-Dimensional, Two-Phase Flow in Oil and Gas Reservoirs," *Soc. Pet. Eng. J.* (Dec. 1967) 377–88; *Trans.*, AIME, **240**.

48. Spillette, A.G., Hillestad, J.G., and Stone, H.L.: "A High Stability Sequential Solution Approach to Reservoir Simulation," paper SPE 4542 presented at the 1973 SPE Annual Meeting, Las Vegas, Sept. 30–Oct. 3.

49. Lantz, R.B.: "Rigorous Calculation of Miscible Displacement Using Immiscible Reservoir Simulators," *Soc. Pet. Eng. J.* (June 1970) 192–202.

50. Todd, M.R. and Longstaff, W.J.: "The Development, Testing, and Application of a Numerical Simulator for Predicting Miscible Flood Performance," *J. Pet. Tech.* (July 1972) 874–82.

51. Bilhartz, H.L. *et al.*: "A Method for Projecting Full-Scale Performance of CO_2 Flooding in the Willard Unit," paper SPE 7051 presented at the 1978 SPE Symposium on Improved Methods for Oil Recovery, Tulsa, April 16–19.

52. Youngren, G.K. and Charlson, G.S.: "History Match Analysis of the Little Creek CO_2 Pilot Test," *J. Pet. Tech.* (Nov. 1980) 2042–52.

53. Todd, M.R., O'Dell, P.M., and Hirasaki, G.J.: "Methods for Increased Accuracy in Numerical Reservoir Simulators," *Soc. Pet. Eng. J.* (Dec. 1972) 515–30; *Trans.*, AIME, **253**.

54. Settari, A., Price, H.S., and Dupont, T.: "Development and Testing of Variational Methods for Simulation of Miscible Displacement in Porous Media," *Soc. Pet. Eng. J.* (June 1977) 228–44.

55. Vinsome, P.K.W. and Au, A.D.K.: "One Approach to the Grid Orientation Problem in Reservoir Simulation," *Soc. Pet. Eng. J.* (April 1981) 160–61.

56. Yanosik, J.L. and McCracken, T.A.: "A Nine-Point Finite Difference Reservoir Simulator for Realistic Prediction of Unfavorable Mobility Ratio Displacements," *Soc. Pet. Eng. J.* (Aug. 1979) 253–62.

57. Ko, S.C.M. and Au, A.D.K.: "A Weighted Nine-Point Finite-Difference Scheme for Eliminating the Grid Orientation Effect in Numerical Reservoir Simulation," paper SPE 8248 presented at the 1979 SPE Annual Technical Conference and Exhibition, Las Vegas, Sept. 23–26.

58. Coats, K.H.: "Reservoir Simulation: State of the Art," *J. Pet. Tech.* (Aug. 1982) 1633–42.

59. Watkins, R.L.: "The Development and Testing of a Sequential, Semi-Implicit Four Component Reservoir Simulator," paper SPE 10513 presented at the 1982 Symposium on Reservoir Simulation, New Orleans, Jan. 31–Feb. 3.

60. Todd, M.R. *et al.*: "Numerical Simulation of Competing Chemical Flood Designs," paper SPE 7077 presented at the 1978 SPE Symposium on Improved Methods for Oil Recovery, Tulsa, April 16–19.

61. Price, H.S. and Donohue, D.A.T.: "Isothermal Displacement Processes With Interphase Mass Transfer," *Soc. Pet. Eng. J.* (June 1967) 205–20.

62. Roebuck, I.F. Jr. *et al.*: "The Compositional Reservoir Simulator: Case 1—The Linear Model," *Soc. Pet. Eng. J.* (March 1969) 115–30.

63. Culham, W.E., Farouq Ali, S.M., and Stahl, C.D.: "Experimental and Numerical Simulation of Two-Phase Flow with Interphase Mass Transfer in One and Two Dimensions," *Soc. Pet. Eng. J.* (Sept. 1969) 323–37.

64. Kniazeff, V.J. and Naville, S.A.: "Two-Phase Flow of Volatile Hydrocarbons," *Soc. Pet. Eng. J.* (March 1965) 37–44; *Trans.*, AIME, **234**.

65. Nolen, J.S.: "Numerical Simulation of Compositional Phenomena in Petroleum Reservoirs," *Numerical Simulation*, Reprint Series, SPE, Dallas (1973) **11**, 269–84.

66. Kazemi, H.A., Vestal, C.R., and Shank, G.D.: "An Efficient Multicomponent Numerical Simulator," *Soc. Pet. Eng. J.* (Oct. 1978) 355–68.

67. Besserer, G.J., Serra, J.W., and Best, D.D.: "An Efficient Phase Behavior Package for Use in Compositional Reservoir Simulation Studies," paper SPE 8288 presented at the 1979 SPE Annual Technical Conference and Exhibition, Las Vegas, Sept. 23–26.

68. Rowe, A.M. Jr.: "Internally Consistent Correlations for Predicting Phase Compositions for Use in Reservoir Composition Simulators," paper SPE 7545 presented at the 1978 SPE Annual Technical Conference and Exhibition, Houston, Oct. 2–4.

69. Van Quy, N., Simandoux, P., and Corteville, J.: "A Numerical Study of Diphasic Multicomponent Flow," *Soc. Pet. Eng. J.* (April 1972) 171–84; *Trans.*, AIME, **253**.

70. Corteville, J., Van Quy, N., and Simandoux, P.: "A Numerical and Experimental Study of Miscible or Immiscible Fluid Flow in Porous Media with Interphase Mass Transfer," paper SPE 3481 presented at the 1971 SPE Annual Meeting, New Orleans, Oct. 3–6.

71. Coats, K.H.: "An Equation of State Compositional Model," *Soc. Pet. Eng. J.* (Oct. 1980) 363–76.

72. Fussell, L.T. and Fussell, D.D.: "An Iterative Technique for Compositional Reservoir Models," *Soc. Pet. Eng. J.* (Aug. 1979) 211–20.

73. Fussell, D.D. and Yanosik, J.L.: "An Iterative Sequence for Phase Equilibria Calculations Incorporating the Redlich-Kwong Equation of State," *Soc. Pet. Eng. J.* (June 1978) 173–82.

74. Metcalfe, R.S., Fussell, D.D., and Shelton, J.L.: "A Multicell Equilibrium Separation Model for the Study of Multiple Contact Miscibility in Rich-Gas Drives," *Soc. Pet. Eng. J.* (June 1973) 147–55.

75. Fussell, D.D., Shelton, J.L., and Griffith, J.D.: "Effect of 'Rich' Gas Composition on Multiple-Contact Miscible Displacement — A Cell-to-Cell Flash Model Study," *Soc. Pet. Eng. J.* (Dec. 1976) 310–16.

76. Nghiem, L.X., Fong, D.K., and Aziz, K.: "Compositional Modeling with an Equation of State," *Soc. Pet. Eng. J.* (Dec. 1981) 687–98.

77. Sigmund, P.M. *et al.*: Laboratory CO_2 Floods and Their Computer Simulation," *Proc.*, 10th World Pet. Cong., Bucharest (1979) **3**, 243–50.

78. Spivak, A., Perryman, T.L., and Norris, R.A.: "A Compositional Simulation Study of the Sacroc Unit CO_2 Project," *Proc.*, Ninth World Pet. Cong., **4**, Tokyo (1975) 187.

79. Warner, H.R. Jr., Hardy, J.H., and Davidson, C.D.: "University Block 31 Middle Devonian Reservoir Study—Part 2: Projections," *J. Pet. Tech.* (Aug. 1979) 971–78.

80. Graue, D.J. and Zana, E.: "Study of a Possible CO_2 Flood in the Rangely Field, Colorado," *J. Pet. Tech.* (July 1981) 1312–18.

81. Fussell, L.T.: "A Technique for Calculating Multiphase Equilibria," *Soc. Pet. Eng. J.* (Aug. 1979) 203–08.

82. Zudkevitch, D. and Joffe, J.: "Correlation and Prediction of Vapor-Liquid Equilibrium with the Redlich-Kwong Equation of State," *AIChE J.* (Jan. 1970) **16**, 112–19.

83. Soave, G.: "Equilibrium Constants from a Modified Redlich-Kwong Equation of State," *Chem. Eng. Sci.* (1977) **27**, 1197–1203.

84. Peng, D.Y. and Robinson, D.B.: "A New Two-Constant Equation of State," *Ind. Eng. Chem. Fund.* (1976) **15**, 59–64.

85. Yarborough, L.: "Application of a Generalized Equation of State to Petroleum Reservoir Fluids," paper presented at the 1978 I&EC Symposium on Equations of State in Engineering and Research, 176th ACS Natl. Meeting, Miami Beach, Sept. 10–15.

86. Young, L.C. and Stephenson, R.E.: "A Generalized Compositional Approach for Reservoir Simulation," paper SPE 10516 presented at the 1982 SPE Symposium on Reservoir Simulation, New Orleans, Jan. 31–Feb. 3.

87. Camy, J.P. and Emanual, A.S.: "Effect of Grid Size in the Compositional Simulation of CO_2 Injection," paper SPE 6894 presented at the 1977 SPE Annual Technical Conference and Exhibition, Denver, Oct. 9–12.

88. Chase, C.A. and Todd, M.R.: "Numerical Simulation of CO_2 Flood Performance," paper SPE 10514 presented at the 1982 SPE Symposium on Reservoir Simulation, New Orleans, Jan. 31–Feb. 6.

89. Mrosovsky, I., Wong, J.Y., and Lampe, H.W.: "Construction of a Large Field Simulator on a Vector Computer," *J. Pet. Tech.* (Dec. 1980) 2253–64.

90. Jacks, H.H., Smith, O.J.E., and Mattax, C.C.: "The Modeling of a Three-Dimensional Reservoir with a Two-Dimensional Simulator — The Use of Dynamic Pseudo Functions," *Soc. Pet. Eng. J.* (June 1973) 175–85.

91. Christian, L.D. *et al.*: "Planning A Tertiary Oil Recovery Project for Jay-Little Escambia Creek Fields Unit," *J. Pet. Tech.* (Aug. 1981) 1535–44.

92. Gillund, G.N. and Patel, C.: "Depletion Studies of Two Contrasting D-2 Reefs," paper 80-31-37 presented at the 1980 Annual Technical Meeting of the Petroleum Society of CIM, Calgary, May 25–28.

93. Brigham, W.E.: "Mixing Equations in Various Geometries," paper SPE 4585 presented at the 1973 SPE Annual Meeting, Las Vegas, Sept. 30–Oct. 3.

94. Kane, A.V.: "Performance of a Large-Scale CO_2-Wag Project, SACROC Unit—Kelly Snyder Field," *J. Pet. Tech.* (Feb. 1979) 217–31.

95. Warner, H.R. Jr. *et al.*: "University Block 31 Middle Devonian Reservoir Study—Part 1: History Match," *J. Pet. Tech.* (Aug. 1979) 962–70.

96. Simlote, V.N. and Withjack, E.M.: "Estimation of Tertiary Recovery by CO_2 Injection — Springer A Sand, Northeast Purdy Unit," *J. Pet. Tech.* (May 1981) 808–18.

97. DesBrisay, C.L. *et al.*: Review of Miscible Flood Performance, Intisar 'D' Field, Socialist People's Libyan Arab Jamairiya," *J. Pet. Tech.* (Aug. 1982) 1651–60.

98. Henderson, L.E.: "The Use of Numerical Simulation to Design a Carbon Dioxide Miscible Displacement Project," *J. Pet. Tech.* (Dec. 1974) 1327–34.

99. Griffith, J.D. and Horne, A.L.: "South Swan Hills Solvent Flood," *Proc.*, Ninth World Pet. Cong., Tokyo (1975) **4**, 269–78.

100. DesBrisay, C.L., Gray, J.W., and Spivak, A.: "Miscible Flood Performance of the Intisar 'D' Field, Libyan Arab Republic," *J. Pet. Tech.* (Aug. 1975) 935–43.

101. Graue, D.J. and Blevins, T.R.: "SACROC Tertiary CO_2 Pilot Project," paper SPE 7090 presented at the 1978 SPE Symposium on Improved Oil Recovery Methods, Tulsa, April 16–19.

102. Higgins, R.V. and Leighton, A.J.: "A Computer Method to Calculate Two-Phase Flow in Any Irregularly Bounded Porous Medium," *Trans.*, AIME (1962) **225**, 679–83.

103. Higgins, R.V. and Leighton, A.J.: "Computer Prediction of Water Drive of Oil and Gas Mixtures Through Irregularly Bounded Porous Media — Three-Phase Flow," *Trans.*, AIME (1962) **22**, 1048–54.

104. Faulkner, B.L.: "Reservoir Engineering Design of a Tertiary Miscible Gas Drive Pilot Project," paper SPE 5539 presented at the 1975 SPE Annual Meeting, Dallas, Sept. 28–Oct. 1.

105. Drennon, M.D., Kelm, C.H., and Whittington, H.M.: "A Method for Appraising the Feasibility of Field-Scale CO_2 Miscible Flooding," paper SPE 9323 presented at the 1980 SPE Annual Technical Conference and Exhibition, Dallas, Sept. 21–24.

106. Hall, H.N. and Geffen, T.M.: "A Laboratory Study of Solvent Flooding," *Trans.*, AIME (1957) **210**, 48–57; *Miscible Processes*, Reprint Series, SPE, Dallas (1965) **8**, 133–141.

107. Price, H.S., Cavendish, J.C., and Varga, R.S.: "Numerical Methods of Higher-Order Accuracy for Diffusion-Convection Equations," *Soc. Pet. Eng. J.* (Sept. 1968) 293–303; *Trans.*, AIME, **243**.

108. Shum, Y.M.: "Use of the Finite-Element Method in the Solution of Diffusion-Convection Equations," *Soc. Pet. Eng. J.* (June 1971) 139–44.

109. Young, L.C.: "A Finite-Element Method for Reservoir Simulation," *Soc. Pet. Eng. J.* (Feb. 1981) 115–28.

110. Russell, T.F.: "Finite Elements with Characteristics for Two-Component Incompressible Miscible Displacement," paper SPE 10500 presented at the 1982 SPE Symposium on Reservoir Simulation, New Orleans, Jan. 31–Feb. 3.

111. Darlow, B.L., Ewing, R.E., and Wheeler, M.F.: "Mixed Finite Element Methods for Miscible Displacement Problems in Porous Media," paper SPE 10501 presented at the 1982 SPE Symposium on Reservoir Simulation, New Orleans, Jan. 31–Feb. 3.

112. Croes, G.A., Geertsma, J., and Swartz, N.: "Theory of Dimensionally Scaled Models of Petroleum Reservoirs," *Trans.*, AIME (1956) **207**, 118–25.

113. Pozzi, A.L. and Blackwell, R.J.: "Design of Laboratory Models for Study of Miscible Displacement," *Soc. Pet. Eng. J.* (March 1963) 28–40.

114. Craig, F.F. *et al.*: "A Laboratory Study of Gravity Segregation in Frontal Drives," *Trans.*, AIME (1957) **210**, 275–82.

115. Mahaffey, J.L., Rutherford, W.M., and Matthews, C.W.: "Sweep Efficiency by Miscible Displacement in a Five-Spot," *Soc. Pet. Eng. J.* (March 1966) 73–80; *Trans.*, AIME, **237**.

116. Craig, F.F.: *The Reservoir Engineering Aspects of Waterflooding*, Monograph Series, SPE, Dallas (1971) **3**.

117. Suder, F.E. and Calhoun, J.C. Jr.: "Waterflood Calculations," *Drill. and Prod. Prac.*, API (1949) 260.

118. Prats, M. *et al.*: "Prediction of Injection Rate and Production History for Multifluid Five-Spot Floods," *Trans.*, AIME (1959) **216**, 98–105.

119. Stiles, W.E.: "Use of Permeability Distribution in Waterflood Calculations," *Trans.*, AIME (1949) **186**, 9–13.

120. Dykstra, H. and Parsons, H.L.: "The Prediction of Oil Recovery by Waterflooding," *Secondary Recovery of Oil in the United States* (second edition), API, New York City (1950) 160.

121. Fitch, R.A. and Griffith, J.D.: "Experimental and Calculated Performance of Miscible Floods in Stratified Reservoirs," *J. Pet. Tech.* (Nov. 1964) 1289–98.

122. Agan, J.B. and Fernandes, R.J.: "Performance of a Miscible-Slug Process in a Highly Stratified Reservoir," *J. Pet. Tech.* (Jan. 1962) 81–86; *Trans.*, AIME, **225**.

123. Doepel, G.W. and Sibley, W.P.: "Miscible Displacement — Multilayer Technique for Predicting Reservoir Performance," *J. Pet. Tech.* (Jan. 1962) 73–80; *Trans.*, AIME, **225**.

124. Deppe, J.C.: "Injection Rates — The Effect of Mobility Ratio, Area Swept and Pattern," *Soc. Pet. Eng. J.* (June 1961) 81–91; *Trans.*, AIME, **213**.

125. Muskat, M.: *Physical Principles of Oil Production*, McGraw-Hill Book Co. Inc., New York City (1949).

126. Hansen, P.W.: "A CO_2 Tertiary Recovery Pilot, Little Creek Field, Mississippi," paper SPE 6747 presented at the 1977 SPE Annual Technical Conference and Exhibition, Denver, Oct. 9–12.

127. Claridge, E.L.: "A Trapping Hele-Shaw Model for Miscible-Immiscible Flooding Studies," *Soc. Pet. Eng. J.* (Oct. 1973) 255–70.

128. Claridge, E.L.: "Prediction of Recovery in Unstable Miscible Flooding," *Soc. Pet. Eng. J.* (April 1972) 144–55.

129. Lacey, J.W., Faris, J.E., and Brinkman, F.H.: "Effect of Bank Size on Oil Recovery of the High Pressure Gas-Driven LPG-Bank Process," *J. Pet. Tech.* (Aug. 1961) 806–12; *Trans.*, AIME, **222**; *Miscible Processes*, Reprint Series, SPE, Dallas (1965) **8**, 215–21.

130. Dietz, D.N.: "A Theoretical Approach to the Problem of Encroaching and By-Passing Edge Water," Akademi Van Wetershappen, Amsterdam, *Proc.*, Ser. B Physical Science (1953) **56**, 83.

131. Hawthorne, R.G.: "Two-Phase Flow in Two-Dimensional Systems — Effects of Rate, Viscosity, and Density on Fluid Displacement in Porous Media," *Trans.*, AIME (1960) **219**, 81–87.

132. Pearson, D.G., Bray, J.A., and Strom, N.A.: "Simplified Approach for Assessing Gravity-Viscous Effects in Horizontal Steady-State Solvent Floods," paper 6823 presented at the 1968 Annual Technical Meeting of the Petroleum Soc. of CIM, Calgary, May 7–10.

Chapter 5
The First-Contact Miscible Process

5.1 Estimation of Miscibility Conditions

LPG's such as ethane, propane, butane, or their mixtures are the common solvents for first-contact miscible flooding. Propane has been the most frequently used solvent in field trials.

LPG slugs must be designed to achieve first-contact miscibility with oil at the leading edge of the slug and with driving gas at the trailing edge. To accomplish this, pressure must be above the cricondenbar of the pressure/composition diagrams for both the solvent/oil system and the drive gas/solvent system. At this condition, all mixtures of the displaced and displacing fluids are single phase, and no mixtures lie within the multiphase region.

The pseudoternary diagrams in Fig. 5.1 illustrate phase behavior considerations for achieving first-contact miscibility at both the leading and trailing edges of an LPG slug. In these diagrams, the LPG is represented by the lower right corner of the triangle. For first-contact miscibility between LPG and reservoir oil, the straight line connecting reservoir oil composition and LPG must not pass through a multiphase region. Similarly, for first-contact miscibility between LPG and drive gas, the line connecting drive-gas composition and LPG must not pass through a multiphase region. This latter line is represented by the right side of the triangle.

Figs. 5.1a and 5.1b illustrate two situations that would cause solvent and oil not to be first-contact miscible: (1) the two-phase region intersects the bottom side of the pseudoternary diagram—i.e., the side representing all mixtures of LPG and the heavy pseudocomponent, and (2) the LPG is diluted sufficiently with a gas such as methane that some mixtures of solvent and oil fall within the two-phase region. The phase behavior shown in Fig. 5.1a can occur either when pressure is below the solvent vapor pressure or at some pressures when reservoir temperature is above the solvent critical temperature. For example, at 180°F the reservoir pressure must be greater than about 475 psia for propane to be above its vapor pressure. At 250°F, which is higher than the critical temperature of either propane or ethane, pressure must be greater than about 720 psia for first-contact miscibility between propane and n-heptane and higher than about 1,620 psia for first-contact miscibility between ethane and n-decane. Presumably, at this temperature, even higher pressures would be required for first-contact miscibility with higher molecular weight oils. Typically, the pressure required for first-contact miscibility between LPG and reservoir oil is low enough that it is not restrictive.

Fig. 5.1b shows that LPG can be diluted with methane only up to a limiting Composition "A" and still be first-contact miscible with Oil A.* According to the phase behavior concepts discussed in Chap. 2, this maximum permissible methane concentration in the slug for a given pressure decreases with increasing temperature because the size of the two-phase region increases, which causes the tangent line through the oil composition to intersect the right side of the pseudoternary diagram at lower methane concentrations. Conversely, the pressure required to achieve first-contact miscibility with a given mixture of LPG and light components increases with increasing reservoir temperature.

At temperatures below the critical temperature of LPG, the two-phase region will intersect the right side of the pseudoternary diagram at pressures that are above the vapor pressure of LPG but below the cricondenbar of the drive gas/solvent system. At this condition, some mixtures of drive gas and solvent will be two-phase. This situation is illustrated in Fig. 5.1c. For first-contact miscibility between solvent and drive gas, pressure must be high enough that phase behavior of this type does not occur.

---

*Section 2.5 discusses how even higher concentrations of the light components, between Compositions "a" and "b," will develop condensing-gas drive miscibility with Oil A.

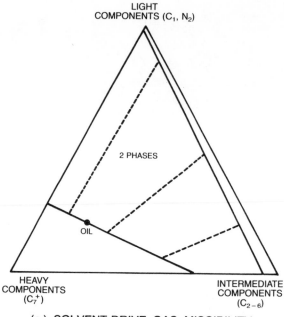

(a.) SOLVENT-DRIVE GAS MISCIBILITY;
SOLVENT-OIL IMMISCIBILITY

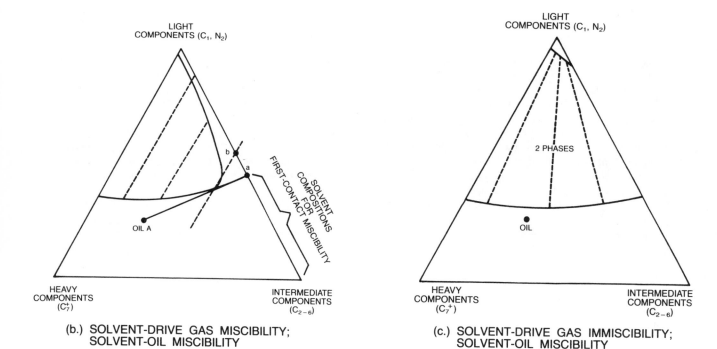

(b.) SOLVENT-DRIVE GAS MISCIBILITY;
SOLVENT-OIL MISCIBILITY

(c.) SOLVENT-DRIVE GAS IMMISCIBILITY;
SOLVENT-OIL MISCIBILITY

Fig. 5.1—Phase behavior considerations for first-contact miscibility; pseudoternary diagrams.

These phase behavior concepts can be seen in a different perspective by examining the conceptual pressure/temperature diagrams for drive gas/solvent and solvent/oil shown in Fig. 5.2. This figure shows how the cricondenbars for drive gas/solvent and solvent/oil might vary with temperature (the figure is modeled after pure component behavior). At temperature T_1, which is below the critical temperature of solvent, pressure P_1 is below the solvent vapor pressure and falls within the two-phase region for solvent/oil. This situation also is depicted by Fig. 5.1a. Pressure P_2 is above the vapor pressure of solvent but falls within the two-phase region for solvent/drive gas. This situation is depicted by Fig. 5.1c. Pressure P_3 is sufficiently high that all drive gas/solvent mixtures are single-phase and all solvent/oil mixtures are single-phase (Fig. 5.1b). However, P_3 is still below the cricondenbar for drive gas and oil, and these fluids are immiscible. Fig. 5.2 shows that at temperatures below the solvent critical temperature, miscibility between solvent and driving gas determines

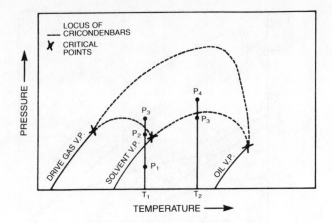

Fig. 5.2—Phase behavior considerations for first-contact miscibility; pressure-temperature diagram.

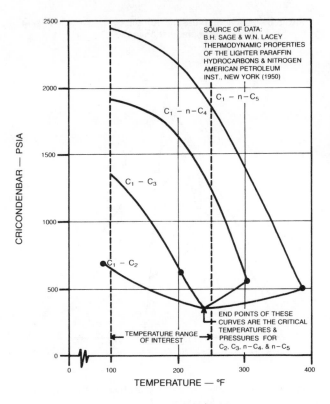

Fig. 5.3—Loci of cricondenbars for methane and potential slug materials (C_2, C_3, nC_4, nC_5) (after Ref. 86).

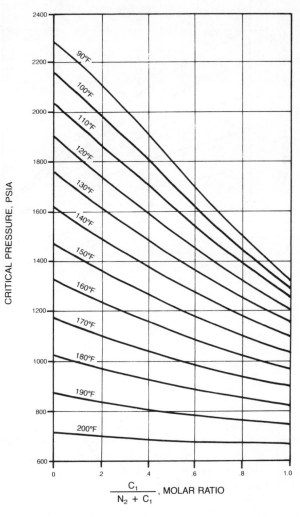

Fig. 5.4—Critical loci for the nitrogen/methane/propane system (after Ref. 7).

the minimum pressure required for first-contact miscible displacement at the leading and trailing edges of an LPG slug.

At temperature T_2 (Fig. 5.2), which is above the solvent critical temperature, two-phase mixtures of solvent and oil are possible at pressure P_3, although all drive gas/solvent mixtures are first-contact miscible. This situation is also depicted by Fig. 5.1a. At the higher pressure P_4, solvent and oil are first-contact miscible at temperature T_2.

The pressure required for first-contact miscibility between solvent and oil can be found experimentally by determining the P-X diagram. The maximum permissible concentration of methane and nitrogen in an LPG slug for first-contact miscibility also can be estimated by

calculating pseudoternary phase behavior using the techniques described in Refs. 1 through 5. In addition to estimating the limiting concentration for first-contact miscibility, calculation of a pseudoternary diagram is useful for estimating the minimum concentration to which the solvent slug can be diluted by reservoir mixing with oil and drive gas.

Ref. 6 also suggests a method for estimating the maximum quantity of natural gas that can be added to LPG without losing first-contact miscibility. It does not involve phase behavior calculations. However, the method was derived from phase behavior data on only two oils—an aromatic, asphaltic oil and a paraffinic, nonasphaltic oil—and should be used with caution.

Figs. 5.3 through 5.6 are useful for estimating the

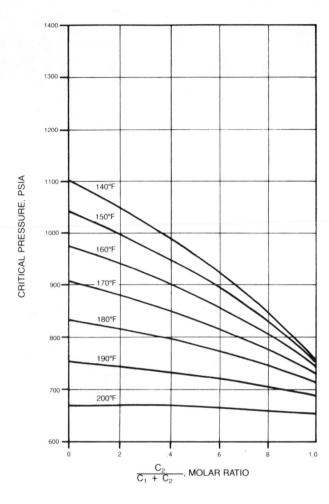

Fig. 5.5—Critical loci for the methane/ethane/propane system, 140 to 200°F (after Ref. 7).

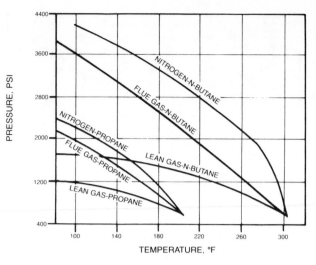

Fig. 5.6—Minimum pressure for miscibility, miscible slug/driving gas combinations (after Ref. 8).

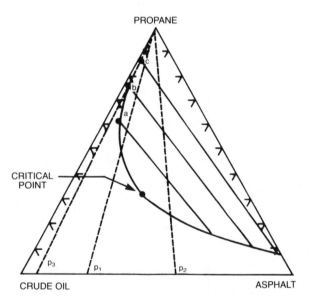

Fig. 5.7—Phase diagram for a California crude/propane/asphalt system (after Ref. 10).

pressure required for first-contact miscibility between drive gas and solvent. Fig. 5.3 shows cricondenbars for methane and several pure-component slug materials. Above the cricondenbar, methane is directly miscible with these intermediate-molecular-weight hydrocarbons. Miscibility pressure decreases with increasing temperature and increases with increasing molecular weight of the slug material. For a propane slug, the pressure required for miscibility with methane decreases from about 1,350 psia at 100°F to 1,000 psia at 160°F. Required pressure with butane is 1,900 psia at 100°F and 1,800 psia at 160°F.

Volatile hydrocarbons in methane, such as ethane and propane, as well as nonhydrocarbons, such as nitrogen and CO_2, can affect miscibility with first-contact miscible solvents. Figs. 5.4 and 5.5 show the miscibility of methane/nitrogen and methane/ethane mixtures with propane. Miscibility pressure increases with increasing nitrogen content in the drive gas and decreases with increasing ethane content. Estimates from these figures tend to be too low when in error.[7] The figures are useful in preliminary design but experimental determination is recommended for final design.

Fig. 5.6 shows miscibility pressure of a natural gas with propane and with butane.[8] Miscibility pressure is lower with natural gas than with pure methane because of the intermediate-molecular-weight hydrocarbons in natural gas. A full-page version of Fig. 5.6 can be found in Appendix E as Fig. E6.

Flue gas or nitrogen also can be used as drive fluids in miscible flooding. Fig. 5.6 shows miscibility of these gases with propane and butane. CO_2 in the flue gas causes miscibility pressure to be lower than with pure nitrogen.

Instead of being a pure component, the LPG slug may be a mixture containing various percentages of ethane, propane, butanes, and pentanes. In this situation, dry gas/solvent phase behavior can be estimated with fair accuracy by any sufficiently accurate equation of state, such as the Benedict-Webb-Rubin equation of state,[1,7] the Redlich-Kwong equation of state,[2,4,5] the Peng-Robinson equation of state,[9] or by the Rowe and

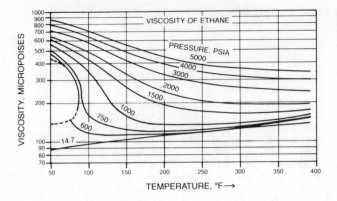

Fig. 5.8—Viscosity of ethane (after Ref. 13).

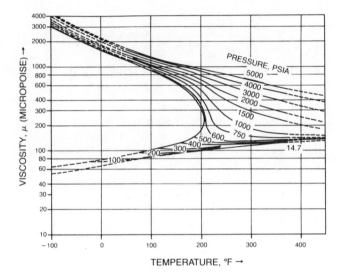

Fig. 5.9—Viscosity of propane (after Ref. 13.).

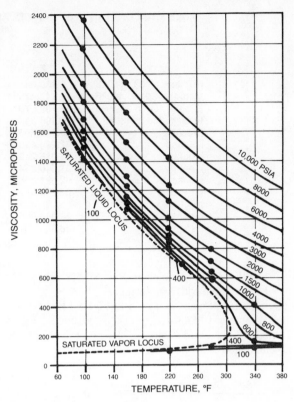

Fig. 5.10—n-butane viscosity vs. temperaure (after Ref. 12).

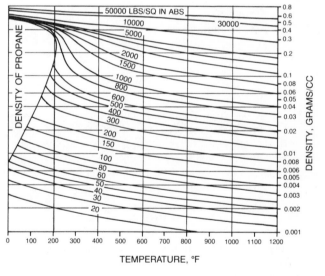

Fig. 5.11—Density of propane (after Ref. 13).

Silberberg convergence pressure method.[3]

Lower-molecular-weight hydrocarbons such as ethane through hexane will precipitate some of the asphalt from asphaltic crudes. The severity of precipitation decreases as molecular weight of the solvent increases. Bossler and Crawford[10] discuss the phenomenon in detail.

Fig. 5.7 shows a pseudoternary diagram for a California crude oil/propane/asphalt system. Precipitation occurs only over a certain range of propane concentration. For example, with an initial asphalt concentration p_1 in the oil, precipitation begins at about 50% propane and continues to high propane concentrations. Strictly speaking, a first-contact miscible displacement does not occur, since some compositions fall within a two-phase region. However, the mixing of propane and original crude oil forms a transition zone of fluids within the partially deasphalted phase such that all compositions in this transition zone are contiguously miscible—e.g., the transition zone described by compositions along Path P_1-a-b-c-propane in Fig. 5.7. Miscible displacement is achieved within this transition zone, although precipitation of asphalt occurs within part of the mixed zone. Severe asphalt precipitation may reduce permeability and affect well injectivities and productivities. It also may cause plugging in producing well tubulars.

5.2 Viscosity and Density of LPG Solvents

Figs. 5.8, 5.9, and 5.10 show the influence of temperature and pressure on the viscosity of ethane, propane, and n-butane.[11-13] For most floods, ethane viscosity will be in the range of 0.02 to 0.07 cp, propane viscosity will be in the range of 0.06 to 0.12 cp, and n-butane viscosity will be in the range of 0.1 to 0.18 cp. Often, LPG solvent slugs will be mixtures of LPG hydrocarbons. Viscosity of these mixtures can be estimated by the method recommended by Lohrenz et

TABLE 5.1—CONSTANTS OF INDIVIDUAL PURE HYDROCARBONS FOR EQS. 5.2 AND 5.3
(after Ref. 16)

	Substance			
	K	n	$m \times 10^4$	C
Methane, 70 to 300°F	9,160.6413	61.893223	3.3162472	0.50874303
Methane, 301 to 460°F	147.47333	3,247.4533	− 14.072637	1.8326695
Ethane, 100 to 249°F	46,709.573	− 404.48844	5.1520981	0.52239654
Ethane, 250 to 460°F	17,495.343	34.163551	2.8201736	0.62309877
Propane	20,247.757	190.24420	2.1586448	0.90832519
i-butane	32,204.420	131.63171	3.3862284	1.1013834
n-butane	33,016.212	146.15445	2.902157	1.1168144
n-pentane	37,046.234	299.62630	2.1954785	1.4364289
n-hexane	52,093.006	254.56097	3.6961858	1.5929406
n-heptane	82,295.457	64.380112	5.2577968	1.7299902
n-octane	89,185.432	149.39026	5.9897530	1.9310993
n-nonane	124,062.650	37.917238	6.7299934	2.1519973
Decane	146,643.830	26.524103	7.8561789	2.3329874

al.,[14] although experimental viscosity measurement may be advisable for final designs.

Density for propane is shown in Fig. 5.11.[13,15] Densities of other LPG solvents can be estimated by the method of Alani and Kennedy.[16] First, molal volumes are calculated from the following equation.

$$V^3 - \left(\frac{10.73T}{p} + b \right) V^2 + \frac{aV}{p} - \frac{ab}{p} = 0, \quad \ldots\ldots (5.1)$$

where

T = absolute temperature, °R,
p = absolute pressure, psia, and
V = molal volume, cu ft/lbm-mole,

and constants a and b are defined as

$$a = \sum_{i}^{n} \left(K_i e^{n_i/T} \right) x_i \quad \ldots\ldots\ldots\ldots\ldots (5.2)$$

and

$$b = \sum_{i}^{n} (m_i T + C_i) x_i. \quad \ldots\ldots\ldots\ldots\ldots (5.3)$$

In Eqs. 5.2 and 5.3, the subscript i refers to component i in the LPG mixture and x_i signifies the mole fraction of that component in the mixture. Constants K_i, n_i, m_i, and C_i for pure hydrocarbons are given in Table 5.1. Eq. 5.1 has three roots, at least one of which is real under all temperatures and pressures. In cases where Eq. 5.1 has more than one real root, the minimum value represents the real liquid volume.

Density in pounds mass per cubic foot is calculated from

$$\rho = \frac{\sum_{i}^{n} (MW_i) x_i}{V},$$

where MW_i is the molecular weight of component i.

Methods for calculating viscosities and densities of drive gases are discussed in Secs. 6.2 and 7.2.

5.3 Sources of LPG and Drive Gases

Most LPG for first-contact miscible slugs has come from nearby natural gasoline plants. Gases from oil and gas fields often contain enough liquefiable hydrocarbons to justify plants for their recovery.[13] In these plants, the liquefiable hydrocarbons are stripped from raw field gas in an absorber with a gas oil as the absorbent. This oil then is stripped of propane, butane, and natural gasoline constituents, which in turn are separated by fractional distillation in distillation towers. Overhead products from the debutanizer and depropanizer towers are suitable materials for first-contact miscible slugs.

Low-temperature separation, either by expansion or by refrigeration, is occasionally used for liquid recovery.[13] These facilities are also potential sources for LPG mixtures.

Sources of natural gas, flue gas, or nitrogen for driving miscible slugs are discussed in Sec. 7.3.

5.4 Field Experience

The first-contact miscible process has received the most field testing of all the miscible processes. There have been more than 50 field tests of this method, the majority being run in the 1950's and 1960's with 1957–58 as peak years for initiation of new floods. Some early projects, however, amounted to little more than LPG storage, and most were small pilot tests involving one or at most a few solvent injection wells. Test size typically varied from several tens to several hundred acres, although a few projects were field- or unit-wide in scope and involved several thousand acres. Details about many of the first-contact miscible field tests are documented in Tables 5.2 and 5.3. Table 5.2 shows secondary recovery floods that were conducted in reservoirs that had not been waterflooded. Table 5.3 shows field tests that were conducted in reservoirs that either had been waterflooded or produced by natural water drive before the test.

The process has been tested in both sandstone and carbonate reservoirs, although the majority of tests were in sandstones. Most were secondary recovery floods, although at least seven were tertiary recovery tests, all in sandstones.[74,75,77-80] In early project designs, the LPG slug was followed by continuous injection of lean

TABLE 5.2—FIRST-CONTACT MISCIBLE PROJECTS

Field	Operator	Year Started	Project Type	Oil Gravity (°API)	Oil Viscosity (cp)	Viscosity Ratio μ_o/μ_s	Depth (ft)	Thickness (ft)	Acreage	Slug Size (% HCPV)	Recovery (% OOIP)	Status	Reference
San Pedro Block D (Peru)	Intl. Petroleum Co.	1951	LPG	39	10		2,300	164	20	12	10 incremental	term., succ.	17*
West Panhandle (TX)	Diamond Shamrock Corp.	1952	LPG storage; oil recovery						600			term., good oil response	18*
Seminole City (OK)	Carter Oil Co.	1953	LPG→G	37	3	30	3,500	22		8.5	>7.5 incremental	term., oil response	18*,19
Chicon Lake (TX)	Winthrop Corp.		LPG storage; oil recovery tertiary									term.	18*
Eram (OK)	Sunray		LPG→G										18
Sand Creek (WY)	Mobil Oil Corp.		LPG→G										18
South Leonard (OK)	Amoco Production Co.		LPG→G									term., unsucc.	18*
Spraberry Trend Area (TX)	Phillips Petroleum Co.		LPG									term.	18*
Camp Hill (TX)	Frank Buttram		LPG									term., poor response	18*
Killan (TX)	Petrocel Corp.		LPG storage; oil recovery									term., poor response	18*
West Panhandle (TX)	Diamond Shamrock Corp.	1955	LPG storage; oil recovery										18
Taylor-Ina (TX)	Wherry & Green	1956	LPG→G									term., unsucc., channelling	18*
Millican Reef (TX)	Sun Oil Co.	1956	LPG→G				6,000		300	1.5			18,20–23
North Pembina Cardium Unit (Alberta)	Mobil Oil Corp.	1957	LPG→G	37		14	5,000	16	10	7.1	72 ultimate	term., succ.	24*
Newhall-Potrero (CA)	Sun Oil Co.	1957	LPG→G										18,25,26
Bisti (NM)	Sun Oil Co.	1957	LPG→G		0.9	10	4,900	130	40-pilot 1,600	5.5	55 Pilot ultimate	term., unsucc., prof.	18*,27–37
Short Junction (OK)	Conoco Inc.	1957	LPG→G						640				18,38
Elk City P1 (OK)	Shell Development Co.	1957	LPG→G						pilot				18

(continued on next page)

TABLE 5.2 (CONTINUED)

Field	Operator	Year Started	Project Type	Oil Gravity (°API)	Oil Viscosity (cp)	Viscosity Ratio μ_o/μ_s	Depth (ft)	Thickness (ft)	Acreage	Slug Size (% HCPV)	Recovery (% OOIP)	Status	Reference
Panhandle (TX)	Cities Service Co.	1957	LPG→G						320				18
Willbarger Co. Regular (TX)	Cities Service Co.	1957	LPG→G						26				18
Three Bar (TX)	Amoco Production Co.	1957	LPG→G		0.38				160			term., unsucc.	18*,39
Parks Pennsylvanian (TX)	Mobil Oil Corp.	1957	LPG→G	45		5	10,400	21	6,400	4	14 incremental	NC, succ., prof.	18,39,40, 41*–48
Tijerina-Canales-Blucher (TX)	Texaco Inc.	1957	LPG→G						pilot				18,49
Meadow Creek Lakota B (WY)	Conoco Inc.	1957	LPG→G →G/W		0.42	10	7,300	9.5	400	7		term., disc., prof.	18,50,51*
Slaughter Boyd Lease (TX)	ARCO Oil & Gas Co.	1958	LPG→G ←W	30	1.3	13	5,000	40	320	0.6	4 incremental		18,52,53
Rincon (TX)	Conoco Inc.	1958	LPG→G						120				
Midway-Sunset 29-D (CA)	Standard Oil Co. of California	1958	LPG→G	31	0.84	13	5,300	20	160	10	51 ultimate		57
Rio Bravo (CA)	Superior Oil Corp.	1958	LPG→G	38	0.28	5	11,200	60	300	12			58
Guijairal Hills (CA)	Union Oil Co. of California	1959	LPG→G										18
Paloma (CA)	Western Oil Corp.	1959	LPG→G										18
Elk City M Zone (OK)	Shell Development Co.	1959	LPG→G						pilot				18
Sunflower (OK)	Tekoil Corp.	1959	LPG→G						40				18,59
East Washunga (OK)	Tekoil Corp.	1959	LPG→G						40				18
La Gloria (TX)	Mobil Oil Corp.	1959	LPG→G						59				18
Panhandle (TX)	Skelly	1959	LPG→G						160				18
Baskington (LA)	Sun Oil Co.	1962	LPG→G		0.4	4	3,800	20	3 wells			term., gravity seg. disc.	60*
Pembina Lobstick Cardium (Alberta)	Amoco Production Co.	1964	LPG→G/W				5,000		2,240				61,62*
West Ranch Howard Glasscock (TX)	Mobil Oil Corp.	1964	High-relief LPG→G	31	1.4		5,500		480			75% succ., prof.	41*
North Coles Levee (CA)	ARCO Oil & Gas Co.	1967	LPG→G	36	0.41		9,500		624			term., prof.	41*

Field	Operator	Year Started	Project Type	Oil Gravity (°API)	Oil Viscosity (cp)	Viscosity Ratio μ_o/μ_s	Depth (ft)	Thickness (ft)	Acreage	Slug Size* (% HCPV)	Incremental Recovery* (% OOIP)	Status	Reference
Pembina Cardium (Alberta)	Texaco Inc.	1972	LPG←G←W	32	2	12	5,000	250	5,540			term., partly succ., unprof.	51,63*
Wasson (TX)	Shell Oil Co.	1959	LPG←G	30	1.2	14	5,000	16	160				64–66
North Pembina Cardium Unit (Alberta)	Mobil Oil Corp.	1959	LPG←G	37	0.5		5,000	650	13,000	1–1.7		term., disc.	49,67*–69
Wizard Lake (Alberta)	Texaco Inc.	1969	vertical LPG←G	39			6,200		4,800	7.4	84 ultimate projected	HF, succ., prof.	63*,70–71
Rainbow area (Alberta)	various		vertical LPG←G										

*Primary reference reporting status.

Symbol key:
LPG = liquified petroleum gas
G = lean gas
LPG←G = LPG-slug driven by gas
LPG←G/W = LPG-slug displaced by alternate gas/water injection
NC = nearing completion
disc. = discouraging
term. = terminated
HF = half finished
prof. = profitable
succ. = successful.

TABLE 5.3—FIRST-CONTACT MISCIBLE TERTIARY RECOVERY TESTS

Field	Operator	Year Started	Project Type	Oil Gravity (°API)	Oil Viscosity (cp)	Viscosity Ratio μ_o/μ_s	Depth (ft)	Thickness (ft)	Acreage	Slug Size* (% HCPV)	Incremental Recovery* (% OOIP)	Oil/Gross Slug (STB/RB)	Oil/Net Slug (STB/RB)	Reference
Burkett Unit (KS)	Phillips Petroleum Co.	1958	LPG←G←W	42	—	25	2,100	30	10	10	7.0	0.67		18,74
Johnson (NE)	Ohio Oil Co.	1958	LPG←G	37	1.1	18	4,600	10	164	5.5	34†	~7	2.2	18,75,76
South Ward (TX)	ARCO Oil & Gas Co.	1959	LPG←G←W	35	4	40	2,400	32	10	7.5	11.4	1.5		18,77
Adena Clar A (CO)	Union Oil Co. of California	1962	LPG←G	44	0.42	6.5	5,500	28	1 injector 2 producers	~17‡		0.36	1.7	78
Adena Hough A (CO)	Union Oil Co. of California	1963	LPG←G/W	44	0.42	6.5	5,500	28	80	~9.0‡		0.47	0.93	78
Hibberd Pool South Cuyama (CA)	ARCO Oil & Gas Co.	1963	LPG←G←W	35	1.7	23	4,300	60	80	7.4	13.5	1.6	>3	79
Phegly Unit (CO)	Mobil Oil Corp.	1964	LPG←G/W	37	3	30	4,900	8	785	3.3	3.5	0.9		80

*Treated area.
**Ultimate estimated from decline curve analysis.
†Nineteen percent OOIP in entire field.
‡Treated area difficult to define because of nature of test.

Symbol key:
LPG←G = LPG-slug driven by gas
LPG←G/W = LPG-slug driven by alternate gas/water injection.

hydrocarbon gas. Later, in an effort to improve the overall mobility ratio and to counteract the effect of permeability stratification, water was injected alternately with lean hydrocarbon gas in some projects after injection of a buffer slug of lean gas. In the majority of floods, reservoir pressure was sufficiently high, either initially or as a result of repressuring, that miscibility was achieved at both the leading and trailing edges of the slug. In some projects, however, reservoir pressure was below the pressure required for first-contact miscibility between LPG and lean hydrocarbon gas, and these two fluids were not miscible. The LPG in these latter projects was transported by vaporization rather than by miscible displacement. [17,19]

Solvent slug sizes have ranged from less than 1% HCPV [52,53] to as high as 15 to 22% HCPV. [17] Slug size for most projects was in the 1 to 6% HCPV range. Oil gravities have ranged from 30 to 51°API with the majority between 36 and 42°API.

Information on miscible field projects is found in technical papers, news reports in trade journals, and records of regulatory body hearings. Technical papers containing performance histories and engineering evaluations are the most valuable. Unfortunately, although a large number of first-contact miscible tests have been completed, relatively few were published in enough detail for a meaningful analysis, and many of these were ongoing rather than completed projects. Some of the better documented tests are described in the remainder of this section.

Secondary Recovery Field Tests

San Pedro, Block D. [17] This project, conducted by Intl. Petroleum Company in Peru, was one of the earliest LPG-slug tests. The flood was in a small, 20-acre isolated fault block. Properties of the reservoir included 16.5% average porosity, 155-md average permeability, 164-ft gross thickness, and 82-ft net thickness. Depth was 2,300 ft subsea. Viscosity ratio of butane with the 39°API oil was about 10.

This was essentially a one-well, one-producer project with the wells located on opposite sides of the fault block. In 1951, a 12% HCPV slug of butane was followed by natural gas injection. Reservoir pressure was below 900 psig, and miscibility between butane and drive gas was not achieved. Although butane appeared in the producer before the first significant increase in oil production, overall response was characterized by a decrease in GOR and an increase in oil production. Incremental recovery from the miscible flood, compared with expected primary recovery, was 10% OOIP.

Seminole City Pool. [19] In this flood conducted by Carter Oil Co. in Oklahoma, the low reservoir pressure also prevented attainment of miscibility between solvent and drive gas. Injection was into the Booch sandstone of the McAlester formation. This reservoir had an average porosity of 18% and an average permeability of 30 md. Reservoir depth was 3,500 ft, and average thickness varied from 12 to 32 ft. Mobility ratio of propane with the 36 to 38°API oil was about 30.

The reservoir had been depleted by solution gas drive but was repressured with gas before propane-slug injec-

tion, which began through two wells in 1953. BHP during lean gas injection following the 8.5% HCPV slug was slightly more than 700 psig. GOR's in the surrounding producers decreased shortly after propane injection. Breakthrough of propane into all five test wells was observed almost immediately after the start of gas drive, but the quantities were not excessive. Incremental recovery over primary production was in excess of 7.5% OOIP in the test area when a third of the propane had been produced and the only case history was published.

Millican Field. [20-23] In 1956, Sun Oil Co. tested LPG flooding in the Strawn limestone of this west Texas field. Average porosity of the 6,000-ft-deep reef was 6.6% and average permeability was 10 md.

Produced gas had been returned since early in the life of the field. A small propane/butane slug, amounting to only 1.5% of the sweepable HCPV in the project area, was injected into one well. The slug was driven by lean hydrocarbon gas at a pressure sufficient for miscibility between solvent and drive gas. GOR's declined almost immediately after slug injection but began increasing again after about 6 months. About 15 to 20% of the pilot area was swept before LPG breakthrough. Project engineers attributed the premature breakthrough of gas injected behind the slug to dilution of the slug and loss of miscibility caused by an adverse permeability distribution and an insufficient amount of solvent. Nevertheless, the operator estimated that 18% OOIP in the test area had been displaced by the solvent and drive gas.

H.T. Boyd Lease, Slaughter Field. [52,53] This west Texas carbonate reservoir flood was conducted by ARCO Oil and Gas Co. Properties of the 5,000-ft-deep San Andres reservoir included 12% average porosity, 10-md average permeability, and 40-ft average thickness. Viscosity ratio between the 30°API oil and propane was about 13.

The reservoir had been produced to well below the bubble point by 1958 when a 0.6% HCPV propane slug was injected into one inverted five-spot and two inverted nine-spots encompassing about 360 acres. The propane slugs were followed by 11% HCPV lean gas slugs and then by continuous water injection. Reservoir pressure was below the solvent/dry gas miscibility pressure at the start of the project, and miscibility at the trailing edge of the slug was attained by injecting at rates sufficiently high to establish the required pressure at the rear of the propane slug through a pressure gradient. The project was designed for only partial miscible sweepout because migration of fluids to surrounding low-pressure leases was anticipated at the high injection pressures required for sustained miscible displacement. Drive water was injected at reduced pressure. All patterns responded to the flood as evidenced by a decrease in GOR and an increase in oil rate. The five-spot pattern behaved in an almost textbook manner with some slight channeling of fluids. In the nine-spot patterns, breakthrough and production of miscible fluids was much more severe, apparently because of more severe permeability stratification and loss of miscibility. Water was injected alternately with dry gas in an attempt to control channeling in the nine-spots. Overall incremental recovery over waterflooding was estimated to be 4% OOIP.

North Pembina Cardium Unit. [24,62,67,68] To minimize pilot test costs some operators used inverted five-spot patterns of one injector with four producers. Mobil Oil Corp. conducted such a test in a 10-acre area of this Alberta, Canada, field.

The 5,000-ft-deep reservoir was a stratified sand of 16-ft gross thickness. Shale breaks separated sand intervals to the extent that only about 8 ft of the gross interval had good permeability and could be flooded. Other reservoir properties included 20% porosity and approximately 80-md permeability. Viscosity ratio between propane and the 37°API oil was 14.

In 1957, a propane/butane slug amounting to 7.1% of the floodable pattern volume was injected. This LPG slug was followed by gas, by a water slug, by gas again, and finally by water. Wells were operated to have no net withdrawal. Although early solvent breakthrough occurred, ultimate recovery was estimated by the operator to be 72% of the floodable OIP.

Pilot flood behavior was so encouraging that the project was expanded in 1959 to five 41-spot patterns encompassing 13,000 acres in the North Pembina Cardium Unit. This was the largest project by far of all the field trials of LPG-slug flooding. The gas-driven slugs in the expanded flood varied from 1 to 1.7% HCPV. Performance of the expanded flood was discouraging because of many premature breakthroughs of solvent and drive gas and subsequent production of injection fluids. In 1967, the Alberta Oil and Gas Conservation Board found that miscibility no longer existed and reduced the maximum permissible oil rate for the flood. The project subsequently was converted to a waterflood.

Bisti. [27,28] Another small, inverted pattern pilot test was conducted by Sunray in this New Mexico field. Injection in 1957 was into the 4,900-ft-deep Gallup sandstone. The 130-ft-thick reservoir contained alternating layers of sandstone, sandy shale, and shale. Only about 20% of the total thickness had permeability greater than 1 md. Porosity of the most permeable section was about 12%. Viscosity ratio between oil and the LPG slug was about 10.

Pattern size of the pilot test was 40 acres. A 5.5% HCPV propane/butane slug was driven with dry gas. Despite premature breakthrough, the ultimate recovery from this test was 55% OOIP, and the flood was expanded in 1959 to ten 160-acre five-spots. Channeling was severe in the expanded flood, however, and it was not expanded further.

In retrospect, the results from the small pilot tests at Pembina and Bisti apparently were misleading because of the inverted nature of the flooding pattern. In this type of pattern, swept volume could have exceeded the volume of the pilot area.

Parks Field. [39,40,42-48] Mobil Oil Corp. conducted a large propane-slug flood on approximately 6,400 acres in this west Texas field. Target formation was the 10,400-ft-deep Pennsylvania limestone—a thin, finely crystalline limestone interspersed throughout with shale stringers. Average net thickness was 21 ft, average porosity was 6.8%, and average permeability was 2.6 md. Viscosity ratio between propane and the 45°API oil was five.

Beginning in 1957, a 4% HCPV propane slug was injected into 23 wells to flood patterns that were approximately five-spots in shape. The propane slug was displaced by lean gas and ultimately by water. The field had been produced by solution gas drive, and GOR decreased and oil rate increased in response to miscible flooding. Breakthrough behavior was about as anticipated and, overall, the flood performed successfully. The operator felt that incremental recovery by miscible flooding compared with projected waterflood recovery was 14% OOIP.

29-D Pool, Midway-Sunset Field. [57] In 1958, Mobil Oil Corp. tested propane slug flooding in about 160 acres of this California field. Injection was into a 5,300-ft-deep, dipping, stratigraphic Miocene sand consisting of three major productive intervals of sand fingers interbedded with shale. Individual sand thickness averaged between 10 and 30 ft. Average permeability and porosity were 100 md and 24%. Viscosity ratio between the 31°API oil and solvent was 13.

Injection into two updip wells flooded two of the sands, while injection into another two updip wells flooded the third sand. The flood was conducted downdip across three rows of producing wells. Displacement rate was above the critical rate for gravity-stabilized flow.

The reservoir had been produced by solution gas drive with some secondary gas cap formation, and there had been some immiscible gas injection prior to miscible flooding. GOR's in producing wells decreased and oil rates increased in response to the miscible flood. Although early breakthroughs were experienced, large quantities of oil were produced after breakthrough. The original plan called for injection of a 10% HCPV slug of propane followed by dry gas. Because of production of the slug with oil in the first row wells, the plan was modified to include recycling of produced propane. Ultimate recovery was projected to be 51% OOIP compared with 25% by primary depletion and 31% by immiscible gas injection. There has been no subsequent published confirmation of this projection.

Meadow Creek Unit, Lakota B Reservoir. [50] Two sand intervals were flooded with a propane slug by Continental Oil Co. in this 7,300-ft-deep reservoir. The sands were separated by a 10-ft shale break and had about 9.5 ft of net pay combined. Other properties included 14% porosity and 100-md average permeability. Viscosity ratio was 10.

The reservoir had been immiscibly flooded with about 4% HCPV of propane before pressure was raised to 790 psi in 1958 by injection of water alternately with an additional 7% HCPV of propane. To achieve miscible displacement at the trailing edge of the slug, an ethane-rich gas was injected before switching to lean gas and water. Miscible fluids were injected into four wells in this 350-acre reservoir to displace oil into four producers. Injection of lean gas and water had to be increased from the design ratio of 4 RB/RB to 16 RB/RB to control water cut. The operator concluded that water injection prevented early gas breakthrough. From response to the miscible flood, the operator also concluded that 13% more oil would be recovered compared with recovery from waterfloods in similar reservoirs in the area. No confirmation of this projection has been published.

Wizard Lake, D3A Pool. [71-73] This project is a gravity-stable flood in a Canadian pinnacle reef. The flood was started in late 1969 when Texaco Inc. began injection of a 7.4% HCPV slug. At this time, primary recovery was about 27% OOIP. The 6,200-ft-deep reservoir had a maximum thickness of about 650 ft, covered about 4,800 surface acres, and originally contained 385 million bbl of oil. An LPG slug composition of 73% ethane-and-heavier hydrocarbons achieved miscibility at only 2,100 psi with the 39°API oil.

The solvent bank was sized to withstand diffusion of solvent and oil into and out of dead-end pores, which the operator believed was demonstrated to occur from core displacements. This resulted in a slug-size requirement double that estimated for a design considering longitudinal dispersion only.

In this project, lean drive-gas injection was begun before the entire LPG slug had been injected. Tubing in the solvent injectors was lowered periodically to maintain injection as close as possible to the gas/solvent interface. Water was injected into the underlying aquifer to raise the water/oil contact. This was necessary since the reef flared at the bottom, and, if allowed to reach this region of the reservoir, the solvent/oil interface would spread, causing the slug to decrease in thickness. Advance of both the oil/water and solvent/oil interfaces are being monitored with observation wells.

An ultimate recovery of 66% OOIP was expected for continued primary depletion by a combination gas expansion, water drive, and gravity segregation. By Dec. 1980, oil recovery from the miscible flood had reached 61% OOIP, and an ultimate recovery of 84% OOIP was expected.

Field Tests in Waterflooded Reservoirs

Field tests of LPG flooding for tertiary recovery began in the late 1950's. Published information is available on the seven tests discussed in the following subsections.

Johnson Field. [75,76,81] Marathon Oil Co. tested tertiary recovery LPG flooding in the Muddy J sand of this Nebraska field. The reservoir had a net pay thickness of 10 ft and contained a 37°API crude. Viscosity ratio between propane and the oil was 18. Other properties of the 4,600-ft-deep reservoir included an average porosity of 23% and an average permeability of 220 md.

A propane slug equivalent to 5.5% of the treated HCPV was injected through a single well in 1958 to flood the 164 acres under test. Water influx had partially maintained pressure; and at the start of miscible flooding, oil recovery had reached 31% OOIP, and the average producing water cut was 67%. Dry gas followed the propane slug, and premature breakthrough of both propane and dry gas was attributed to the wide permeability variation in the reservoir. To control gas production, water was injected first as a slug and then simultaneously with the injected gas. Ultimate recovery for the entire field was almost 64% OOIP,* which was

very close to the value predicted by the operator at the time recovery had reached 50% OOIP. Predicted recovery for a continued water drive was 45% OOIP. All of the field was not affected by the miscible flood. Incremental recovery from that part of the field that was affected by the miscible project was estimated to be as high as 34% OOIP.*

Burkett Unit. [74] Another tertiary recovery test was operated by Phillips Petroleum Co. in this Kansas field. Injection was into the 2,100-ft-deep Bartlesville sand which had an average porosity of 16% and an average permeability of 11 md in the test area. Reservoir thickness averaged 22 ft in the injection well. Viscosity ratio between the 42°API oil and propane was 25.

In 1958, a propane slug amounting to 10% HCPV was injected into a 10-acre, inverted five-spot pattern. The propane was followed by lean-gas injection and finally by water injection. Before the miscible test began, the reservoir had been produced by primary production, gas injection, and waterflooding for a recovery of 58% OOIP. At the start of miscible flooding, residual oil saturation in the test area was estimated to be 22% PV. Gas production began soon after gas injection and increased to 25% of the input rate before water injection began. Gravity of the produced oil declined during gas production but increased back to normal as the gas production declined during water injection. The operator thought that this behavior indicated that miscibility may not have been achieved at the trailing edge of the slug and that stripping was occurring. Because injection pressure was well above the propane/dry gas miscibility pressure, dry gas may have penetrated the propane slug and come into immiscible contact with oil. Oil recovery amounted to 6.7% of the oil initially in place in the pattern at discovery. Approximately 95% of the LPG and 72% of the injected gas remained within the reservoir.

W.D. Johnson Lease, South Ward Field. [77] A normal five-spot pattern was used by ARCO Oil and Gas Co. to test propane slug flooding. Injection was into the Penn Bennet sand of the Yates formation at a depth of 2,400 ft. The 10-acre pattern had a net thickness of 32 ft, an average porosity of 22.8%, and air permeabilities generally of 100 to 200 md. Crude oil gravity was 35°API, and oil/propane viscosity ratio was 40.

The sand had been waterflooded to a high water cut before the miscible-flood test began in 1961. The test pattern was ringed with water injectors to confine the miscible fluids. Waterflood residual oil saturation was thought to be 20% PV. Following a propane slug of 7.5% HCPV, a small band of lean gas was injected and followed by alternate injection of gas and water. The central producer responded significantly to propane injection, and oil cut peaked between 30 and 40%. Gas breakthrough occurred only 11 days after gas injection started, but oil recovery from the test reached 10.5% OOIP and was expected to reach 11.4% OOIP based on decline curve analysis.

Adena Field. [78] These small pilot tests were run in watered-out areas of the field by Union Oil Co. Depth of this sandstone was 5,500 ft. Average thickness was 28 ft, average porosity was 19.7%, and average permeability was 356 md. The estimated oil saturation after waterflooding was 30% PV. Viscosity ratio was 6.5 between the 44°API oil and propane.

The Clar A test involved a single injector and two producers in a multiple-line drive arrangement. In 1962, a propane slug was followed by alternate gas/water injection. Test results indicated that the volumetric conformance was low, but most of the oil contacted was displaced. Oil recovery in this test was 0.46 STB per surface barrel of propane injected (0.36 STB/RB) or 2.2 bbl oil recovered per surface barrel of propane left in the reservoir (1.7 STB/RB).

A second test was run in the Hough A lease, again involving a single injector and several producers aligned in the direction of flow caused by the pressure gradient established during waterflooding. The propane slug was followed by injection of a small buffer slug of lean gas and then by alternate gas/water injection. Oil recovery was 0.6 STB per surface barrel of propane injected (0.47 STB/RB) or 1.2 bbl per surface barrel of propane left in the reservoir (0.93 STB/RB).

Hibberd Pool.[79] ARCO Oil and Gas Co. tried a propane slug flood in this 200-acre pool located on the crest of the South Cuyama anticline in California. The sandstone reservoir was at a depth of 4,300 ft, had a gross thickness of 60 ft, and was interbedded with hard, dense sandstone shells. Other properties included an average porosity of 23% and an average permeability of 272 md. The 35.5°API oil had a viscosity of 1.7 cp, and its viscosity ratio with propane was 23.

Primary production was by solution gas drive with a partial water drive. This was followed by a peripheral-type waterflood in downstructure injection wells to give a total recovery of 46.4% OOIP. In April 1963, all but two producers on the crest of the structure were watered-out, and these were producing at high water cuts. Propane injection began into two wells located approximately on the 73% water saturation contour. This saturation was thought to be the highest at which oil was still mobile in the reservoir. A producer located between the injectors was placed on production to draw propane from both injectors and to form a bank of solvent along the strike of the field. When this was accomplished, dry gas was injected into all three wells to drive the solvent bank toward the two crestal producers. Water was injected into downstructure wells to confine miscible fluids to the top of the structure. The propane slug amounted to 5.4% PV upstructure of the injection wells.

Incremental recovery from the miscible flood amounted to approximately 13% OOIP in the reservoir that was upstructure from the injection wells.* The miscible flood recovered 1.6 STB of incremental oil per reservoir barrel of propane injected.* About 70% of the injected propane ultimately was produced, and more than 3 STB of incremental oil were recovered per reservoir barrel of propane left in the reservoir.*

Phegly Unit.[80] The miscible flood conducted by Mobil Oil Corp. in this Colorado field was the largest propane-

slug tertiary recovery test. The entire 785-acre unit was flooded through 11 injection wells, and tertiary oil was captured by 9 producers.

The Muddy D sandstone was found at a 4,900-ft depth and had an average thickness of about 8 ft in the Phegly Unit. Average porosity was 18%, and average permeability was 168 md. Viscosity ratio of the 37°API oil with propane was 30.

The reservoir was waterflooded with five-spot pattern development for an ultimate recovery of 41.4% OOIP. Residual oil saturation from flood pot tests was 25% PV. At the start of miscible flooding in 1964, the average WOR was 13. A propane slug amounting to about 3.3% HCPV of the expected floodable reservoir volume was followed by lean gas alternated with water and finally by water only. The lean gas slug was sized to leave a 25% residual gas saturation in the reservoir. Response of the oil production in timing and amount was approximately as anticipated. Breakthrough of the lean gas was more rapid than expected. Production of tertiary oil amounted to 3.5% OOIP in the flood area. Volumetric sweepout of the miscible fluid was estimated to be 28%.

5.5 Assessment of Process

Secondary Recovery Tests

Field experience has shown that first-contact miscible solvents will displace oil in reservoirs containing an initial gas saturation and will bank the oil into a secondary oil bank of increased saturation. This is true whether the gas saturation occurred from solution gas drive, from gas injected for pressure maintenance, or from gas injected for repressuring.[17,19,20] Response to the miscible flood resembles response to waterflooding—decreasing GOR's and increasing oil productivity when the secondary oil bank arrives at producing wells.

In most projects, rapid breakthrough of both solvent and lean hydrocarbon gas occurred. There were exceptions, such as the flood in the Parks field where only 7 out of 32 producing wells had experienced premature breakthrough of solvent after injection of 0.2 HCPV of solvent and lean gas.[40] More typical were the Pembina and Bisti tests, where solvent breakthrough was detected after injection of about 0.1 HCPV. Most often, solvent breakthrough occurred practically coincident with or soon after arrival of the secondary oil bank at producing wells.[17,19,20,24,28,52] Similarly, lean-gas breakthrough often has occurred almost simultaneously with solvent breakthrough or soon thereafter. In most floods, the bulk of the secondary oil bank was produced concurrently with solvent and lean gas.

Mobility ratio was unfavorable at both the leading and trailing edges of the LPG slugs in all field tests, and the early breakthroughs experienced in oilfield floods are consistent with behavior found in laboratory models for unfavorable-mobility-ratio displacements. This laboratory research showed that early breakthrough of injected fluids could result from viscous fingering or gravity tonguing (see Chap. 3). At high values of the viscous-to-gravity force ratio, viscous fingering prevails, and early breakthrough of injected fluids is caused by solvent fingering into the oil bank and from lean gas fingering into both solvent and oil bank. At low values of the

*Previously unpublished results, ARCO Oil & Gas Co. (1982).

viscous/gravity force ratio, gravity segregation dominates, and early breakthrough of injected fluid is caused by gravity tonguing. Vertical sweepout becomes poorer as the mobility ratio becomes more unfavorable.

Gravity undoubtedly was an important factor in some, or perhaps many, field tests. In the Baskington field, sampling at an observation well located 565 ft from a solvent injector verified that the propane slug had assumed a nearly horizontal rather than vertical orientation at this location.[60]

Permeability stratification, in addition to viscous fingering and/or gravity segregation, was responsible for early solvent breakthrough and relatively low miscible sweepout in some projects.[17,20,52] This was particularly evident in the Boyd lease flood in Slaughter field where two patterns in a more stratified part of the field had early breakthrough of solvent and significant production of injection gases, while a third pattern had considerably different performance characteristic of a more homogeneous reservoir.[52] When permeability stratification was recognized to be an important factor before solvent injection was begun, the solvent slug was sized to sweep miscibly only the more permeable strata.[24] This maximized efficient use of solvent and minimized nonproductive solvent flow through previously swept high-permeability strata.

Injection of water to improve mobility ratio and to moderate the influence of stratification was tried in several projects. Alternate injection of water slugs with lean gas slugs slowed the rate of GOR increase after solvent and lean-gas breakthrough in the Meadow Creek[50] and Slaughter Boyd lease[52] tests. Operators of these projects tried to achieve the equal-velocity ratio of Caudle and Dyes[82] (see Sec. 3.8.) Although this ratio often is about 1 res bbl of lean gas per reservoir barrel of water, a ratio of lean gas to water as high as 16:1 was necessary in the Meadow Creek project to prevent water from advancing faster than lean gas.

Presolvent water injection was tried in the Pembina Lobstick Cardium Unit flood to distribute the solvent slug among strata more uniformly.[61] Although the operator felt that pre-water injection was beneficial and calculations support this view, there are no comparisons of field performance with and without pre-water injection to demonstrate its effectiveness.

In many projects, a substantial fraction of injected LPG was produced (as much as 65% in the Millican field test).[20] This is important for project economics because produced solvent can be recovered and either be reinjected or sold. In one project, the operator had plans to refrigerate produced gas and recover the LPG for reinjection.[57]

Total oil recovery alone is not a completely adequate measure of success for miscible flood tests. Incremental recovery may be a better measure of miscible flooding effectiveness than total recovery, because there are usually alternative methods for producing the reservoir, such as waterflooding or immiscible dry-gas injection.

Both total oil recovery and incremental recovery can be misleading in some pilot tests. For example, in some single-injector pilot tests, the producing wells may capture only a portion of the oil displaced from the pilot area.[83] Oil recovery from such a test could be conservative, if a higher capture efficiency would be achieved in a larger project with repeating patterns. In other single-injector tests, miscible fluids may sweep outside the test area, and recovery from these tests can be overly optimistic.

In secondary recovery floods, of course, there is no direct measure of incremental recovery, and any estimate of it must be made by estimating performance of the alternative recovery method and comparing with the observed performance of the miscible flood. There is necessarily a considerable uncertainty in such an estimate of incremental recovery. Craig[84] compared recoveries from both LPG and condensing-gas drive slug projects with recovery expected from conventional dry gas injection, and of 31 projects analyzed, he concluded that about 22% had recovered an incremental oil volume greater than four times the volume of the solvent slug, while in about 26%, incremental oil recovery did not even equal the solvent slug volume. No difference in behavior was noted between LPG or condensing-gas drive slug processes.

There are only a few published attempts to compare first-contact miscible flood secondary recovery performance with estimated waterflood performance.[40,50,52] In these projects, estimates of incremental recovery ranged from 8 to 35% more oil than would have been realized under waterflood.

The relationship between oil recovery and slug size depends on conditions specific to a given reservoir such as permeability stratification, vertical/horizontal permeability ratio, oil viscosity, phase behavior, well spacing, and flooding pattern. Greater oil recovery should be expected as larger solvent slugs are injected, although conceptually the incremental recovery per barrel of slug injected eventually should decrease with increasing slug size. Field test conditions for first-contact miscible flooding were too varied to establish even a rough correlation of slug size with oil recovery. A solvent slug as small as 0.6% HCPV was partially effective in the Slaughter Boyd lease test but lost miscibility before reaching producing wells.[52] Floods in the Pembina field performed poorly with slug sizes of 1 to 1.7% HCPV[62,85] and 3% HCPV,[62] and a test in the Millican field failed with a slug size of 1.5% HCPV.[20] From the bulk of field test experience, a slug size of 4 to 5% HCPV seems to be about the minimum for maintaining miscibility over much of the floodable pore volume.

There are several gravity-stable, first-contact miscible floods in pinnacle reef reservoirs,[63,70,73] but performance details have been published only for the project at Wizard Lake, where a 7.4% HCPV LPG slug was displaced by lean hydrocarbon gas.[73] This project appears to be performing successfully and about as anticipated. By late 1980, oil recovery had reached 61% OOIP, about equal to the recovery expected for continued primary depletion by immiscible gas expansion and water drive. Ultimate recovery for the miscible flood was expected to be 84% OOIP.

There have been relatively few additional operating problems caused by LPG slug flooding. Cooling caused by gas expansion has caused gas hydrates to form in orifice plates and chokes of injection wells;[28] cooling has also caused low separator temperature[50] and some

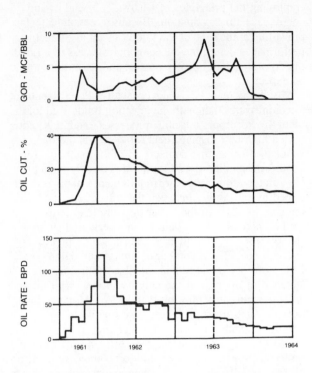

Fig. 5.12—Pilot producer performance, W.D. Johnson Well 33 (after Ref. 77).

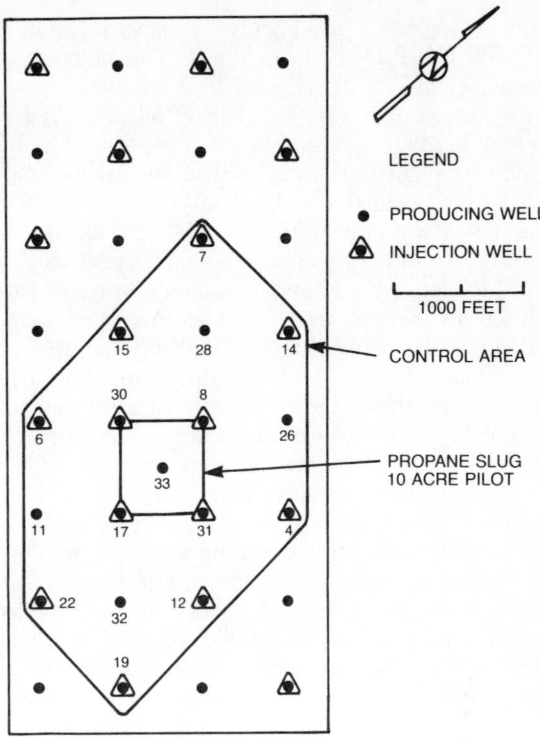

Fig. 5.13—Location of pilot on 320-acre W.D. Johnson lease (after Ref. 77).

paraffin deposition problems in flow lines.[50] Downhole pump freeze-ups were experienced, which required installation of downhole separators.[50] Unexpectedly high gas production from early solvent and dry gas breakthrough has overloaded recycle compressors, requiring a reduction in oil production;[28] and some difficulty was experienced in maintaining BHP because of solvent-supply interruption and the seasonal nature of the solvent supply.[28] Asphalt deposition was anticipated in some tests,[60] and, to prevent asphalt plugging in the immediate vicinity of injection wells, one operator injected a small slug of refinery oil to displace crude away from the wells before injection of solvent.[80]

Tests in Waterflooded Reservoirs

Field tests in reservoirs that have been waterflooded give the best measure of how much more oil can be recovered by miscible flooding than by waterflooding. Table 5.3 shows results of first-contact miscible floods that were conducted in waterflooded reservoirs. In these tests, solvent slug sizes generally were in the 3-to-10% HCPV range. Oil gravities were high, in the 35-to-44°API range, and swept-zone waterflood residual oil saturations were in the range of 20 to 30% PV.

All the tests definitely showed that miscible solvent slugs could displace waterflood residual oil and move it to producing wells. Significant quantities of water were produced before arrival of tertiary oil at the producing wells caused a decrease in WOR.

Breakthrough behavior for solvent and lean gas was generally similar to the behavior found in secondary recovery floods. Early breakthroughs were typical; often both solvent and dry gas appeared at producing wells

almost simultaneously with first response of tertiary oil or at least soon thereafter. As in secondary recovery floods, this behavior could have been caused by viscous fingering, gravity tonguing, permeability stratification, or some combination of these factors.

To illustrate these remarks, performance for the pilot test in the W.D. Johnson lease, South Ward field is shown in Fig. 5.12.[77] Fig. 5.13 is a map of the lease showing arrangement of wells. Water was injected into wells surrounding the normal five-spot test pattern to confine miscible fluids to the test area. Calculations showed that 30% of the propane slug and lean-gas drive fluid went into the test pattern. After injection of the 7.5% HCPV propane slug, a buffer slug of lean gas was injected. The buffer slug was followed by alternate injection of small slugs of lean gas and water at the equal-velocity ratio. There was a slight trend of increasing injection pressure with time as slugs of lean gas and water were alternated; gas and water injection rates were relatively constant. After injection of roughly 0.1 HCPV of total hydrocarbons, gas breakthrough was observed in the central producing well at about the same time that oil cut began to increase (see Fig. 5.12). After breakthrough, the GOR quickly rose to about 4,000 scf/STB. However, alternate gas/water injection was effective in moderating gas production, and GOR began to decline after injection of the gas band was completed and alternate gas/water injection began. This was followed by a subsequent slow increase in GOR and a slow decline in oil cut. The operator attributed early breakthrough of injected fluids to viscous fingering or gravity tonguing rather than to stratification.

Incremental oil recovery in the Burkett,[74] South

Ward,[77] Hibberd,[79] and Phegly[80] tests varied from 3.5 to 13.5% OOIP in the treated area. This amounted to between 9 and 37% of the waterflood residual oil saturation in the treated area. Incremental oil recovered per reservoir barrel of LPG injected in these tests as well as in the two small tests completed in the Adena field[78] varied from about 0.4 to 1.6 STB/RB. Although data were not always available on LPG production and recovery, the ratio of incremental oil recovered per barrel of LPG left in the reservoir was in the range of 1 to 2 STB/RB for several tests,[77,78] and was more than 3 STB/RB in the Hibberd test.[79] Probably the effect on flood behavior caused by reservoir heterogeneity overshadowed the effects caused by differences in slug sizes and mobility ratio; however, the lowest oil recovery (but not the lowest ratio of oil recovered per barrel of LPG injected) resulted for the smallest slug size tested.[80]

Results from the test in the Johnson field are anomalously favorable when compared with results of first-contact miscible floods in other waterflooded reservoirs (and also compared with condensing-gas drive and CO_2-miscible floods in waterflooded reservoirs, Chaps. 6 and 8). This 164-acre test was in a reservoir that had been produced by natural water drive until the start of miscible flooding, and the average producing water cut at that time was 67%. The propane slug was only about 5.5% of the contacted pore volume, but incremental recovery was almost 34% OOIP in the treated area.* The ratio of incremental oil recovered per reservoir barrel of slug injected was more than 7 STB/RB. There was no explanation offered by the operator for the cause of this exceptionally favorable performance.

The project in the Phegly Unit was the largest first-contact miscible flood in a waterflooded reservoir.[80] The entire 785-acre unit was flooded through 11 injection wells. Because of its size, this test may have the greatest validity. Incremental recovery was about 3.5% OOIP in the treated volume for injection of a 3.3% HCPV LPG slug driven by alternate lean gas/water injection. Incremental recovery per reservoir barrel of solvent injected was about 0.9 STB/RB.

Although tests in waterflooded reservoirs give the best available measure of incremental recovery over waterflooding, even in these tests there is still substantial uncertainty that the incremental recovery from the relatively small tests adequately characterizes the recovery to expect in large-scale floods with regular, repeating patterns. In unconfined tests or tests with irregular well arrangement, some of the tertiary oil actually displaced may not be captured by producing wells. In the Phegly Unit, for example, the operator estimated that 550,000 bbl of oil were displaced by the propane slug, although only 160,000 bbl actually were produced.[80]

5.6 Screening Criteria

First-contact miscible flooding cannot be practiced indiscriminately. It is best suited only for certain types of reservoirs and oils. The remainder of this section summarizes advantages, disadvantages, and criteria for practicing this method of miscible flooding. The criteria should be considered only as general guidelines that are

the starting point for a sound engineering evaluation employing the principles of Chaps. 2, 3, and 4, and do not replace this evaluation. Because economic factors prevailing at the time of the flood as well as oil recovery performance determine the economic success of a flood, the criteria cannot be very precise and exceptions are to be expected.

Miscibility can be achieved with LPG solvents at lower pressures than with the injection fluids used for the other hydrocarbon miscible processes, and this along with a high unit displacement efficiency is the main advantage of the first-contact miscible method. First-contact miscibility of LPG with oil is usually not a problem, and minimum pressure requirement usually is determined by miscibility between the solvent slug and driving gas. Because of the lower pressure requirement, the method potentially can be used in reservoirs that are too shallow for the other hydrocarbon miscible processes.

The process has a number of limitations. High cost of LPG solvents dictates that small solvent slugs be injected. But dilution of the slug by mixing with oil and drive gas can cause miscibility to be lost. Viscous fingering can cause drive gas to penetrate small slugs and come into direct (and immiscible) contact with oil. Proper sizing of the slug to account for the influence of fluid and reservoir properties on slug dilution is uncertain. Miscible sweep efficiencies are often poor because of the small amount of solvent injected, unfavorable viscosity ratios at front and rear of the slug, and density contrasts between oil, solvent, and drive gas.

Guidelines for judging the suitability of reservoirs for first-contact miscible flooding are as follows.

1. Low oil viscosity is desirable to minimize the effects of unfavorable viscosity ratio. Less than 1 cp is preferred for horizontal floods. An upper viscosity limit is not well defined but may be approximately 5 cp. The upper limit for gravity-controlled floods depends on reservoir permeability (see Sec. 3.6). The viscosity limitation generally excludes oils with API gravity less than 30° from being flooding prospects.

2. With a propane slug driven by lean gas, miscibility pressure between propane and the lean gas will be in the range of 1,100 to 1,300 psi. Depending on the reservoir parting pressure, a minimum reservoir depth in the range of 1,500 to 2,500 ft is required to operate above the miscibility pressure. Lower miscibility pressures, about 900 psi, are required with ethane-rich slugs. There is no maximum depth limitation, but at depths where reservoir temperature exceeds the solvent critical temperature or when the LPG is diluted with light gases, reservoir pressure should be sufficiently high to ensure solvent/oil miscibility.

3. In horizontal reservoirs, restricted vertical permeability, which might result from scattered shale lenses and laminations or from dense streaks, is preferred to minimize gravity segregation. Reservoirs with substantial structural relief are preferred if permeability is high enough for gravity-stabilized flooding.

4. Extensive fracturing, a gas cap, a strong water drive, and a high permeability contrast increase project risk.

5. Oil saturation in the reservoir volume that will be swept miscibly is important for project feasibility. Acceptable saturation depends on injection fluid costs, oil

---

*Predicted by extrapolation of miscible flood and water-drive decline curves and later confirmed by ultimate production (personal communication, F.S. Cordiner, Marathon Oil Co. 1982).

price, reservoir properties, and fluid properties, but 25% PV is approximately a minimum value.

References

1. Benedict, M., Webb, G.B., and Rubin, L.C.: "An Empirical Equation for Thermodynamic Properties of Light Hydrocarbons and Their Mixtures—Constants for Twelve Hydrocarbons," *Chem. Eng. Prog.* (1951) **47**, 419.

2. Zudkevitch, D. and Joffe, J.: "Correlation and Prediction of Vapor-Liquid Equilibria with the Redlich-Kwong Equation of State," *AIChE J.* (Jan. 1970) 112–19.

3. Rowe, A.M. and Silberberg, I.H.: "Prediction of Phase Behavior Generated by the Enriched-Gas Drive Process," *Soc. Pet. Eng. J.* (June 1965) 160–66.

4. Fussel, D.D. and Yanosik, J.L.: "An Iterative Sequence for Phase Equilibrium Calculations Incorporating the Redlich-Kwong Equation of State," *Soc. Pet. Eng. J.* (June 1978) 173–82.

5. Fussell, L.T.: "A Technique for Calculating Multiphase Equilibria," *Soc. Pet. Eng. J.* (Aug. 1979) 203–08.

6. Meldau, R.F. and Simon, R.: "Phase Behavior in Miscible Systems of Solvent, Dry Gas, and Oil," paper SPE 226 presented at the 1961 SPE California Regional Meeting, Bakersfield, Nov. 2–3.

7. Yarborough, L. and Smith, L.R.: "Solvent and Driving Gas Compositions for Miscible Displacement," *Soc. Pet. Eng. J.* (Sept. 1970) 298–310.

8. Herbeck, E.F., Heintz, R.C., and Hastings, J.R.: *Fundamentals of Tertiary Oil Recovery*, Energy Communications, Dallas (1977).

9. Peng, D.Y. and Robinson, D.B.: "A New Two-Constant Equation of State," *Ind. Eng. Chem. Fund.* (1976) **15**, 59.

10. Bossler, R.B. and Crawford, P.B.: "Miscible-Phase Floods May Precipitate Asphalt," *Oil and Gas J.* (Feb. 23, 1959) 137–45.

11. Smith, A.S. and Brown, G.G.: "Viscosity of Ethane and Propane," *Ind. Eng. Chem.* (1943) **35**, 705.

12. Dolan, J.P. *et al*: "Liquid, Gas, and Dense Fluid Viscosity of n-Butane," *J. Chem. Eng. Data* (July 1963) **8**, 396–99.

13. Katz, D.L. *et al.*: *Handbook of Natural Gas Engineering*, McGraw-Hill Book Co. Inc., New York City (1959).

14. Lohrenz, J., Bray, B.G., and Clark, C.R.: "Calculating Viscosities of Reservoir Fluids From Their Compositions," *J. Pet. Tech.* (Oct. 1964) 1171–76.

15. Brown, G.G. *et al.*: *Natural Gasoline and the Volatile Hydrocarbons*, Natural Gas Assn. of America, Tulsa (1948).

16. Alani, G.H. and Kennedy, H.T.: "Volumes of Liquid Hydrocarbons at High Temperatures and Pressures," *Trans.*, AIME (1960) **219**, 288–92.

17. Moscrip, R. III: "Butane Injection in San Pedro Block D," *Pet. Eng.* (Sept. 1958) B-10–B-24.

18. "Miscible Drives: A Growing Tool for Oil Recovery," *Oil and Gas J.* (March 23, 1959) 64–69.

19. Jenks, L.H., Campbell, J.B., and Binder, G.G. Jr.: "A Field Test of the Gas-Driven Liquid Propane Method of Oil Recovery," *Trans.*, AIME (1957) **210**, 34–39.

20. Sturdivant, W.C.: "Pilot Propane Project Completed in West Texas Reef," *J. Pet. Tech.* (May 1959) 27–30.

21. "Early LPG Flood Runs Into Trouble," *Oil and Gas J.* (Dec. 2, 1957) 59.

22. "New Recovery Idea Looks Like a Champ," *Oil and Gas J.* (March 11, 1957) 96.

23. "LPG Injection Test Proves Successful," *Petroleum Week* (Sept. 25, 1958) 22.

24. Justen, J.J. *et al.*: "The Pembina Miscible Displacement Pilot and Analysis of Its Performance," *Trans.*, AIME (1960) **219**, 38–45.

25. "LPG Flood Combined with Storage in California Project," *Oil and Gas J.* (July 8, 1957) 74.

26. Schnake, C.: "New Repressuring Method Promises to Triple Output," *World Oil* (April 1958) **145**, 201.

27. Brinkley, T.W.: "Sunray's LPG Pilot Test is Paying Off," *World Oil* (Aug. 1, 1960) 85–93.

28. Ramsey, E.H.: "Operating a Miscible Flood, Central Bisti Unit," *Prod. Monthly* (Sept. 1961) 12–15.

29. "Bisti Looks to LPG," *Oil and Gas J.* (July 22, 1957) 51.

30. Brooks, R.E.: "Miscible Displacement in Bisti Pool, San Juan Co., New Mexico," paper 1037-G presented at the 1958 AIME Mid-Continent Section Petroleum Conference on Production and Reservoir Engineering, Tulsa, March 20–21.

31. "LPG Flood in Bisti Pool is Succeeding," *Petroleum Week* (May 27, 1960) 27.

32. "LPG Spurs Bisti Production," *Oil and Gas J.* (July 21, 1958) 56.

33. "Sunray Begins Big Miscible Phase Flood at Bisti," *Oil and Gas J.* (July 6, 1957) 72.

34. "Busy Days Ahead for Bisti Operations," *Oil and Gas J.* (Feb. 23, 1959) 78.

35. "Recovery Mechanisms Used Simultaneously in the Bisti Unit," *Oil and Gas J.* (July 11, 1960) 106.

36. "Greater Production Thought for Bisti Miscible Phase Unit," *Oil and Gas J.* (May 16, 1960) 108.

37. Brooks, R.E.: "Miscible Drive Field Application in Bisti Pool," *J. Pet. Tech.* (May 1958) 22–24.

38. "Conoco Readies First LPG Flood," *Oil and Gas J.* (July 22, 1957) 50.

39. Secondary Recovery Texas Railroad Commission Hearing Summaries as extracted from the Texas State House Reporter, R.W. Bynum and Co., Austin (1961).

40. Marrs, D.G.: "Field Results of Miscible-Displacement Using Liquid Propane Driven by Gas, Parks Field Unit, Midland County, Texas," *J. Pet. Tech.* (April 1961) 327–32.

41. Bleakley, W.B.: "Journal Survey Shows Recovery Projects Up," *Oil and Gas J.* (March 25, 1974) 69–78.

42. "Record Miscible Phase Project Planned," *Oil and Gas J.* (June 3, 1957) 64.

43. "Parks Propane Flood Gets First Start," *Oil and Gas J.* (July 8, 1952) 76.

44. "Magnolia Plans Big Miscible Displacement Recovery Project," *Pet. Eng.* (Aug. 1957) **29**, B-84.

45. "Biggest Propane Flood Hits Its Stride," *Oil and Gas J.* (Sept. 9, 1957) 58.

46. Marrs, D.G.: "Field Wide Miscible Displacement Project, Parks Field Unit, Midland Co., Texas," paper 1042-G presented at the 1958 AIME Mid-Continent Section Petroleum Conference on Production and Reservoir Engineering, Tulsa, March 20–21.

47. Marrs, D.G.: "Miscible Drive in Parks Field," *J. Pet. Tech.* (May 1958) 20–21.

48. "Largest U.S. Miscible Drive Enters Fourth Year a Spectacular Success in Parks Field, West Texas," *Oil and Gas J.* (Sept. 19, 1960) 66.

49. "Another Miscible Job Starts," *Oil and Gas J.* (Oct. 14, 1957) 117.

50. Gernet, J.M. and Brigham, W.E.: "Meadow Creek Unit Lakota "B" Combination Water-Miscible Flood," *J. Pet. Tech.* (Sept. 1964) 993–97.

51. "U.S. Pushes Enhanced-Recovery Effort," *International Petroleum Encyclopedia*, Petroleum Publishing Co., Tulsa (1977) 282–94.

52. Sessions, R.E.: "How Atlantic Operates the Slaughter Flood," *Oil and Gas J.* (July 4, 1960) 91–98.

53. Sessions, R.E.: "Small Propane Slug Proving Success in Slaughter Field Lease, *J. Pet. Tech.* (Jan. 1963) 31–36.

54. "Three Texas Floods Approved," *Oil and Gas J.* (May 19, 1958) 103.

55. "Conoco Plans Miscible Flood In Texas Rincon Field," *Oil and Gas J.* (Sept. 19, 1960) 79.

56. Secondary Recovery Railroad Commission Hearing Summaries as extracted from the Texas State Reports, R.W. Bynum and Co., Austin (1960) 69.

57. Block, W.E. and Donovan, R.W.: "An Economically Successful Miscible Phase Displacement Project," *J. Pet. Tech.* (Jan. 1961) 35–40.

58. Glasser, S.R.: "History and Evaluation of an Experimental Miscible Flood in the Rio Bravo Field," *Prod. Monthly* (Jan. 1964) 17–20.

59. "Tekoil May Use LPG Flood on Oklahoma Lease," *Oil and Gas J.* (June 10, 1957) 65.

60. Lane, L.C., Teubner, W.G., and Campbell, A.W.: "Gravity Segregation in a Propane Slug-Miscible Displacement Project—Baskington Field," *J. Pet. Tech.* (June 1965) 661–63.

61. Kloepfer, C.V. and Griffith, J.D.: "Solvent Placement Improvement by Pre-Injection of Water, Lobstick Cardium Unit Pembina Field," paper SPE 948 presented at the SPE 1964 Annual Meeting, Houston, Oct. 11–14.

62. Craig, D.R. and Bray, J.A.: "Gas Injection and Miscible Flooding," *Proc.*, Eighth World Pet. Cong., Moscow (1971) **3**, 275–85.

63. Matheny, S.L. Jr.: "EOR Methods Help Ultimate Recovery," *Oil and Gas J.* (March 31, 1980) 91–92.

64. "Shell's Miscible Pilot in Wasson Field Looking Good," *Oil and Gas J.* (Feb. 1, 1960) 66.

65. "Shell Will Test Miscible Displacement in Big Wasson Field," *Oil and Gas J.* (June 29, 1959) 48.

66. "Pilot Miscible Drive Started in Big Wasson Field," *Oil and Gas J.* (July 13, 1959) 79.

67. "Plans for Biggest Miscible Flood Abandoned," *Oil and Gas J.* (March 27, 1961) 71.

68. "Biggest Miscible Flood Turned to Water at Pembina," *Oilweek* (August 1961).

69. "Mobil Starts Biggest Miscible Phase Job in Pembina," *Oil and Gas J.* (Sept. 7, 1959) 100.

70. "Canadian Enhanced-Recovery Activity is Moderate, Centers on Thermal Projects," *Oil and Gas J.* (April 5, 1976) 128–29.

71. "Texaco Miscible-Flood Program Begins for Alberta Pool Soon," *Oil and Gas J.* (Dec. 22, 1969) 48.

72. "Wizard Lake D-3A Pool—Engineering Study of Primary Performance and Miscible Flooding," Texaco Exploration Co. application to Alberta Oil and Gas Conservation Board, Calgary (Dec. 1968).

73. Martin, W.E.: "The Wizard Lake D-3A Pool Miscible Flood," paper SPE 10026 presented at the 1982 SPE Intl. Petroleum Exhibition and Technical Symposium, Beijing, China, March 19–22.

74. Weimer, R.F.: "LPG-Gas Injection Recovery Process, Burkett Unit, Greenwood County, Kansas," *J. Pet. Tech.* (Oct. 1963) 1067–72.

75. Cordiner, F.S. and Popp, M.P.: "Application of Miscible Phase Displacement Techniques to a Water-Drive Reservoir, Johnson Field, Cheyenne Co., Nebraska," *Drill. and Prod. Prac.*, API (1962) 144.

76. "Nebraska LPG Flood Begins," *Oil and Gas J.* (June 9, 1958) 76.

77. Blanton, J.R., McCaskill, N., and Herbeck, E.F.: "Performance of a Propane Slug Pilot in a Watered-Out Sand—South Ward Field," *J. Pet. Tech.* (Oct. 1970) 1209–14.

78. Holm, L.W.: "Propane-Gas-Water Miscible Floods in Watered-Out Areas of the Adena Field, Colorado," *J. Pet. Tech.* (Oct. 1972) 1264–70.

79. Chambers, F.T.: "Tertiary Oil Recovery Combination Water-Gas Miscible Flood—Hibberd Pool," *Prod. Monthly* (Jan. 1968) 17.

80. Connally, C.A. Jr.: "Tertiary Miscible Flood in Phegly Unit, Washington, County, Colorado," paper SPE 3775 presented at the 1972 SPE Improved Oil Recovery Symposium, Tulsa, April 16–19.

81. Rhodes, H.N.: "Secondary Recovery Booms in Western Nebraska," *Pet. Eng.* (May 1960) **32**, B-34.

82. Caudle, B.H. and Dyes, A.B.: "Improving Miscible Displacement by Gas-Water Injection," *Trans.*, AIME (1958) **213**, 281–84; *Miscible Processes*, Reprint Series, SPE, Dallas (1965) **8**, 111–14.

83. Craig, F.F. Jr.: "A Laboratory Model Study of Single Five-Spot and Single Injection Well Pilot Waterflooding," *Trans.*, AIME (1965) **234**, 1454–67.

84. Craig, F.F. Jr.: "A Current Appraisal of Field Miscible Slug Projects," *J. Pet. Tech.* (May 1970) 529–36.

85. *Report and Decision on Review of Certain Solvent Flood Operations in the Pembina Cardium Pool*, Alberta Oil and Gas Conservation Board, Calgary (1967).

86. Hutchinson, C.A. and Braun, P.H.: "Phase Relations of Miscible Displacement in Oil Recovery," *AIChE J.* (March 1961) **7**, 64–72.

Chapter 6
The Condensing-Gas Drive Process

6.1 Estimation of Miscibility Conditions

Miscibility between reservoir oil and injected gas in the condensing-gas drive process is achieved by in-situ mass transfer of intermediate-molecular-weight hydrocarbons such as ethane, propane, and butane from an injected gas containing these materials into the reservoir oil (i.e., the intermediate-molecular-weight hydrocarbons "condense" into the oil). For suitable conditions of pressure and gas composition, the reservoir oil can become so enriched with these materials that miscibility results between the injected gas and the enriched oil. In the remainder of this section, an injection gas that contains a relatively high concentration of intermediate-molecular-weight hydrocarbons will be called a "rich gas."

The required phase behavior for condensing-gas drive miscibility is illustrated conceptually by the pseudoternary diagram shown in Fig. 2.14. This figure shows that condensing-gas drive miscibility can be achieved by adding intermediate-molecular-weight hydrocarbons to a lean injection gas until a rich gas of Composition B is attained. Mixtures of methane and intermediate-molecular-weight hydrocarbons between Compositions A and B will miscibly displace the example oil in Fig. 2.14 by the condensing-gas drive process. First-contact miscibility occurs with gases richer in intermediate-molecular-weight hydrocarbons than Composition A. Lean drive gas is directly miscible with the rich-gas slug at pressures where condensing-gas drive miscibility is attained between oil and the rich-gas slug.

The concentration of intermediate-molecular-weight hydrocarbons required in the injection gas for condensing-gas drive miscibility depends on reservoir pressure. An increase in pressure reduces the size of the two-phase region. This decreases the concentration requirement because the limiting tie line intersects the right side of the pseudoternary diagram (i.e., the side representing methane/intermediate-molecular-weight hydrocarbon mixtures) at a lower concentration of the intermediate hydrocarbons, and Point B in Fig. 2.14 moves toward the methane apex of the diagram.

Pressure and injection gas concentration requirements for condensing-gas drive miscibility depend on (1) oil composition, (2) composition of the intermediate-molecular-weight hydrocarbons contained in the rich gas, and (3) reservoir temperature. Benham et al.[1] derived a correlation to estimate miscibility requirements. These authors mathematically combined reservoir fluids in various proportions with different mixtures of methane and intermediate-molecular-weight hydrocarbons and calculated the critical temperature and pressure for the resulting overall mixtures. They assumed that a rich gas having an intermediate-molecular-weight hydrocarbon concentration equal to or greater than that of the critical mixture would miscibly displace the reservoir fluid at a reservoir temperature equal to the calculated critical temperature if pressure were higher than the calculated critical pressure. They felt that such an assumption would be conservative because the minimum intermediate-molecular-weight hydrocarbon concentration necessary for condensing-gas drive miscibility is lower than the concentration of intermediates in the critical mixture or plait-point fluid (see Fig. 2.14). They recognized, however, that the correlation derived from such calculations was heavily dependent on the accuracy of the critical temperature and pressure correlations used in the calculations.

Figs. 6.1, 6.2, and 6.3 are crossplots at 100, 150, and 200°F of the original Benham et al.[1] correlations. Full-page versions of these figures can be found in Appendix E as Figs. E-1, E-2, and E-3. Oil composition is characterized by the molecular weight of the pentanes-

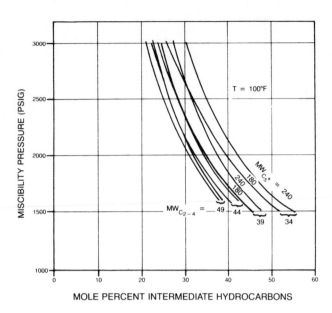

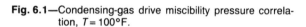

Fig. 6.1—Condensing-gas drive miscibility pressure correlation, $T = 100°F$.

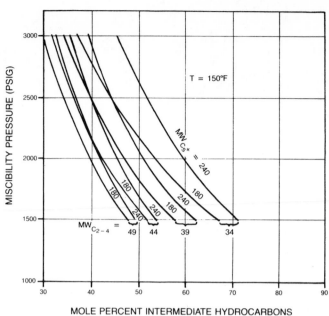

Fig. 6.2—Condensing-gas drive miscibility pressure correlation, $T = 150°F$.

and-heavier fraction, and composition of the intermediate-molecular-weight hydrocarbon fraction is characterized by the average molecular weight of the ethane-through-butane components. These figures give the minimum concentration of intermediate-molecular-weight hydrocarbons required in the rich gas to achieve miscibility at a given pressure, or, conversely, the minimum pressure required for miscibility with a given rich-gas composition. Either a higher pressure or a greater concentration of intermediate-molecular-weight hydrocarbons in the rich gas is required for miscibility as (1) molecular weight of the pentane-and-heavier fraction increases, (2) average molecular weight of the intermediate hydrocarbon fraction decreases, and (3) reservoir temperature increases. The original correlation was developed for the following conditions.

Pentanes-and-heavier fraction molecular weight	180 to 240
Average molecular weight of ethane-through-butane fraction	34 to 58
Reservoir temperature, °F	70 to 260
Pressure, psig	1,500 to 3,000

Yarborough and Smith[2] compared predictions from the Benham *et al.* correlation with experimentally determined miscibility conditions for 10 oils. Predictions for eight of the oils required extrapolating the correlation. Differences between experimental and predicted intermediate-molecular-weight hydrocarbon concentrations ranged from 1.5 to 9 mol% units and averaged 4 units. Except for two oils, the correlation predicted conservatively—i.e., predicted concentration was higher than the experimentally determined concentration. This author tested the correlation with 10 other experimentally determined miscibility conditions and found that at fixed gas composition the average difference between predicted and experimental miscibility pressures was 340

psi, although the maximum difference was slightly more than 800 psi. The correlation predicted conservatively (i.e., too high) for only five of the experiments. The Benham *et al.* correlation appears to be useful for preliminary designs, but slim-tube experiments should be run for final design conditions.

In principle, the pressure and rich-gas composition required for miscibility can be estimated with mathematical models by simulating the multiple contacting of oil by rich gas and calculating the resulting phase behavior. In practice, calibration of equations or correlations used in the phase behavior calculations with experimental PVT data usually is necessary to achieve the accuracy required for miscibility prediction. This can require as much or more effort than running slim-tube experiments.

Rowe and Silberberg[3] described a method for predicting condensing-gas drive phase behavior that mathematically simulated a batch multicontact approach to miscibility between rich gas and oil. After each contact, flash calculations were made employing *K*-values. Metcalfe *et al.*,[4] on the other hand, calculated condensing-gas drive phase behavior with the Redlich-Kwong equation of state.[5] Metcalfe *et al.* and Corteville *et al.*[6] used compositional models to calculate condensing-gas drive displacements, approximating both flow and phase behavior in their calculations.

The calculation of miscibility conditions with mathematical models is no more accurate than the ability of the model to represent phase behavior in the region near the plait point. A method for obtaining greater accuracy near the plait point in phase-behavior calculations was described by Fussell and Yanosik[7] and by Fussell and Fussell.[8] Accuracy of model phase-behavior predictions in the critical region should be checked with experimental data in this region.

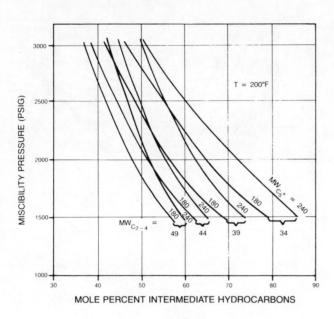

Fig. 6.3—Condensing-gas drive miscibility pressure correlation, $T = 200°F$.

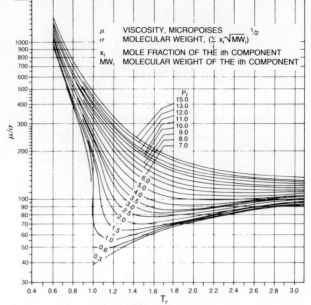

Fig. 6.4—Viscosity of hydrocarbon systems as a function of reduced pressure (after Ref. 9).

Example Calculation

Estimation of Miscibility Conditions Using the Benham *et al.*[1] Correlation.

Problem. Estimate the minimum concentration of intermediate-molecular-weight hydrocarbons for condensing-gas drive miscibility at 2,500 psig, 125°F, with a reservoir fluid that has a pentanes-and-heavier fraction molecular weight of 200. The intermediate-molecular-weight hydrocarbon fraction of the gas is 80% propane and 20% butane.

Solution. Molecular weight of intermediate hydrocarbon fraction $= 44(0.8) + 56(0.2) = 46.4$. After interpolating Figs. 6.1 and 6.2 for temperature and for the intermediate-hydrocarbon-fraction molecular weight, the minimum permissible concentration of intermediate hydrocarbons is estimated to be 32 mol%.

6.2 Viscosity and Density of Hydrocarbon Gases

Viscosity of hydrocarbon gases can be estimated from Fig. 6.4, which relates viscosity to the reduced temperature and pressure of the gas mixture.[9] Reduced temperature and pressure are defined by

$$T_r = \frac{T}{T_c}, \quad p_r = \frac{p}{p_c}, \quad \ldots\ldots\ldots\ldots\ldots\ldots(6.1)$$

where

T = absolute temperature, °R,
T_c = critical temperature, °R,
p = absolute pressure, psia, and
p_c = critical pressure, psia.

In practice, molecular-average or pseudocritical temperatures and pressures are used when calculating the reduced temperature and pressure of mixtures: the pseudocritical temperature is given by

$$\sum_i^n T_{ci} x_i$$

and the pseudocritical pressure by

$$\sum_i^n p_{ci} x_i$$

where

T_{ci}, p_{ci} = critical temperature and critical pressure of component i in the gas mixture, and

x_i = mole fraction of component i in the gas mixture.

Fig. 6.5 gives an additive correction, σ, to the molecular weight calculated by the formula shown on Fig. 6.4 for CO_2 concentrations up to 80 mol%. Gonzales and Lee[9] compared viscosities predicted by Fig. 6.4 with experimental viscosities of eight natural gases over a temperature range of 40 to 460°F and a pressure range of 14.7 to 10,000 psia. Methane content in these gases ranged from 71.7 to 97.9 mol%. The maximum deviation of predicted and experimental viscosities was 5%.

Density of hydrocarbon gases can be calculated from the following equation.

$$\rho_g = \frac{p \sum_i^n (MW_i x_i)}{10.73zT}, \quad \ldots\ldots\ldots\ldots\ldots\ldots(6.2)$$

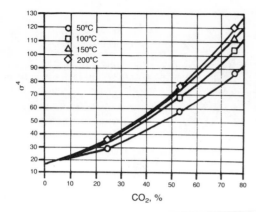

Fig. 6.5—Correlation chart for mixtures containing CO_2 (after Ref. 9).

where

ρ_g = gas density, lbm/cu ft,
p = absolute pressure, psia,
z = compressibility factor, and
T = absolute temperature, °R.

Fig. 6.6 relates compressibility factor to reduced temperature and pressure. [10]

Full-page versions of Figs. 6.4 and 6.6 can be found in Appendix E as Figs. E-7 and E-8.

6.3 Sources of Rich Gas and Drive Gases

Rich-gas slugs most often are blended by enriching field separator gas or residue gas from a gasoline plant with intermediate-molecular-weight hydrocarbons. In some instances, field separator conditions can be adjusted to give a separator gas with a composition sufficient for achieving miscibility. In most projects, however, the natural gas that is available for injection must be enriched with intermediate-molecular-weight hydrocarbons to achieve miscible displacement at the expected reservoir pressure.

Sources of LPG for enriching injection gas were discussed in Sec. 5.3. They include natural gasoline plants located in the field of interest or in nearby fields, low-temperature separation facilities in the field, or oil refineries.

Field separator gas and residue gas from a gasoline plant are also potential sources of drive gas. Flue gas or nitrogen are alternative gases for driving miscible slugs. Sources of these gases are discussed in Sec. 7.3.

6.4 Field Experience

Table 6.1 gives locations and other details of 18 selected condensing-gas drive projects. Some of the early projects began in the mid-1950's, but others started during the 1960's and early 1970's. The Ante Creek, South Swan Hills, Levelland, Central Mallet, Intisar Reef, and several Rainbow area projects were still active in 1982. Most of these projects were 8 to 14 years old at that time.

Most of the condensing-gas drive floods were larger in scope than the majority of first-contact miscible process tests, often involving hundreds or occasionally thousands of acres. Sometimes the entire field or unit was flooded. All but two were secondary-recovery

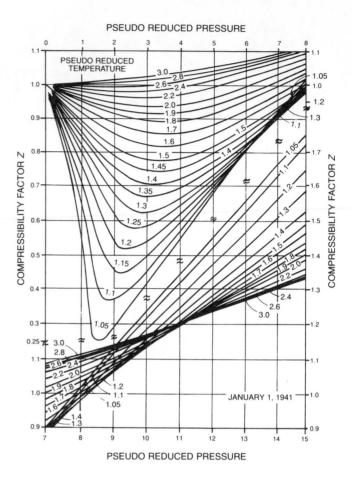

Fig. 6.6—Compressibility factor for natural gases (after Ref. 10).

floods. Oil gravities have been in the 30-to-50°API range, and oil viscosities generally have been low, less than about 2 cp. Projects have been conducted in both sandstone and carbonate reservoirs. In addition to floods in low-relief, nearly horizontal reservoirs, gravity-stabilized displacements in pinnacle reef reservoirs have been an important aspect of condensing-gas drive field experience.

Floods in Horizontal Reservoirs

Seeligson Field, Zone 20B-07. [12-21] This project was initiated in 1957 by Humble Oil & Refining Co. and is the most thoroughly documented condensing-gas drive miscible flood. The reservoir was a stratified Frio sand at 6,000 ft and originally contained 7.4 million STB of a 40°API oil. Average reservoir characteristics included 150-md permeability, 24% porosity, 12-ft thickness, and 4° dip. The surface area of the test encompassed 877 acres. Oil/rich-gas viscosity ratio was about 12.

Primary production resulted primarily from solution gas drive, and oil recovery was only 10% OOIP out of an expected 22.3% by primary depletion when rich-gas injection began. Initially a rich-gas slug of 50% propane and 50% residue gas was injected through two upstructure wells to flood across three main rows of some 19 producers on 40-acre spacing. After injection of 0.2 HCPV of rich gas, ethane was used as enriching material until a total slug of 0.52 HCPV was injected. This was followed by residue gas and shortly thereafter by alternate injection of residue gas and water.

First-row producing wells responded quickly with reduced GOR and increased pressure. Average sweepout at breakthrough into each producing row was about 23%. Well tests in first- and third-row producers showed that vertical conformance was influenced by gravity segregation and stratification. The flood had recovered 50% OOIP at last report,[14] and an ultimate recovery of 54% was expected. Waterflooding would have recovered only 42 to 48% OOIP according to projections.

Midland Farms.[24,25] In contrast to the large rich-gas slug of the Seeligson field, only a 2% HCPV slug was injected in the Midland Farms project completed by Amoco Production Co. The test was conducted in the Wolfcamp formation, a fine-to-medium crystalline organic limestone reef with some oolitic deposits occurring at a depth of 8,350 ft. Average permeability was 30 md and average porosity was 12.9%. Gravity of the oil was 41°API, and surface area of the test was about 600 acres.

The Wolfcamp reservoir was produced by solution gas drive to 700 psi below the oil's bubble point before injection of rich gas began in 1960 into four wells. The injectors were arranged in a peripheral manner around 10 interior producers on 40-acre spacing to give a flooding pattern with converging radial flow. The rich-gas slug was followed by a 5% HCPV buffer slug of lean gas and then by alternate injection of lean gas and water. The operator reported that performance of the flood was discouraging, although alternate gas/water injection was beneficial in reducing the GOR and propane production in breakthrough wells.

Ante Creek.[31,33] In this project operated by Amoco Canada, rich-gas slugs were alternated with water slugs after injection of a small rich-gas buffer slug. Injection is into the 11,000-ft deep Beaverhill Lake reservoir, a porous reefal development with an average net pay thickness of 29.5 ft that originally contained nearly 38 million bbl of a 0.13-cp oil. The reservoir has an average porosity of 6.2% and an average permeability of 9.3 md. Oil/gas viscosity ratio was about five, and with alternate water injection, mobility ratio was estimated to be less than one.

In 1968, injection of rich gas was begun through two wells to flood the 6000-acre, 19-well field. The reservoir had been produced by solution gas drive up to the time of rich-gas injection, although GOR's had not risen appreciably by the time the project was started. A gas enrichment of 35% ethanes-through-butanes was needed for miscibility at the 4,290-psi reservoir pressure. Calculations showed that a water/gas injection ratio in the range of 0.74 to 1.6 RB/RB was necessary to both replace voidage and maintain the buffer slug. For operational reasons, water/gas injection ratios have been less than planned, and the overall mobility ratio may have become unfavorable. As of April 1982 with the project about half finished, the operator had not published an overall project evaluation but reported the project was performing successfully.[40]

Levelland Unit.[31,38] This project in the Levelland field of west Texas was started by Amoco Production Co. in 1972. Injection is in the San Andres formation, a 250-ft thick, highly stratified dolomite at 4,900-ft depth. Porous, productive zones alternate with dense zones of limited permeability. Average reservoir permeability is only 2 md, and average porosity is 10.2%. Net thickness is 100 ft. The reservoir crude is a 30°API gravity oil with a viscosity of 1.9 cp at reservoir conditions. Oil/gas viscosity ratio is approximately 68.

The reservoir was depleted below the oil's bubble point to 940 psig before miscible flooding began. MMP was found to be 1,100 psi with a gas containing 60 to 65% ethane and propane. Rich gas was injected into twenty-eight 42-acre five-spot patterns. After injection of a 5.8% HCPV slug of water to raise pressure and to improve the solvent's injection profile, a 1.8% HCPV slug of rich gas was followed by alternate injection of rich gas and water in a ratio of 4 RB/RB. The operator injected a 14% HCPV total rich-gas slug before switching injection to lean gas alternated with water. Cycle sizes have been about 0.3% HCPV water and 1.4% HCPV rich gas. After all wells responded to rich-gas injection, the GOR decreased to 800 scf/STB compared with a ratio of 4,000 at the start of flooding. After injection of 7.5% HCPV of rich gas, there had been some gas breakthroughs, but cumulative rich-gas production was only 3% of the total rich gas injected. A decrease in water injectivity was experienced after the first gas cycle, which persisted through other cycles. The operator reports that overall flood behavior has been promising.[40]

South Swan Hills.[39,43] Another flood of a carbonate formation is being conducted by Amoco Canada in this Alberta field. Injection is in the 8,600-ft-deep Beaverhill Lake formation, a stratified reef with reef front, lagoonal, and forereef development. The project area had an average thickness of 73 ft and originally contained 540 million bbl of a 42°API, 0.39-cp oil. Average porosity and permeability are 8.2% and 49 md. Viscosity ratio of reservoir oil and the rich gas was about 18.

Rich-gas injection began in 1973 into twenty-one 640-acre inverted nine-spot patterns. With a gas enrichment of 50% ethane-and-heavier hydrocarbons, miscibility pressure was 2,350 psi. Some of the wells had received previous water injection, but in most patterns the rich gas was a secondary-recovery injection fluid. In these latter patterns, a 2% HCPV slug of water was injected before the injection of rich gas.

Original plans called for a total rich-gas slug of 10% HCPV injected alternately with water and followed by a total residue gas slug of 30% HCPV, also injected alternately with water. Residue gas injection was to be followed by continuous water injection. In practice, about 15% HCPV of solvent had been injected as of the last published report.[43] Rich-gas/water injection ratio began at 1.25 RB/RB but decreased to about 1.0 RB/RB.

Significant solvent breakthrough occurred within 1 year and after about 3.4% HCPV of rich-gas injection. After 6 years' injection, the GOR had increased to about 1,300 scf/STB from an original 700 scf/STB. Injection and production profiles showed that some wells had a poor fluid distribution, which was correlatable with early gas breakthrough. The operator tried to improve problem

TABLE 6.1—CONDENSING-GAS DRIVE FIELD PROJECTS

Field	Operator	Year Started	Project Type	Oil Gravity (°API)	Oil Viscosity (cp)	Viscosity Ratio μ_o/μ_s	Depth (ft)	Thickness (ft)	Acreage	Slug Size (% HCPV)	Recovery (% OOIP)	Status	References
Haynesville (LA)	Haynesville Operators Comm.	1953	LPG→RG →G →W						3,200				11
Bronte (TX)	Exxon U.S.A.	1956	RG						1,300				11
Bronte (TX)	Exxon U.S.A.	1956	RG						460				11
Seeligson Zone 20B-07 (TX)	Exxon U.S.A.	1957	RG→G/W	40		12	6,000		877	52	5 to 10 (incremental) 50 to 54 (ultimate)	NC, disc.	11-21*
Stratton (TX)	Southern Minerals Corp.	1957	RG										11
South Coles Levee (CA)	Ohio	1957	RG										11,22
Elk Basin (WY)	Amoco Production Co.												11
Midland Farms (TX)	Amoco Production Co.	1960	RG→G →G/W	41	0.3		8,350		600	2		term., disc., prof.	23*-25
Neale Lilliedoll (LA)	ARCO Oil & Gas Co.	1962	tertiary RG→G/W			11	10,100	13.5	640	15		term.	26,†
Golden Spike (Alberta)	Imperial Oil Co.	1964	vertical RG→G	37	0.8		5,672	550	1,305	7	3.1 (incremental)	comp., disc., prof.	27-31*
Pembina Bear Lake Cardium (Alberta)	Cities Service Co.	1968	RG→G →G/W	37.6	1.5	52	4,857	50	4,160	3			32
Ante Creek (Alberta)	Amoco Production Co.	1968	RG/W	42	0.13	4.7	11,000	29.5	6,000	cont. rich gas		HF, NE	27,31,33
Intisar D Reef (Libya)	Occidental Petroleum Corp.	1969	vertical RG	40	0.46	20	8,950	888	3,300		70** (ultimate)	HF, succ., prof.	34-37*
Levelland (TX)	Amoco Production Co.	1972	RG→RG/W →G/W	30	1.9	86	4,900	250	1,190	14	>27** (incremental)	HF, succ., prof.	31*,37,38

(continued on next page)

Field	Operator	Year Started	Project Type	Oil Gravity (°API)	Oil Viscosity (cp)	Viscosity Ratio μ_o/μ_s	Depth (ft)	Thickness (ft)	Acreage	Slug Size (% HCPV)	Recovery (% OOIP)	Status	References
South Swan Hills (Alberta)	Amoco Production Co.	1973	RG/W ←G/W ←W	42	0.39	18	8,500	73	830	15	20** (incremental) 65** (ultimate)	HF, prom., prof.	27,31*,39
Central Mallett Slaughter (TX)	Amoco Production Co.	1972		30	2		4,800		12	20 and increasing	6.0 (incremental) and increasing	JS TETT unprof.	31*,41
Camp Sand Haynesville (TX)	Marathon Oil Co.	1966	RG←G←W	46	1.05		8,000	5	~2,200	~10	42		27,31
Rainbow area (Alberta)	various		vertical										

Symbol key:
LPG = liquified petroleum gas
RG = rich gas
G = lean gas
RG←G/W = rich gas displaced by alternate gas-water
NC = nearing completion
disc. = discouraging

term. = terminated
NE = not elevated
TETT = too early to tell
JS = just started
HF = half finished
prof. = profitable

* Primary reference reporting status.
** Estimated with simulators.
† Unpublished ARCO Oil & Gas Co. results.

zones with workovers. A pressure core showed rich gas was reducing the oil saturation to about 7.9% PV average. This compares with a waterflood residual oil saturation of about 35% PV. Miscible flood ultimate oil recovery was projected with a numerical simulator to be 65% OOIP compared with a projected waterflood recovery of 45% OOIP.

Neale Field. [26,*] ARCO Oil and Gas Co. tested tertiary recovery in the Lilliedoll sand of the Louisiana field. The thin, 13.5-ft sand occurred at a depth of 10,100 ft and contained shale lenses scattered discontinuously throughout the continuous main sand body. Viscosity ratio of the reservoir oil and rich gas was about 11. The reservoir initially was produced by combined solution gas drive/partial water drive mechanisms. At the start of rich-gas injection, all wells were producing at high water cuts, and average oil saturation was estimated to be 35% PV. Rich-gas injection began in the first 160-acre inverted five-spot pattern in 1962. A 15% HCPV slug of rich gas was followed by a small dry-gas buffer slug and then by alternate injection of dry gas and water. Response was encouraging; three producers responded to the tertiary oil bank with decreased water cut and increased oil rate. One well did not respond. Approximately 1.6 STB of oil were recovered per reservoir barrel of rich gas injected. In April 1965, the flood was expanded to two more patterns followed shortly by a fourth pattern. No buffer lean gas was injected and alternate injection of lean gas and water immediately followed the rich-gas slugs. Flue gas and water also were injected in the latter two patterns because of a shortage of lean gas. All the expansion patterns performed more poorly than the original test pattern, with oil recovery ranging from 0.23 to 0.71 STB oil per reservoir barrel of rich gas injected.

Floods in Pinnacle Reef Reservoirs

Several condensing-gas drive miscible floods are in progress in pinnacle reef fields. [28-31,34-36,44] In most of these projects, solvent slugs were injected near the crest of the reefs and displaced vertically downward with lean gas, and the vertical displacement rate was designed to be below the gravity-stable critical rate to prevent viscous fingering of solvent into oil or of lean gas into the rich-gas solvent. In at least one project the rich-gas solvent is being injected continuously. [36] Some of the reefs contained gas caps, and the solvent slug had to be placed between the oil and gas cap gas before downward displacement began.

Golden Spike. [28-30] This reservoir initially contained about 320 million bbl of oil. It had a maximum thickness of 624 ft, and was about 1,400 acres in area. Average horizontal permeability was 320 md. Viscosity of the 37°API oil was 0.8 cp. A dense seal at the base of the reservoir isolated it from the extensive underlying Cooking Lake aquifer.

Starting in 1964, Imperial Oil Co. injected a rich-gas solvent bank of 7.8% HCPV between the secondary gas cap and the oil zone through centrally located wells. Tracers in the rich-gas solvent showed the solvent bank had covered the gas/oil surface after about half the sol-

*ARCO Oil and Gas Co., unpublished results (1971).

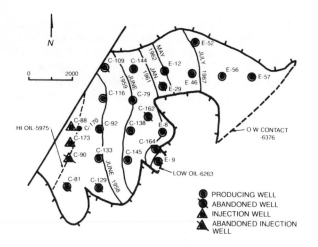

Fig. 6.7—Enriched gas breakthrough experience, Zone 20B-07 (after Ref. 14).

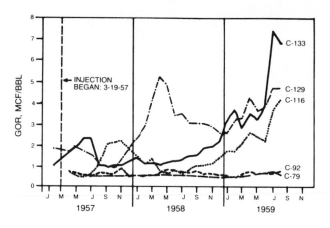

Fig. 6.8—GOR's for enriched-gas-drive project, Seeligson field, Zone 20B (after Ref. 12).

vent slug was injected. Injection of the solvent bank was essentially completed by 1972. A unique feature of the Golden Spike project was the generation of solvent by producing the field at a high rate, separating oil and solution gas, extracting intermediate-molecular-weight hydrocarbons for enriching the slug, and reinjecting all crude in excess of market requirements. Injection gas contained 60% C_2-through-C_4 hydrocarbons and was dynamically miscible with reservoir oil at a pressure of 1,750 psig.

Rich gas was injected through three wells and lean gas through two injectors. Crude was returned through two former producers, and oil initially was produced through four high-volume producers. By 1972, well productivities had declined to the point where rising allowables could not be met by the existing wells. Extensive infill drilling began at this time, and by the end of 1975, infill development was completed.

Original projections showed that ultimate recovery under gravity-stable displacement could be as high as 94.7% OOIP. Initial performance was promising, although one well showed coning at high oil rates of 6,000 to 8,000 B/D, and a second well was drilled to reduce the drawdown.[45] Logging of wells in 1971 disclosed a distortion of the oil surface by as much as 77 ft from the horizontal.[30] This distortion had grown to 130 ft by 1973. Fluid samples in 1974 showed that a band of pure solvent no longer existed. Reservoir analysis using new data from the infill wells showed that the rich-gas/oil interface did not remain horizontal during its downward displacement because of previously unsuspected barriers to vertical flow. In 1977, the operator revised the projection of ultimate recovery downward to 67% OOIP and reported that even though the project had been profitable, overall performance was discouraging. The project has been completed.

Intisar D Reef.[34-35] Another interesting application of miscible displacement by the condensing-gas drive method is the Intisar D Reef project in Libya. The reservoir initially contained about 1.8 billion bbl of oil and is approximately circular in areal extent with an outside diameter of about 3 miles. Porous calcarenite, together with some scattered calcilutite and biomicrite of lower porosity, comprise the majority of the reservoir rock.

The gross section is about 888 ft thick and has good vertical permeability. Individual permeability values from core data range from 4 to 500 md. Depth to the top of the formation is 8,946 ft. The reservoir crude has an API gravity of 40° and an in-situ viscosity of 0.46 cp. Oil/gas viscosity ratio is about 20. With a bubble point of 2,224 psi, the oil was highly undersaturated at the 4,257-psi initial reservoir pressure.

After discovery by Occidental Petroleum Corp. in 1967, water was injected into the bottom of the reservoir through 15 wells in 1968. This was followed by continuous rich-gas injection above the 4,000-psi miscibility pressure through seven wells in the crest of the structure in 1969. Some wells were produced at first through completions at the top of the reservoir to draw gas over the top of the structure. After the gas blanket was in place, wells were produced to move the gas/oil interface vertically downward. There was only one gas breakthrough after injection of 386 Bcf of rich gas, and the GOR has increased gradually during the life of the project. After injection of 1,420 Bcf of rich gas, the GOR had climbed from 650 to 3,000 scf/STB. Injection rate has generally varied between 300,000 and 400,000 Mcf/D and reached a high of 600,000 Mcf/D in 1979. Oil producing rate generally has been in the range of 100,000 to 200,000 STB/D.

Neutron logging surveys showed that an oil saturation was left behind the advancing gas front and that this saturation decreased slowly with time. By March 1981, ultimate recovery had reached 56% OOIP, and performance forecasts with a numerical simulator indicated that ultimate recovery would reach 67.5% OOIP.

6.5 Assessment of Process
Horizontal Floods
Response in condensing-gas drive projects has been similar to the response observed in first-contact miscible floods. In secondary recovery projects, response generally has been characterized by reduced GOR's and increased reservoir pressure as the oil bank arrived at producing wells. Limited tertiary recovery testing has demonstrated that the process will displace waterflood residual oil to producing wells.[26,41,*]

*ARCO Oil and Gas Co., unpublished results for Lilliedoll test (1971).

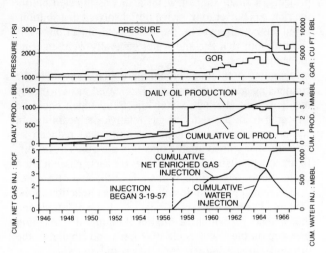

Fig. 6.9—Production data, Zone 20B-07 (after Ref. 14).

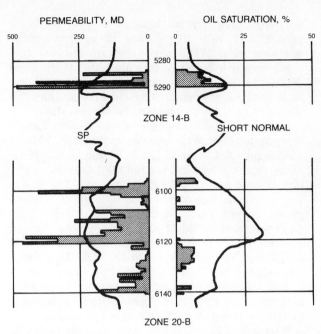

Fig. 6.10—Core analysis, K.R. Seeligson Well C-170 (after Ref. 13).

As in first-contact miscible floods, solvent breakthrough occurred relatively early. Breakthrough was observed after only 0.05 HCPV of rich-gas injection in the South Swan Hills[39] and Levelland[38] floods. Average breakthrough volumetric sweepout for each row of producers at Seeligson was about 23%.[14] Gas production after breakthrough at South Swan Hills and Levelland was moderated by the alternate injection of water and was not excessive. Despite the early breakthrough at Levelland, only 3% of the cumulative gas injected had been produced after injection of 0.075 PV of rich gas.

The Seeligson project is a well-documented example of condensing-gas drive miscible flooding for secondary recovery in a sandstone reservoir.[12-14,21] Fig. 6.7 shows the location of wells in this test and also shows contours that illustrate the approximate advance of the rich-gas front with time. Fig. 6.8 shows the GOR behavior of first-row producers, and Fig. 6.9 shows the overall producing behavior of the project.

Reservoir oil in Zone 20B-07 at Seeligson was saturated at the original reservoir pressure of 3,010 psi and had an initial dissolved GOR of approximately 600 cu ft/bbl. Reservoir pressure had declined to less than 2,400 psi from solution gas drive production by the time rich-gas injection began in 1957. GOR declined very soon after rich-gas injection, indicating that the oil saturation had increased because of formation of an oil bank. The rapid response in this project indicated that the free-gas saturation was small at the start of rich-gas injection. In many wells, the GOR had decreased to near that of solution gas before rich-gas breakthrough occurred. The rate of increase in GOR after rich-gas breakthrough was moderate, and a large volume of the oil recovered in the Seeligson project was produced after breakthrough. More than 87% of the rich gas was produced eventually.

The operator of the Seeligson test thought that flood performance was dominated by a combination of gravity segregation and permeability stratification. Production

tests on a first-row producer showed rich gas was present in only the upper part of the sand, whereas similar tests in a third-row producer showed 75% of produced fluid was coming from 1 or 2 ft of the best-developed sand at the bottom of the interval.

Unit displacement efficiency at Seeligson was very good. A core taken 100 ft from an injection well showed that zones with good permeability were swept almost clean of oil. Fig. 6.10 shows the oil saturations found in this core along with a permeability profile for the well. For comparison, data from a well cored in another Seeligson zone (14-B) that had been flooded with water only are shown in Fig. 6.10 also. Forty-six samples in the core swept by rich gas had no oil, and 25 samples had saturations no higher than 1% PV.

Evidence of good unit displacement efficiency in condensing-gas drive miscible flooding was found also in the South Swan Hills project.[39,43] Oil saturations in floods of laboratory cores with restored connate water saturations averaged 3.3% PV. Water saturation seemed to have little effect on the oil saturation left to miscible flooding in this instance. The average saturation found in preserved cores from a similar reservoir that were laboratory flooded at various water/solvent injection ratios was only 5.3% PV. This finding was significant because it meant that precise control of the water and solvent advance through different portions of the reservoir was not critical to unit displacement efficiency. A pressure core taken behind the rich-gas front revealed a 7.9% PV average oil saturation.[43]

Rich-gas slug sizes have ranged from 2% HCPV in the Midland Farms project to 52% HCPV in the Seeligson project. Slug sizes in general have been larger than in first-contact miscible projects because the enriching material is less concentrated in rich-gas slugs and the slug cannot be diluted as much as a first-contact miscible slug before miscibility is lost. In the majority of horizontal floods, slug size has been 10 to 15% HCPV or greater, and 10% HCPV may be an approximate minimum value. Field testing has not resolved conclusively which design better withstands fingering and dilution to the minimum permissible LPG concentration—small, high-concentration, first-contact miscible slugs or larger, lower-concentration rich-gas slugs.

Results also are available from several tertiary recovery tests. Rich-gas injection in the Central Mallet 12-acre pilot test began in late 1978, and by mid-1981, incremental recovery had reached 6% OOIP after injection of more than 20% HCPV of rich gas. The project was still under way at that time although oil rate appeared to have peaked and started to decline.[41] Incremental recovery varied considerably among the four patterns ultimately flooded at Lilliedoll with a 15% HCPV slug, ranging from 0.2 to 1.6 STB/RB. Some of this variation may have been caused by different effective oil saturations in the different patterns. Downstructure patterns may have been waterflooded more thoroughly than patterns higher on the structure.

Estimates of incremental recovery have been reported for several of the secondary recovery projects. Ultimate recovery for the 52% HCPV slug injected at Seeligson was expected to reach 53 to 54% OOIP before the zone was abandoned. Waterflood recoveries from similar reservoirs in the area were 42 to 48% OOIP; thus, the incremental recovery from condensing-gas drive flooding at Seeligson should be on the order of 5 to 10% OOIP.[14] A low waterflood residual oil saturation may have been a major factor contributing to the low incremental recovery for such a large slug size.

The operators of the South Swan Hills and Levelland projects used reservoir simulators to project incremental recoveries of 20 and 27% OOIP for slug sizes of approximately 15% HCPV.[38,39] A later comparison of South Swan Hills performance with waterfloods in similar fields supported the original projections.[43] A somewhat lower recovery is anticipated currently for Levelland as a result of recent unpublished studies. Both floods are no more than half completed, and although it remains to be seen whether the projections will be achieved, reports so far have been encouraging.[31,40]

Discouraging results were reported for the Midland Farms project, which has been completed.[23] The small rich-gas slug size may have contributed to the disappointing performance.

Alternate gas/water injection was used in many of the condensing-gas drive tests. Generally, the project operators felt that water injection improved performance.[24,38,39] In early projects, alternate injection of water was started after injection of the rich-gas slug was completed. For example, in the Midland Farms test, alternate lean-gas/water injection followed a 2% HCPV slug of rich gas and a 5% HCPV buffer slug of lean gas.[24] Breakthrough occurred in three wells before injection of the lean gas buffer was completed, causing both GOR and propane content in the gas to increase in these wells. The subsequent injection of alternate slugs of water with the lean gas caused both GOR and propane content in the gas to decrease.

In later floods at Ante Creek, Levelland, South Swan Hills, and Central Mallet, water was injected in alternate slugs with the rich gas to achieve the benefits of water in-

jection earlier in the displacement. In these projects, alternate water injection was begun after injection of a small buffer slug of rich gas. In the secondary recovery projects, a small slug of water also was injected before the rich-gas buffer to improve initial solvent distribution. Fluid entry surveys in the Levelland injection wells showed both rich gas and water generally were entering the same zones.

Operating problems were not particularly severe in the condensing-gas drive projects. Decreased gas injectivity resulted from alternate water injection, which could be expected from the reduced gas relative permeability. In some cases, gas relative permeability was reduced so severely that with available injection pressures it was difficult to establish initial gas flow on each gas-injection cycle. Backflowing injection wells at the end of a water cycle seemed to promote subsequent gas injection in the Midland Farms project. BHP data were not reported, and the extent to which a higher bottomhole injection pressure on the water cycle may have contributed to the difficulty in starting gas injection is unknown.

A severe decrease in water injectivity following the first gas cycle was experienced in the Levelland project, and it persisted through remaining cycles. Complete cause of the reduced injectivity never was established definitely, although laboratory testing showed that a greater-than-expected reduction in water relative permeability from the trapped gas saturation may have been responsible. Acid stimulations helped restore injectivity to some extent. Severely decreased water injectivity was not experienced in three other condensing-gas projects examined by Harvey et al.[46] High wellhead pressure at the end of gas cycles required in one instance that the well be killed with an auxiliary high-pressure water pump before using the main water pump.

Hydrate formation was reported as a problem in several projects employing alternate gas/water injection.[24,33] If the problem is severe, separate injection lines for gas and water could be required.

Increased gas production caused foaming and pumping problems on occasion.[38]

Vertical Floods

The vertical flood in the Intisar D reef appears to be performing successfully. This project has been characterized by continuous injection of the rich gas, whereas in most vertical floods the rich-gas slug is driven by natural gas. Ultimate recovery was projected at the start of the flood to be as high as 80% OOIP,[35] but recent projections based on a history match of production data indicate a lower, but still respectable, ultimate recovery of 70% or higher.[36] A rough estimate of the ultimate recovery that might have been achieved by waterflooding indicated a value of approximately 40% OOIP, whereas a more detailed waterflooding study for a nearby and similar reef indicated that the ultimate waterflood recovery for that reef might be as high as 46% OOIP.[34] Thus, the incremental recovery over waterflooding for the Intisar D reef condensing-gas drive flood may be on the order of 20 to 30% OOIP.

Analyses have not been published for the vertical floods in Canadian pinnacle reefs except for the Golden Spike project. However, qualitative operator evaluations have ranged from "promising" to "successful."[31,40] Many of these floods, however, are less than half completed at this writing.

The Golden Spike flood was not successful.[30] Performance was dominated by permeability barriers to vertical flow that were completely unsuspected when the project was begun. The following paragraphs discuss the Golden Spike flood in some detail to illustrate the great importance of reservoir heterogeneity and the need for an adequate reservoir description when miscible floods are engineered.

At the start of the Golden Spike flood in 1964, project engineers considered the vertical cross section relatively homogeneous and, as a result of critical rate calculations for this view of reservoir description, expected the solvent/oil interface to remain more or less horizontal as the flood progressed. Early performance seemed to verify these expectations. However, extensive infill drilling beginning in 1972 and subsequent geological work disclosed the existence of a major permeability barrier separating the Middle and Upper Leduc members and located about midway down in the cross section. The areal extent of this barrier was large, and it effectively blocked the downward flow of fluids over a significant area of the reservoir. Many smaller permeability barriers of more limited areal extent also were identified. Because of these barriers, the solvent/oil interface did not advance downward uniformly. Some oil and solvent were held up on the barriers and were by-passed by the remainder of the advancing solvent front. Oil and solvent then drained off the sides of the barriers and mixed into the lean gas driving the slug. In 1973, a survey based on bottomhole samples from 8 wells and logs from 30 wells showed that the gas cap had partially underrun the main barrier, although about half the injected solvent was still identified as being above this barrier. The survey indicated that the remainder of the solvent had dispersed into the driving gas. Another survey in 1974 showed that drive gas had underrun the main barrier completely, and a maximum difference of 126 ft existed in the elevation of the solvent/oil interface. There was no pure solvent detected at this time. The operator concluded that miscibility had been lost and that the flood had reverted to an immiscible-gas drive. A subsequent analysis of the project using a reservoir simulator showed that the 7.8% HCPV rich-gas slug had increased ultimate recovery only by 3.1% OOIP.

6.6 Screening Criteria

The condensing-gas drive process is best suited for only certain types of reservoirs and oils, and cannot be practiced indiscriminately. Screening criteria must be qualified much as the criteria for first-contact miscible flooding were qualified. They should be considered only as general guidelines that cannot be very precise, and exceptions are to be expected.

Miscibility can be achieved in the condensing-gas drive process over approximately the same pressure range as with the first-contact miscible method, although in practice excessive enrichment may be necessary to achieve the lowest miscible pressures attainable in first-contact miscible flooding. Rich gas usually develops dynamic miscibility at significantly lower pressures than

can be achieved with natural gas, nitrogen, or flue gas in the vaporizing-gas drive method. As a result, the condensing-gas drive process can be applied to much more shallow reservoirs than can the vaporizing-gas drive method. Miscibility between rich gas and oil is controlling rather than miscibility with drive gas.

Flexibility of design and operation is an advantage of the condensing-gas drive method. According to pseudoternary diagram concepts, if reservoir pressure should drop below miscibility pressure during the flood, dynamic miscibility can be re-established by repressuring to design pressure as long as rich gas of design composition is injected. However, experimental data are lacking either to support or to disprove this concept. Miscibility pressure can be adjusted up or down by varying the concentration and composition of the enriching hydrocarbons. Slug compositions can be designed to minimize or to avoid repressuring if desired. An optimal economic balance can be achieved between the benefits of low operating pressure (reduced compression cost, fewer standard cubic feet to occupy a reservoir barrel of pore space) and the benefits of high operating pressure (reduced enriching material requirement, higher producing well productivity).

Cost and supply of the rich-gas slug and drive gas are important considerations in screening reservoirs as potential condensing-gas drive candidates. A favorable situation occurs if separator gas needs only a minimum of enrichment and is available at a relatively low cost because of special contract or market conditions. Location of a gasoline plant or other LPG facility at the field also can be a favorable factor.

Other guidelines for judging the suitability of reservoirs for condensing-gas drive miscible flooding are as follows.

1. An oil viscosity of less than 1 cp is preferred for horizontal floods. An upper viscosity limit is not well defined but may be approximately 5 cp. The upper limit for gravity-stable floods depends on reservoir permeability. Because of the viscosity limitation, oils with gravities above 30°API are most suited to this process.

2. Typical pressure required for miscibility is in the range of 1,500 to 3,000 psi, although higher pressures may be required, depending on oil gravity and reservoir temperature. Depending on reservoir parting pressure, minimum reservoir depth for this process is in the range of 2,000 to 3,000 ft. There is no maximum depth limitation.

3. Reservoirs with substantial structural relief are preferred if permeability is high enough for gravity-stable flooding. In horizontal reservoirs, restricted vertical permeability, which might result from scattered shale lenses and laminations or from dense streaks, is preferred to minimize severe gravity tonguing.

4. Extensive fracturing and reservoirs with a gas cap, a strong water drive, and a high permeability contrast increase project risk.

5. Oil saturation in the reservoir volume that will be miscibly swept should be high. Acceptable saturation depends on injection fluid costs, oil price, reservoir properties, and fluid properties, but 25% PV is an approximate minimum value.

References

1. Benham, A.L., Dowden, W.E., and Kunzman, W.J.: "Miscible Fluid Displacement—Prediction of Miscibility," *Trans.*, AIME (1960) **219**, 229–37; *Miscible Processes*, Reprint Series, SPE, Dallas (1965) **8**, 123–31.

2. Yarborough, L. and Smith, L.R.: "Solvent and Driving Gas Compositions for Miscible Slug Displacement," *Soc. Pet. Eng. J.* (Sept. 1970) 298–310.

3. Rowe, A.M. and Silberberg, I.H.: "Prediction of Phase Behavior Generated by the Enriched-Gas Drive Process," *Soc. Pet. Eng. J.* (June 1965) 160–66.

4. Metcalfe, R.S., Fussel, D.D., and Shelton, J.L.: "A Multicell Equilibrium Separation Model for the Study of Multiple Contact Miscibility in Rich-Gas Drives," *Soc. Pet. Eng. J.* (June 1973) 147–55.

5. Zudkevitch, D. and Jaffe, J.: "Correlation and Prediction of Vapor-Liquid Equilibria with the Redlich-Kwong Equation of State," *AIChE J.* (Jan. 1970) 112–19.

6. Corteville, J., Van Quy, N., and Simandoux, P.: "A Numerical and Experimental Study of Miscible or Immiscible Fluid Flow in Porous Media with Interphase Mass Transfer," paper SPE 3481 presented at the 1971 SPE Annual Meeting, New Orleans, Oct. 3–6.

7. Fussell, D.D. and Yanosik, J.L.: "An Iterative Sequence for Phase Equilibrium Calculations Incorporating the Redlich-Kwong Equation of State," *Soc. Pet. Eng. J.* (June 1978) 173–82.

8. Fussell, L.T. and Fussell, D.D.: "An Iterative Technique for Compositional Reservoir Models," *Soc. Pet. Eng. J.* (Aug. 1979) 211–20.

9. Gonzales, M.H. and Lee, A.L.: "Graphical Viscosity Correlation for Hydrocarbons," *AIChE J.* (March 1968) 242–43.

10. Standing, M.B. and Katz, D.L.: "Density of Natural Gases," *Trans.*, AIME (1942) **146**, 140–49.

11. "Miscible Drives: A Growing Tool for Oil Recovery," *Oil and Gas J.* (March 23, 1959) 64–69.

12. Baugh, E.G.: "Performance of Seeligson Zone 20-B Enriched Gas-Drive Project," *J. Pet. Tech.* (March 1960) 29–33.

13. Grinstead, W.C. Jr. and Barfield, E.C.: "Late Developments in the Seeligson Zone 20-B-07 Enriched Gas-Drive Project," *J. Pet. Tech.* (March 1962) 261–65.

14. Walker, J.W. and Turner, J.L.: "Performance of Seeligson Zone 20-B-07 Enriched Gas-Drive Project," *J. Pet. Tech.* (April 1968) 369–73.

15. Elliot, J.K. and Johnson, H.C.: "Control of Injection Gas Composition in Enriched Gas-Drive Project," *Trans.*, AIME (1959) **216**, 398–99.

16. "Rich-Gas Drive Yields High Recovery in Seeligson Field," *Oil and Gas J.* (Aug. 4, 1958) 80.

17. "New Miscible Method Due Tryout in Texas," *Oil and Gas J.* (April 4, 1960) 74.

18. "Humble Plans Unique Seeligson Flood," *Petroleum Week* (April 1, 1960) 74.

19. "Gas Drive Varied—Special Mixture is Planned by Humble in the Seeligson Field," *Oil and Gas J.* (Aug. 6, 1956) 66.

20. "New Gas Drive Cleared," *Oil and Gas J.* (Sept. 3, 1956) 80.

21. Wright, H.T. and Die, R.R.: "Enriched Gas Drive Project, Zone 20-B, Seeligson Field," paper 1038G presented at the 1958 AIME Mid-Continent Section Petroleum Conference on Production and Reservoir Engineering, Tulsa, March 20–21.

22. "Propane and Ethane Enriched Gas Injected in the 14-12 Sand at the South Coles Levee Field," paper presented at the 1961 API Spring Meeting; Abstract in *World Oil* (May 1961) 135.

23. Bleakley, W.B.: "Journal Survey Shows Recovery Projects Up," *Oil and Gas J.* (March 25, 1974) 69–78.

24. Holloway, H.D. and Fitch, R.A.: "Performance of a Miscible Flood with Alternate Gas-Water Displacements," *J. Pet. Tech.* (April 1964) 372–76.

25. "Texas Railroad Commission Hearing Spurs Five Flood Requests," *Petroleum Week* (June 24, 1960) 36.

26. Coan, S.W.: "Profitable Miscible Flood in a Watered-Out Reservoir," *Oil and Gas J.* (Dec. 12, 1966) 111.

27. "Canadian Enhanced-Recovery Activity is Moderate, Centers on Thermal Projects," *Oil and Gas J.* (April 5, 1976) 128–29.

28. Larson, V.C., Peterson, R.B., and Lacey, J.W.: "Technology's Role in Alberta's Golden Spike Miscible Project," *Proc.*, Seventh World Pet. Cong., Mexico City (1967) **3**, 533–44.

29. "Miscible Flood May Capture 95% of Crude in Place," *Oil and Gas J.* (July 26, 1965) 164.

30. Reitzel, G.A. and Callon, G.O.: "Pool Description and Performance Analysis Leads to Understanding Golden Spike's Miscible Flood," *J. Cdn. Pet. Tech.* (April–June 1977) 39–48.

31. Matheny, S.L.: "EOR Methods Help Ultimate Recovery," *Oil and Gas J.* (March 31, 1980) 79.

32. Bouck, L.S., Hearn, C.L., and Dohy, G.: "Performance of a Miscible Flood in the Bear Lake Cardium Unit, Pembina Field, Alberta, Canada," *J. Pet. Tech.* (June 1975) 672–78.

33. Griffith, J.D., Baiton, N., and Steffesen, R.J.: "Ante Creek—A Miscible Flood Using Separate Gas and Water Injection," *J. Pet. Tech.* (Oct. 1970) 1232–41.

34. Des Brisay, C.L. and Daniel, E.L.: "Supplemental Recovery Development of the Intisar "A" and "D" Reef Fields, Libyan Arab Republic," *J. Pet. Tech.* (July 1972) 785–92.

35. Des Brisay, C.L. *et al.*: "Miscible Flood Performance of the Intisar "D" Reef Field, Libyan Arab Republic," *J. Pet. Tech.* (Aug. 1975) 935–48.

36. Des Brisay, C.L. *et al.*: "Review of Miscible Flood Performance, Intisar "D" Field, Socialist People's Libyan Arab Jamahiriya," paper SPE 10245 presented at the 1981 SPE Annual Technical Conference and Exhibition, San Antonio, Oct. 5–7.

37. "U.S. Pushes Enhanced-Recovery Effort," *International Petroleum Encyclopedia*, Petroleum Publishing Co., Tulsa (1977) 282–94.

38. Brannan, G. and Whittington, H.M. Jr.: "Enriched Gas Miscible Flooding—A Case History of the Levelland Unit Secondary Miscible Project," *J. Pet. Tech.* (Aug. 1977) 919–24.

39. Griffith, J.D. and Horne, A.L.: "South Swan Hills Solvent Flood," *Proc.*, Ninth World Pet. Cong., Tokyo (1975) **4**, 269–78.

40. "Steam Dominates Enhanced Oil Recovery," *Oil and Gas J.* (April 5, 1982) 139.

41. *Improved Oil Recovery Field Reports*, SPE, Dallas (1975–1981).

42. Morgan, J.T. and Thompson, J.E. Jr.: "Reservoir Study of the Camp Sand, Haynesville Field, Louisiana," paper SPE 7553 presented at the 1978 SPE Annual Technical Conference and Exhibition, Houston, Oct. 1–3.

43. Griffith, J.D. and Cyca, L.G.: "Performance of South Swan Hills Miscible Flood," *J. Pet. Tech.* (July 1981) 1319–26.

44. "Banff Shoots for 97% Recovery at Rainbow Key," *Petroleum Engineer* (April 1969) 56–58.

45. Craig, D.R. and Bray, J.A.: "Gas Injection and Miscible Flooding," *Proc.*, Eighth World Pet. Cong., Moscow (1971) **3**, 275–85.

46. Harvey, M.T., Shelton, J.L., and Kelm, C.H.: "Field Injectivity Experiences With Miscible Recovery Projects Using Alternate Rich Gas and Water Injection," *J. Pet. Tech.* (Sept. 1977) 1051–55.

Chapter 7
The Vaporizing-Gas Drive Process

7.1 Estimation of Miscibility Pressure

At sufficiently high pressure, vaporizing-gas drive miscibility can be attained with lean hydrocarbon gas, flue gas, and nitrogen. Vaporizing-gas drive designs usually need be concerned only with achieving and maintaining miscibility between oil and the injection gas. There typically is no further requirement to maintain the miscibility of a solvent slug displaced by a drive gas, as usually is the case in first-contact miscible or condensing-gas drive flooding, because the high-pressure injection gas itself serves as both solvent and drive gas.

Phase behavior required for miscibility with lean hydrocarbon gas is shown conceptually by a pseudoternary diagram (Fig. 7.1). Oil composition must lie to the right of the limiting tie line that passes through the plait point. Stated another way, if vaporizing-gas drive miscibility is to be achieved, no tie line, when extended from the two-phase region, can pass through the oil composition. This condition is not met in Fig. 7.1 at pressure P_1. The two-phase region is so large at this pressure that tie line vl passes through the oil composition. At the higher pressure, P_2, the two-phase region is smaller; no tie lines pass through the oil composition, and lean hydrocarbon gas will miscibly displace the example oil by the vaporizing-gas drive mechanism. MMP is that pressure at which the composition of the oil of interest lies just to the right of the limiting tie line passing through the plait point. Further discussion of the relationship between phase behavior and miscibility in the vaporizing-gas drive method is given in Sec. 2.6.

According to the pseudoternary diagram concept of vaporizing-gas drive miscibility, both oil composition and reservoir temperature should affect miscibility pressure. Discussion of the preceding paragraph illustrated the importance of composition. The requirement that oil composition must lie to the right of the

limiting tie line through the plait point dictates that the most suitable oils for this process should have relatively high concentrations of intermediate-molecular-weight hydrocarbons such as ethane through pentane. Inspection of Fig. 7.1 shows that concentration of methane and nitrogen in the oil is important also. Obviously, vaporizing-gas drive miscible displacement cannot be achieved with a reservoir oil already saturated with the drive gas. Other factors being equal, miscibility pressure with lean hydrocarbon gas decreases with decreasing concentration of methane and nitrogen in the oil. Less obvious from Fig. 7.1 is the influence on miscibility pressure of composition of the intermediate-molecular-weight and heavy hydrocarbon fractions. Size of the two-phase region depends on these compositions. Other factors being equal at the given pressure, as the molecular weight of the intermediate hydrocarbon fraction decreases, a larger two-phase region results, which requires a higher pressure for vaporizing-gas drive miscibility. For example, at a given total concentration of intermediate-molecular-weight hydrocarbons, a higher miscibility pressure will be required for oil rich in ethane compared with oil rich in propane. Size of the two-phase region also increases with decreasing API gravity or increasing molecular weight of the heavy hydrocarbon fraction. Thus, even though two oils may contain the same overall concentrations for the intermediate and heavy hydrocarbon fractions, miscibility pressure will be higher for the oil that has the greater molecular weight for the heavy hydrocarbon fraction.

Miscibility pressure with lean hydrocarbon gas increases with increasing temperature because the solubility of methane in hydrocarbons decreases with increasing temperature, which in turn causes the size of the two-phase region to increase. However, miscibility pressure with nitrogen decreases with increasing temperature at

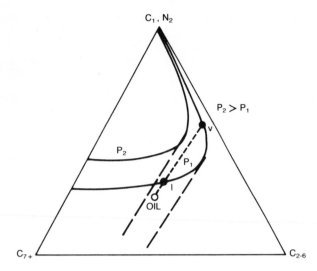

Fig. 7.1—Vaporizing-gas drive miscibility.

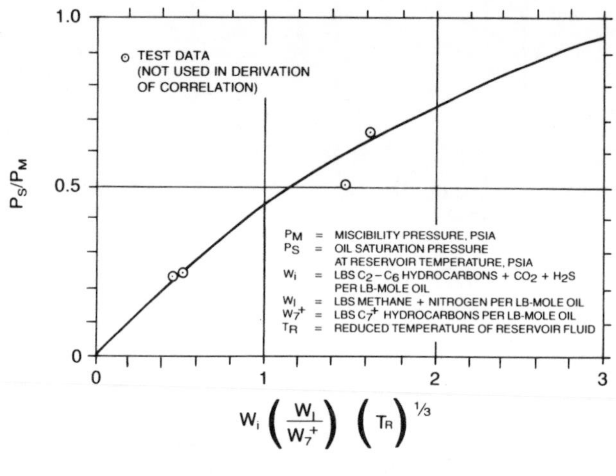

Fig. 7.2—Vaporizing-gas drive miscibility pressure correlation.

temperatures of practical interest in reservoirs[1,2] because the solubility of nitrogen in hydrocarbons increases for temperatures above roughly 100°F.[3-5]

Fig. 7.2 shows a previously unpublished correlation for estimating miscibility pressure with lean hydrocarbon gas. The correlation was derived from experimental miscibility pressures for nine oils covering a temperature range of 140 to 265°F, a saturation pressure range of 596 to 4,035 psi, and a range of heptanes-and-heavier molecular weight from 149 to 216. Experimental miscibility pressures ranged from 3,250 to 4,750 psi. The correlation attempts to account for the concentration of the intermediate and heavy hydrocarbon fractions in the oil as well as for the character of these fractions. It does this through correlating parameters expressed in units of pounds mass of a pseudocomponent per pound-mole of oil. The correlation estimates all the miscibility pressures used in its development within a maximum error of 65 psi.

The correlation was tested with four data points that were not used in its development (shown on Fig. 7.2). Average deviation of estimated from experimental miscibility pressures was 260 psi with a maximum deviation of 640 psi.

The curve of Fig. 7.2 is expressed in Eq. 7.1*:

$$\frac{p_s}{p_m} = \frac{2.252354\, X}{X + 4.09068}, \quad \dotfill \quad (7.1)$$

where

$$X = \frac{W_i (W_1)(T_R^{1/3})}{W_{7+}}$$

and other parameters are defined in Fig. 7.2.

Fig. 7.2 and Eq. 7.1 should be used with caution and for preliminary design or screening only, especially since so few data points were used in its development.

*Blackwell, R.J.: personal communication, Exxon Production Research Co. (1982).

No other vaporizing-gas drive miscibility pressure correlation has been published.

Fig. 2.18 and the discussion of Sec. 2.6 show conceptually how dynamic miscibility is attained with flue gas or nitrogen. Compared with natural gas, both these gases act to enlarge the size of the two-phase region, which in turn increases the pressure required for miscible displacement. The magnitude of difference in miscibility pressure between flue gas or nitrogen and lean hydrocarbon gas is affected by oil composition and temperature — factors that affect the size of the two-phase region and the slope of tie lines.

Koch and Hutchinson[6] speculate that with either flue gas or nitrogen as the displacing gas, methane in the reservoir oil acts to some extent as an intermediate hydrocarbon, and that the difference in miscibility pressure between these gases and lean hydrocarbon gas decreases with increasing methane in the reservoir fluid (assuming the ratio of heavy and intermediate hydrocarbons in the oil remains fixed as methane is added). For example, for the conceptual phase behavior shown in Fig. 2.18, nitrogen cannot displace methane-depleted Oil C miscibly because the composition of this oil lies to the left of the surface of limiting tie lines that pass through the critical locus of the pseudoquaternary diagram. As methane is added to this oil, compositions at some point begin to lie above and to the right of this surface of limiting tie lines such that no tie line when extended passes through the oil composition. For example, Composition C' in the $C_1/C_{2-6}/C_{7+}$ plane results from addition of methane to Oil C and lies to the right of the surface P_1AIH of limiting tie lines. Koch and Hutchinson published some experimental results supporting this hypothesis, but more data are needed to confirm or disprove its validity.

In principle, miscibility pressure with nitrogen should be some higher than with flue gas. The actual difference can be small with some oils, however, because CO_2 is stripped from flue gas by reservoir brine and bypassed oil, and the advancing gas front becomes progressively denuded of CO_2, richer in nitrogen, and similar to the

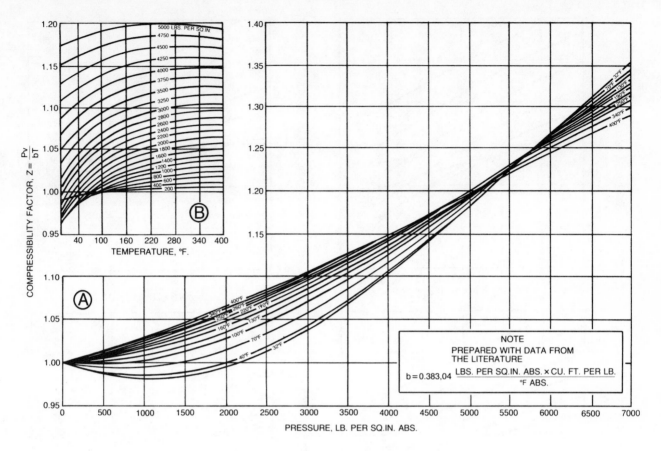

Fig. 7.3—Compressibility factors for nitrogen (after Ref. 10).

composition attained by injecting pure nitrogen. Conceptually, behavior with other oils could be such that the gas front composition becomes miscible with the oil before the front is entirely denuded of CO_2. In this event, a somewhat lower miscibility pressure might result with flue gas.

With some oils, the difference in miscibility pressure between flue gas/nitrogen and lean hydrocarbon gas can be small, perhaps only a few hundred psi.[6,7] Higher temperatures also tend to promote smaller differences between methane and nitrogen miscibility pressures.[8] However, the difference can be substantial in other cases.[9] No correlation is available for miscibility pressure with flue gas or nitrogen. Experimental measurement is necessary.

7.2 Properties of Injection Gases

Compressibility factors for hydrocarbon gases can be estimated from Fig. 6.6. Fig. 7.3 gives compressibility factors for nitrogen,[10] and Fig. 7.4 gives compressibility factors for nitrogen/CO_2 mixtures.[11] Vaporizing-gas drive miscibility occurs at relatively high pressures, which excludes shallow, low-temperature reservoirs. For temperatures from 150 to 250°F and pressures from 3,000 to 7,000 psi, nitrogen compressibility factors range from about 1.08 to 1.32; flue gas (12 mol% CO_2) compressibility factors range from 1.03 to 1.27; and compressibility factors of lean natural gas range from about 0.8 to 1.1.

Viscosity of hydrocarbon gases can be estimated from Fig. 6.4. Fig. 7.5 gives viscosity for nitrogen.[12] Flue gas viscosity can be estimated by the methods of Lohrenz *et al.*[13]

7.3 Injection Gas Sources

Field separator gas and residue gas from natural gasoline plants are potential sources of lean hydrocarbon gas for vaporizing-gas drive injection. In some projects, lean hydrocarbon gas has been purchased from gas transmission pipelines.

Flue gas has been generated in several projects by burning gasoline-plant residue gas in steam boilers.[14-16] In the Block 31 flue gas plant, 1 cu ft of residue gas makes 9 to 11 cu ft of flue gas containing approximately 87 mol% nitrogen, 11 mol% CO_2, and 1 mol% carbon monoxide.[16] Recovery of flue gas may be as low as six volumes of flue gas per volume of residue gas burned, depending on plant design and volume lost from safety equipment venting.[14]

At Block 31, a sulfur treating system sweetens the residue gas prior to burning, which helps reduce fouling and corrosion. The burner is designed to circulate some of the flue-gas product around the periphery of the burner to control flame temperature, because at elevated burner temperatures, highly corrosive nitrogen oxides form. An alternative design could use a catalyst to convert the nitrogen and oxygen compounds.[17,18] Flue gas from the boiler is scrubbed and cooled in a quench tower. Compression from a few inches of water to 1,200 psi is accomplished by a centrifugal gas compressor powered by a steam turbine. Steam generated during residue gas combustion drives the turbine. Five stages of

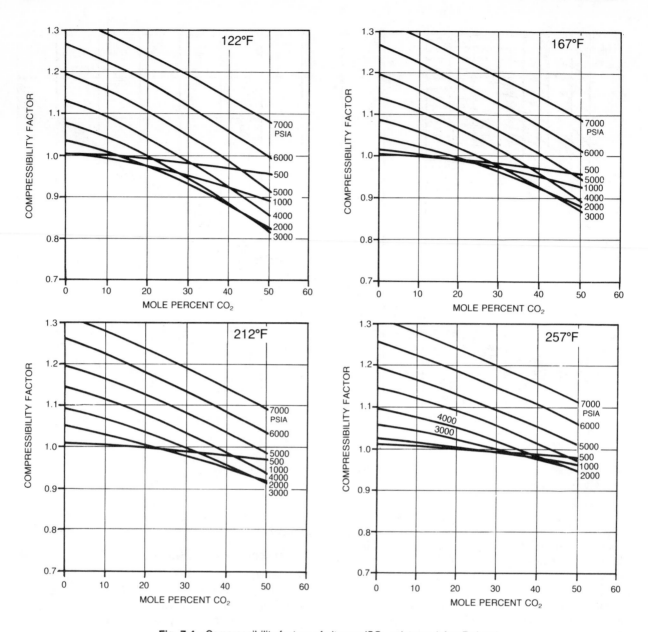

Fig. 7.4—Compressibility factors of nitrogen/CO$_2$ mixtures (after Ref. 11).

centrifugal compression are required to reach 1,200 psi. Compressed gas is cooled by atmospheric coolers, and ammonium hydroxide (NH$_4$OH) is injected to neutralize acid gases and prevent formation of carbonic and nitric acids. The gas is dehydrated at this point by refrigeration and glycol injection. Three 2-stage reciprocating compressors then boost gas pressure to 4,200 psi. Because flue gas manufacture with steam boilers requires the use of turbomachinery, delivery rates below 30 MMscf/D may be impractical.[19]

Gas engine exhaust is a potential source of flue gas for small supply requirements. After catalytic conversion of nitrogen oxides, the exhaust gas is compressed and dried. The largest gas engine exhaust train is about 10 MMscf/D.[19]

Nitrogen for miscible floods can be manufactured on-site also. A cryogenic separation from air typically is used, although separation from boiler flue gas or engine exhaust is an alternative.[20] In the cryogenic process, air

is compressed to about 6 or 7 atm with centrifugal or axial compressors and processed to remove water vapor, CO$_2$, and hydrocarbon contaminants. Two removal systems can be used: (1) a reversing heat exchanger cycle that freezes water and CO$_2$ on the surface of a heat exchanger followed by silica gel adsorption of hydrocarbons at cryogenic temperature and (2) molecular sieve adsorption at ambient temperatures. With either method, the purified air is cooled further before entering the distillation column by heat exchange with products from the distillation column, and it enters this column at about −280°F.

In the distillation column, the air is separated into an oxygen-rich liquid and nitrogen vapor. The oxygen-rich liquid stream from the bottom of the distillation column is boiled in the column's reflux condenser at about −290°F and then expanded in a cryogenic turbine to produce refrigeration. It then is sent to the heat exchangers that cool the incoming air feed to the distillation column. The nitrogen product stream from the distillation column is warmed to ambient temperature by heat exchange and is fed to the suction of the product compressor at about 5 atm. The oxygen-rich stream is vented to air.

This cycle, with reversing heat-exchanger contaminant removal, is generally the preferred cycle for small and medium-sized on-site nitrogen plants of up to approximately 25 Mcf/D. An alternate cycle using a molecular sieve prepurifier and double distillation column has more application as an oxygen plant but may find some use in large nitrogen applications of more than 50 MMcf/D.[20,21]

On-site nitrogen plants will be built and operated by several compressed gas manufacturers.

Ref. 19 compares operating characteristics and economics of flue gas and cryogenic nitrogen manufacture for enhanced oil recovery. Ref. 22 also evaluates the cost of manufacturing cryogenic nitrogen for oilfield use.

7.4 Field Experience

Table 7.1 gives locations and other details of selected vaporizing-gas drive projects. Several of these projects are described in the following subsections.

Block 31 Field[14,16,23,24]

The Block 31 (Devonian) project in west Texas was the earliest recognized field application of the vaporizing-gas drive process. This flood is still in operation today. Produced gas was first reinjected in a few wells into the Middle Devonian formation in 1949 at a pressure greater than what later proved to be the miscibility pressure. Beginning in 1952, hydrocarbon gas was injected into 320-acre inverted nine-spot patterns to maintain gas-front pressure above the miscibility level of 3,500 psi. To limit the volume of purchased hydrocarbon gas, the operator began flue gas injection in about half the field in 1965. Flue gas is generated by burning natural gas in steam boilers and is compressed to injection pressure both with turbine-driven centrifugal compressors and with reciprocating compressors. The injection rate in 1981 was about 135 MMscf/D total gas, of which about 54 MMscf/D was flue gas.

The Middle Devonian formation is a siliceous limestone, occurring at a depth of about 8,400 ft, and is moderately stratified into four major porous zones separated over most of the field by dense intervals. Stratification also occurs within porous zones. Average permeability is about 1 md, with average porosity about 15%. Surface area is 7,840 acres, and the reservoir originally contained about 250 million STB of a 46°API oil. Viscosity of this oil at the 140°F reservoir temperature is 0.2 cp, and viscosity ratio for the displacement is 10.

The Block 31 project has been highly successful. Ultimate recovery is projected at 60% OOIP or greater, and current recovery (1981) already exceeds 46%. Some gas breakthroughs were premature, but others were about as predicted. GOR's did not rise rapidly after breakthrough. Current GOR has risen to about 7,000 scf/STB from an original solution GOR of 1,300. Produced gas is being reinjected. Producing problems have

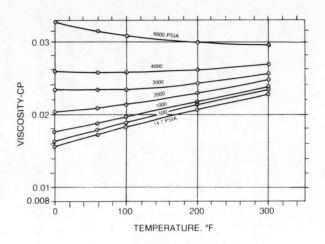

Fig. 7.5—Viscosity of nitrogen.

been minimal, although well plugging from excess compressor lubricant, iron sulfide, and free sulfur has occurred. A production rate of 15,000 STB/D or better has been maintained since 1955; however, infill drilling recently has been necessary to maintain rate. The operator is considering ultimately flooding on 80-acre, five-spot patterns to sweep previously unswept areas.

Fairway Field[31]

In this project, lean gas is being injected alternately with water to improve sweepout. Production is from the 10,000-ft James Lime, a highly stratified reef with three major porous zones. Average porosity of the pay is 12.5%. Total project area is about 23,000 acres. The crude has a gravity of 48°API and had an initial saturation pressure that varied from 3,950 to 4,350 psi. Initial oil in place was more than 400 million STB. Miscibility pressure with lean gas is 4,800 psi. Viscosity ratio in this flood is about five.

The field was discovered in 1960 and produced by solution gas drive with some limited aquifer encroachment until the start of lean gas injection in 1966. Initially lean gas and water were injected in a ratio of 1 RB/RB, but after the first gas breakthroughs this ratio was decreased to 0.7 RB/RB to achieve gas and water breakthroughs at about the same time. The injection rate of lean gas has averaged 100 million scf/D through about 30 injectors, and gas injectivity has been two to three times that for water. GOR behavior in some wells appeared to correlate with gas and water cycles in nearby injectors. Multiple packers and wireline tools were used to isolate zones to improve volumetric conformance, and in some cases breakthrough zones have been shut off mechanically in producers. Oil residual to water has been recovered from areas of the field that had been watered-out by water encroachment. Some wells on the edge of the field that produced at high water cuts showed increased oil production and decreased water cut.

The operator believes that viscous fingers were retarded and GOR was controlled by alternate gas/water injection. Ultimate recovery is expected to be about 50% OOIP, compared with a predicted waterflood recovery of 37%.

TABLE 7.1—SELECTED VAPORIZING-GAS DRIVE FIELD PROJECTS

Field	Operator	Year Started	Project Type	Oil Gravity (API)	Viscosity Ratio μ_o/μ_s	Miscibility Pressure (psi)	Depth (ft)	Thickness (ft)	Acreage	Status	Recovery (% OOIP)	References
Block 31 Devonian (TX)	ARCO Oil & Gas Co.	1949 1952	G,N	46	10	3,500	8,400	125 to 150	7,840	NC, succ., prof.	>60*	14,16,23,24,25†
Neale 10,400-ft (LA)	ARCO Oil & Gas Co.	1956	G	47	4	4,000	10,400	10	—	term.	36**	15,24
University Block 9 (TX)	ARCO Oil & Gas Co.	1960	LPG slug by G by G/W	38	29	4,500	5,300	4 to 45, 10	4,000	—	40	26
Raleigh (MS)	Chevron U.S.A. Inc.	1960	G	45.5	3.5	<5,400	12,600	18	812	—	>52	27
Hassi-Massaoud (Algeria)	SONATRACH	1964	G and G/W	44	7	3,700	10,800	300	18,000	succ., prof.	—	28,29,30†
Fairway (TX)	Hunt Oil Co.	1966	G/W	48	5.5	4,800	10,000	55	22,600	HF, succ., prof.	50*	25†,31
Bridger Lake (WY/UT)	Phillips Petroleum Co.	1970	G	40	19	5,300	16,500	20	3,800	NC, succ., prof.	32*	25†,32
East Binger (OK)	Phillips Petroleum Co.	1977	N	38	—	—	10,000	—	13,000	JS, TETT	—	25†,33
Fordoche (LA)	Sun Gas Co.	1971	G+N	45	5	—	13,500	25 to 35	6,100	HF, succ.	50*	25†,34
Jay-Little (FL)	Exxon U.S.A.	1982	N/W (20% HCPV)	51	6	3,600	15,000	350	14,000	—	54* (6.5* incremental)	8
Lake Barre‡ (LA)	Texaco Inc.	1978	N	40-43	—	5,100 (G) 5,800 (N)	17,500	57	1,194	—	53*	9

*Projected.
**Recovery in the east and west sections of the sand where miscible displacement definitely was achieved was estimated by the operator to be approximately 50%.
†Status as given in the Oil and Gas J. article identified in Reference column.
‡Full miscibility apparently has not been achieved; reservoir pressure is several hundred pounds per square inch below dynamic miscibility pressure.

Symbol key:
G = natural gas
N = flue gas or nitrogen
G/W = alternate gas-water injection
succ. = successful
prof. = profitable
TETT = to early to tell
NC = nearing completion
term. = terminated
HF = half finished
JS = just started

Raleigh Field[27]

The 12,600-ft Hosston sand in the Raleigh field consists of an upper member composed of two thin, tight sand bodies and a lower, more permeable sand member about 18 ft thick. The lower sand has an average porosity of 10.4% and an average permeability of 34 md. The 812-acre reservoir contains a 45°API gravity crude with a viscosity of 0.093 cp at the 257°F reservoir temperature. OOIP was about 4.3 million STB.

The Hosston sand was put on production in 1958. Reservoir pressure dropped quickly under primary production from the original 5,783 psig, and by late 1959 was 3,512 psig—only a few hundred psi above the 3,236-psig bubble point. The severe pressure decline caused a large drop in producing well productivities. Gas injection was selected for pressure maintenance because (1) laboratory slim-tube tests indicated oil recovery characteristic of miscible displacement at 5,400 psig, (2) residual oil saturation to waterflooding was relatively high, (3) gas injection would maintain flowing production longer than would water injection, and (4) sufficient field gas was available.

Gas injection began into one upstructure well in 1960 to flood 27 producing wells in approximately three rows. By late 1961, BHP had increased to more than 5,500 psig. After gas injection began, productivity from wells adjacent to the injection well improved rapidly. The nearest wells went to high GOR soon; but with increasing reservoir pressure, the GOR of the more distant wells increased more gradually. Wells in the last row, most of which were poor producers or noncommercial on original completion, improved to produce low-GOR oil at high rates. By 1965, 52.4% OOIP of the productive sand had been produced, and the project was still producing at a rate of about 5% of the oil in place per year. The operator estimated that ultimate recovery by primary production alone would have been 31.9% OOIP.

Hassi-Messaoud Field[28,29]

This project operated by SONATRACH in Algeria is the largest vaporizing-gas drive flood in a sandstone reservoir. Reserves in this field are estimated to be about 23 billion bbl. Stratification is very severe and the reservoir is complex. Average thickness is about 355 ft. Although some strata are correlatable between wells, others are not. Clay beds sometimes separate strata. This is a horizontal flood, the angle of dip being less than 1°. Viscosity ratio in this flood is seven.

Injection of produced solution gas into one zone began in 1964. Reservoir pressure was maintained at 4,500 psi, well above the miscibility pressure of 3,750 psi. By Jan. 1970, after injection of 330 Bscf, there were only two breakthroughs, and performance was encouraging enough to justify an expansion of gas injection. Sweepout at these first gas breakthroughs was between 10 and 12% with continuous gas injection.

A two-well pilot test was run to evaluate the benefit of alternate gas/water injection. Injector and producer in this small test were 465 ft apart, and analysis showed that breakthrough sweepout was increased to 22% by water injection at a 1.1- to 1.2-RB/RB gas/water ratio. Water injection also helped control GOR increase after gas breakthrough. Alternate slugs of about 3% of the

pore volume in the radius from injector to producer were injected. Sweepout of injected water was good, probably because of crossflow of water between strata. Water injectivity was about as expected, but gas injectivity was only half the expected value.

Neale Field[15]

Another vaporizing-gas drive project in a sandstone reservoir was completed by ARCO Oil & Gas Co. in the 10,400-ft sand reservoir in Neale field, LA. This small, 4-million-bbl reservoir had a partial water drive, which replaced about 25% voidage during primary recovery. It contained a 47°API-gravity, high-shrinkage oil with an original formation volume factor of 2.83 RB/STB. Sand thickness was about 10 ft.

Most wells were producing high water cuts when lean gas injection began in 1956 into three wells spaced along the major axis of the field. Mobility ratio in this flood was only four. At the start of lean gas injection, pressure in the central part of the field was 250 psi below the miscibility pressure of 4,000 psi, and was also below the original saturation pressure by nearly 30 psi. With this high-shrinkage oil, a high gas saturation was present in the center section.

The two ends of the field behaved about as predicted for miscible displacement. However, the GOR continued to rise in the center section despite high injection well pressures, and it is doubtful that a miscible oil bank ever formed. The operator suspected that a high-permeability streak caused the center section behavior and began alternate injection of water with the lean gas to improve gas conformance. Injectivity was impaired unless the alternate cycles were large enough to achieve essentially single-phase flow in the radial flow region around the well. Alternate injection was ineffective in reducing the GOR in center-section wells.

The project has been completed. Overall recovery was about 36% OOIP, compared with a projected 17% OOIP for primary recovery.

Bridger Lake Field[32]

The Bridger Lake flood was started by Phillips Petroleum Co. in 1970. Injection was through a single well into the Dakota sandstone at 16,500 ft. Production came from nine wells. The reservoir is a multisand pay with Sands A and C being the main reservoirs. These two reservoirs are in pressure communication and are the sands being miscibly flooded. The crude has an API gravity of 40° and a viscosity of 0.5 cp. Viscosity ratio of the miscible flood is 19.

Primary recovery was by solution gas drive with some limited water influx. Reservoir pressure at the start of flooding was below the 5,300-psi miscibility pressure, and about 1 year was required for repressuring to this level. Two wells had gas breakthrough within 6 to 12 months, whereas sweepout calculations showed the breakthroughs would not occur for 2 to 3 years. Miscible recovery is expected to be about 32% OOIP in the flooded zones over a project life of 34 years.

University Block 9 Field[26]

A modified vaporizing-gas drive was tried by ARCO Oil & Gas Co. in this field. The Wolfcamp formation was the target reservoir for the miscible flood—a domal, organic limestone structure at about 5,300-ft depth with intercrystalline-to-vugular porosity. Thickness varied between 4 and 45 ft in one productive zone and averaged about 10 ft in another. Average porosity was 10% and average permeability was 14 md.

The reservoir had been produced by fluid expansion and solution gas drive to 1,425 psig when the project was started in 1960. Miscibility pressure with lean gas was 4,500 psig. The operating plan was to inject a 3% HCPV slug of propane to achieve miscible displacement while the reservoir was being repressured, follow the propane with a 10% HCPV slug of lean gas, and then switch to alternate gas/water injection. Viscosity ratio was 7 between oil and propane and 29 between oil and lean gas.

Miscible fluids were injected through 21 wells arranged in a peripheral pattern around the 4,000-acre field. Premature breakthrough of propane and lean gas into the first-row producing wells was caused by viscous fingering and stratification. To reduce GOR, alternate gas/water injection was begun before the full lean-gas slug was injected. Average reservoir pressure reached 4,500 psig in 1963, 900 psig above the original reservoir pressure. Despite increased gas injection, pressure declined after 1965 as fluids leaked to the Wolfcamp aquifer. Analysis with a numerical simulator after 9 years' flooding showed that only regions close to the injection wells were above miscibility pressure. Despite the inability to maintain pressure at a level necessary for full miscibility, recovery was 2¼ times projected primary recovery after 9 years' flooding, and was already close to the ultimate expected for waterflooding. At that time the field still had an expected 10 to 20 years' economic production remaining.

7.5 Assessment of Process

Although there have not been a large number of field trials, good success has been achieved to date in miscible flooding with the vaporizing-gas drive process. This is seen from the status remarks in Table 7.1. For those projects where assessments by the operator are available, the majority are considered to have performed successfully and to have been profitable.

Vaporizing-gas projects generally have been large-scale floods involving thousands of surface acres. Eight of the projects in Table 7.1 have more than 3,000 acres under flood, and as many as 22,600 acres are being flooded in the Fairway field in east Texas. The Hassi-Massaoud field in Algeria is a 22-billion-bbl oil field.

Several projects have been completed, and many of the remaining projects have been operated for a long time. Six of the floods active in 1982 were more than 10 years old, four were more than 14 years old, and the Block 31 flood had been in operation for about 30 years. Because of the size and maturity of these projects, miscible-flood behavior can be assessed and ultimate recovery projected with good confidence.

Exceptionally high ultimate recoveries should be achieved in the majority of Table 7.1 projects. Recovery of about 50% OOIP has been attained in the Block 31 [16,23] and Raleigh floods, [27] and this level of recovery may be met or exceeded in the relatively mature Fairway project. [31] An eventual recovery of more than 60% OOIP

is probable for the Block 31 flood. Recovery already has reached 40% OOIP at the University Block 9 project and is still increasing, [26] and recoveries of more than 50% OOIP have been projected by operators of the floods at Fordoche, [34] Jay-Little Escambia Creek, [8] and Lake Barre. [9]

Floods that were considered successful by the project operator have been conducted in highly stratified carbonate and sandstone reservoirs as well as in less heterogeneous reservoirs. Several factors that have contributed to this success are discussed below.

Continued injection of the miscible solvent without ever switching to a drive gas is an important difference between the vaporizing-gas drive floods and the majority of condensing-gas drive and first-contact miscible floods. This perhaps may be the single most important reason for the relative success of vaporizing-gas drive floods. In the condensing-gas and first-contact miscible methods, a relatively small slug of the solvent is displaced by a drive gas, and the integrity of this solvent slug must be maintained for the displacement to be effective. The vaporizing-gas drive method as practiced in most field projects is not a slug process. The high-pressure gas itself serves as both solvent and drive gas, and miscibility cannot be lost unless pressure at the gas front falls below miscibility pressure. In vaporizing-gas drive floods the amount of miscible solvent injected may be quite large when expressed as a percent of the hydrocarbon pore volume. For example, more than 1 HCPV of lean gas and flue gas have been injected in the Block 31 project. This is contrasted with first-contact miscible and condensing-gas drive floods where the amount of solvent injected typically has been in the range of 4 to 25% HCPV.

Another important difference between vaporizing-gas drive floods and first-contact miscible and condensing-gas drive floods is that the produced gas can be compressed to miscibility pressure and reinjected to maintain miscible sweepout. In condensing-gas drive and first-contact miscible floods, solvent production decreases miscible sweepout. However, breakthrough behavior and subsequent production of gas in vaporizing-gas drive projects have been similar to the behavior experienced in other hydrocarbon miscible processes.

Availability and relatively low cost of lean hydrocarbon gas also have contributed to the favorable economic outcome of past projects. Because of these conditions, it was economical to inject and to cycle the large volumes of lean hydrocarbon gas used in past floods. In future projects, hydrocarbon gas will be more expensive and in shorter supply in the U.S. Flue gas already has been injected in several projects, and both flue gas and nitrogen should be evaluated as injection gases in future projects. In field tests to date, there have been no apparent adverse effects of using flue gas, although miscibility pressure with flue gas or nitrogen generally will be higher than with lean hydrocarbon gas.

Another factor that probably has made an important contribution to the relative success of vaporizing-gas drive floods is a lower overall mobility ratio compared with first-contact miscible and condensing-gas drive floods. Because a high concentration of intermediate-molecular-weight hydrocarbons in the reservoir oil is required to achieve miscibility at attainable reservoir pressures, higher-API-gravity and lower-viscosity oils generally have been displaced in vaporizing-gas drive projects than have been in the other hydrocarbon miscible projects. Because of this, the overall viscosity ratio between oil and driving gas has been more favorable on average in vaporizing-gas projects than in the other hydrocarbon miscible projects.

Many of the vaporizing-gas drive floods were conducted in relatively thin reservoirs or in strata only several tens of feet thick within thicker gross intervals. Considering the large volumes of miscible gas injected, vertical sweepout may have been improved by transverse dispersion, which could have slowed the growth of a gravity tongue and aided recovery of oil bypassed in low-permeability lenses. Also, because of the relatively volatile nature of most oils in vaporizing-gas drive projects and the large throughput of gas, some recovery may have been attributable to vaporization or stripping of hydrocarbons from any oil that was left behind the gas front as a result of the mechanisms discussed in Sec. 3.7 or from oil that was immiscibly swept in regions below the miscibility pressure. The degree to which transverse dispersion and oil stripping contributed to oil recovery in these projects, however, is unknown.

The process has had greatest application in reservoirs that were not particularly good candidates for waterflooding, either because of low permeability and poor water injectivity or because reservoir depth caused problems in producing high-water-cut wells. None of the Table 7.1 projects were in watered-out reservoirs, although the Neale 10,400-ft reservoir had a partial water drive and was producing at a high water cut when natural gas injection began; and the Jay-Little Escambia Creek Unit had been partially waterflooded by the time nitrogen injection began.

Alternate injection of water to reduce mobility ratio was practiced in five projects[8,15,29,31,32] and was reported by the project operators to have had a beneficial effect on flood performance in three of these.[29,31,32] Alternate lean-gas/water injection did not reduce GOR in the Neale 10,400-ft reservoir project. However, of the three injection wells in this project, water was injected only in the center well, and miscibility pressure probably was never attained in the center section of this flood.

Miscibility pressure with natural gas, flue gas, or nitrogen is usually high enough to limit application of the vaporizing-gas drive method to reservoirs with depths greater than about 5,000 ft. Miscibility pressures ranged from 3,500 to 5,800 psi in the Table 7.1 projects. The shallowest reservoir flooded was the University Block 9 field (5,300 ft), and miscibility pressure was not maintained successfully over the life of this project. The majority of floods were conducted at depths of 9,000 ft or greater.

7.6 Screening Criteria

Guidelines for identifying potential vaporizing-gas drive projects are as follows.

1. The reservoir oil must be rich in intermediate-molecular-weight hydrocarbon components. Oil gravity will typically be greater than 40°API.

2. The oil must be undersaturated with respect to the injection gas.

3. A pressure of 3,500 psi is about the lower limit for achieving miscibility. Because of this, the method is restricted to reservoirs that are relatively deep, approximately 5,000 ft or greater.

4. There is no technical upper limit for oil viscosity in horizontal floods. Sweepout becomes progressively poorer as oil viscosity increases. A practical upper limit may be about 3 cp, although oils of 1 cp or less are preferred. The upper oil viscosity limit for gravity-stable floods depends on reservoir permeability.

5. In horizontal floods, reservoirs with restricted vertical permeability are preferred. Thin reservoirs of less than about 10-ft thickness are favorable because transverse dispersion can improve vertical sweepout. Reservoirs with high relief are preferred if permeability is high enough for gravity-stabilized flooding.

6. Extensive fracturing, a gas cap, a strong water drive, and high permeability contrasts all increase project risk.

7. Minimum permissible oil saturation in the reservoir volume to be miscibly swept is determined by project economics. A value of 25% PV is a rough guideline.

References

1. Rushing, M.D. *et al.*: "Miscible Displacement with Nitrogen," *Pet. Eng.* (Nov. 1977) 26–30.
2. Rushing, M.D. *et al.*: "Nitrogen May Be Used for Miscible Displacement in Oil Reservoirs," *J. Pet. Tech.* (Dec. 1978) 1715–16.
3. *Engineering Data Book*, Gas Processors Suppliers Assn., Tulsa (ninth edition, 1972) (fifth revision, 1981).
4. Yarborough, L.: "Vapor-Liquid Equilibrium Data for Multicomponent Mixtures Containing Hydrocarbon and Nonhydrocarbon Components," *J. Chem. Eng. Data* (1972) **17**, No. 2, 129.
5. Jacoby, R.H. and Rzasa, M.J.: "Equilibrium Vaporization Ratios for Nitrogen, Methane, Carbon Dioxide, Ethane, and Hydrogen Sulfide in Absorber Oil-Natural Gas and Crude Oil-Natural Gas Systems," *Trans.*, AIME (1952) **195**, 99–110.
6. Koch, H.A. Jr. and Hutchinson, C.A. Jr.: "Miscible Displacements of Reservoir Oil Using Flue Gas," *Trans.*, AIME (1958) **213**, 7–19.
7. McNeese, C.R.: "The High Pressure Gas Process and the Use of Flue Gas," paper A-67 presented at the 1963 ACS Symposium on Production and Exploration Chemistry, Los Angeles, March 31–April 1.
8. Christian, L.D. *et al.*: "Planning a Tertiary Oil Recovery Project for Jay-Little Escambia Creek Fields Unit," *J. Pet. Tech.* (Aug. 1981) 1535–44.
9. Haag, J.W.: "Analysis and Design of a Deep Reservoir, High Volume Nitrogen Injection Project in the R-1 Sand, Lake Barre Field," paper SPE 10159 presented at the 1981 SPE Annual Technical Conference and Exhibition, San Antonio, Oct. 5–7.
10. Eilerts, C.K., Charlson, H.A., and Mullens, N.B.: "Effect of Added Nitrogen on Compressibility of Natural Gas," *World Oil* (July 1948) 144–60.
11. Haney, R.E.D. and Bliss, H.: "Compressibilities of Nitrogen-Carbon Dioxide Mixtures," *Ind. Eng. Chem.* (Nov. 1944) 985–89.
12. Bicher, L.B. Jr. and Katz, D.L.: "Viscosity of Natural Gases," *Trans.*, AIME (1944) **155**, 246–52.
13. Lohrenz, J., Bray, B.G., and Clark, C.R.: "Calculating Viscosities of Reservoir Fluids From Their Compositions," *J. Pet. Tech.* (Oct. 1964) 1171–76.
14. Caraway, G.E. and Lowrey, L.L.: "Generating Flue Gas for Injection Releases Sales Gas," *Oil and Gas J.* (July 28, 1975) 126–32.
15. Tittle, R.M. and From, K.T.: "Success of Flue Gas Program at Neale Field," paper SPE 1907 presented at the 1960 SPE Annual Meeting, Houston, Oct. 1–4.
16. Hardy, J.H. and Robertson, N.: "Miscible Displacement by High-Pressure Gas at Block 31," *Pet. Eng.* (Nov. 1975) 24–28.
17. Barstow, W.F. and Watt, G.W.: "Fifteen Years of Progress in

Catalytic Treating of Exhaust Gas," paper SPE 5347 presented at the 1975 SPE Rocky Mountain Regional Meeting, Denver, April 7–9.

18. Kuehm, H.G.: "Hawkins Inert Gas Plant: Design and Early Operation," paper SPE 6793 presented at the 1977 SPE Annual Technical Conference and Exhibition, Denver, Oct. 9–12.

19. Wilson, K.: "Enhanced Recovery Inert Gas Processes Compared," *Oil and Gas J.* (July 31, 1978) 161–72.

20. Clancy, J.P., Kroll, D.E., and Gilchrist, R.E.: "How Nitrogen is Produced and Used for Enhanced Recovery," *World Oil* (Oct. 1981) 233–42.

21. Perry, J.H.: *Chemical Engineers' Handbook* (third edition), McGraw-Hill Book Co. Inc., New York City (1950) 1705–11.

22. Donahoe, C.W. and Buchanan, R.D.: "Economic Evaluation of Cycling Gas-Condensate Reservoirs with Nitrogen," *J. Pet. Tech.* (Feb. 1981) 263–70.

23. Herbeck, E.R. and Blanton, J.R.: "Ten Years of Miscible Displacement in Block 31 Field," *J. Pet. Tech.* (June 1961) 543–49.

24. Koch, H.A. Jr.: "High Pressure Gas Injection Is a Success," *World Oil* (Oct. 1956) **143**, 260.

25. Matheny, S.L. Jr.: "EOR Methods Help Ultimate Recovery," *Oil and Gas J.* (March 31, 1980) 84–85.

26. Cone, C.: "Case History of the University Block 9 (Wolfcamp) Field—A Gas-Water Injection Secondary Recovery Project," *J. Pet. Tech.* (Dec. 1970) 1485–91.

27. Meltzer, B.D., Hurdle, J.M., and Cassingham, R.W.: "An Efficient Gas Displacement Project—Raleigh Field, Mississippi," *J. Pet. Tech.* (May 1965) 509–14.

28. Pottier, J. *et al.*: "The High Pressure Injection of Miscible Gas at Hassi-Messaoud," *Proc.*, Seventh World Pet. Cong., Mexico City (1967) **3**, 533–44.

29. Dyes, A.B. *et al.*: "Alternate Injection of HPG and Water—A Two-Well Pilot," paper SPE 4082 presented at the 1972 SPE Annual Meeting, San Antonio, Oct. 8–11.

30. "U.S. Pushes Enhanced Recovery Effort," *International Petroleum Encyclopedia*, Petroleum Publishing Co., Tulsa (1977) 282–94.

31. Lackland, S.D. and Hurford, G.T.: "Advanced Technology Improves Recovery at Fairway," paper SPE 3773 presented at the 1972 SPE Improved Oil Recovery Symposium, Tulsa, April 16–19, 1972.

32. Burt, R.A. Jr.: "High Pressure Miscible Gas Displacement Project, Bridger Lake Unit, Summit County," paper SPE 3487 presented at the 1971 SPE Annual Meeting, New Orleans, Oct. 3–6.

33. "Oklahoma Field Gets Flue Gas Injection," *Oil and Gas J.* (Sept. 12, 1977) 54.

34. Eckles, W.W., Prihoda, C., and Holden, W.W.: "Unique Enhanced Oil and Gas Recovery for Very High-Pressure Wilcox Sands Uses Cryogenic Nitrogen and Methane Mixture," *J. Pet. Tech.* (June 1981) 971–84.

Chapter 8
The CO$_2$-Miscible Process

8.1 Introduction

CO$_2$ is not miscible on first contact with reservoir oils. However, past research shows that at sufficiently high pressure CO$_2$ achieves dynamic miscibility with many reservoir oils. Section 2.7 discusses in some detail a phase behavior concept of how dynamic miscibility is achieved. According to this concept, CO$_2$ vaporizes or extracts hydrocarbons from the crude as heavy as the gasoline and gas/oil fractions. Vaporization occurs at temperatures where the fluid at the displacement front is a CO$_2$-rich gas, and extraction occurs at temperatures where the fluid at the displacement front is a CO$_2$-rich liquid. According to the pseudoternary diagram concept of CO$_2$/reservoir oil phase behavior, no tie lines pass through the reservoir oil composition above the MMP; and because of this, vaporization/extraction can proceed to such an extent and so alter the composition of displacing fluid at the displacement front that dynamic miscibility results after sufficient contacting has occurred between the CO$_2$ and the reservoir oil.

The pressure required for achieving dynamic miscibility with CO$_2$ is usually significantly lower than the pressure required for dynamic miscibility with either natural gas, flue gas, or nitrogen. This is a major advantage of the CO$_2$ miscible process because dynamic miscibility can be achieved at attainable pressures in a broad spectrum of reservoirs.

A disadvantage of CO$_2$ flooding compared with waterflooding results from the low viscosity of CO$_2$ relative to that of oil. For example, at a reservoir temperature of 110°F, CO$_2$ viscosity is about 0.03 cp at 1,500 psi, whereas at 2,500 psi, the viscosity is about 0.06 cp. The low viscosity of CO$_2$ causes the mobility ratio in most CO$_2$ floods to be unfavorable; and, as discussed in Chap. 3, an unfavorable mobility ratio affects sweepout adversely.

The densities of oil and CO$_2$ are similar at many reservoir conditions, which tends to minimize, although not necessarily eliminate, segregation between these fluids in reservoirs that have not been waterflooded. In reservoirs that have been waterflooded or have had water injected with CO$_2$ to counteract the effects of viscosity ratio and permeability stratification, the density contrast between water and CO$_2$ may cause segregation.

The majority of CO$_2$ miscible field projects were initiated after 1970. Most projects to date (1982) have been relatively small-scale pilot tests, although there also have been a few commercial-size projects. Because CO$_2$ is a costly fluid, project designs usually have called for injection of an optimal volume or slug rather than continuous injection. The trend in CO$_2$ flood projects has been to drive relatively large CO$_2$ slugs of 15% HCPV or greater with water rather than with a gas that is miscible with CO$_2$. The objective of doing this is to improve miscible sweepout by achieving a favorable mobility ratio at the trailing edge of the slug. The disadvantage of this method of operation, of course, is that the displacement of CO$_2$ by water is immiscible, and a residual CO$_2$ saturation is left in the reservoir.* In some projects, water has been injected alternately with small slugs of CO$_2$ until the desired total volume of CO$_2$ was injected. As discussed in Sec. 3.8, the objective of alternate water injection is to improve the mobility ratio at the leading edge of the slug. There has been at least one project where the CO$_2$ slug was driven by a miscible gas.[1]

Produced CO$_2$ probably will be reinjected in many future projects to minimize purchased gas requirements. Separation of CO$_2$ from produced hydrocarbon gases may be necessary before reinjection. Reserves and cost of purchased CO$_2$ as well as the cost of reinjection are key elements of this decision.

Ref. 2 is a recent bibliography of CO$_2$ flooding literature.

---

*Some of the residual CO$_2$ is dissolved by and slowly produced with the drive water, although this does not prevent slug dissipation.

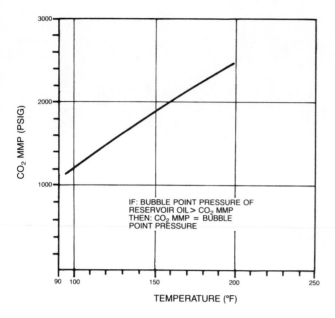

Fig. 8.1—Temperature/bubble-point pressure of CO_2 MMP correlation (after Ref. 5).

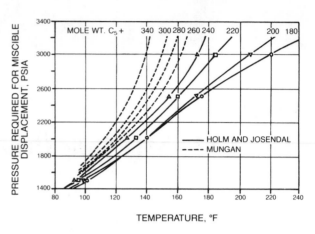

Fig. 8.2—Pressure required for miscible displacement in CO_2 flooding (after Ref. 10).

8.2 Estimation of Miscibility Pressure

From an experimental study of factors affecting CO_2 miscibility pressure, Holm and Josendal[3,4] reached the following conclusions.

1. Dynamic miscibility occurs when the CO_2 density is sufficiently great that the dense gas CO_2 or liquid CO_2 solubilizes the C_5-through-C_{30} hydrocarbons contained in the reservoir oil. For the particular oils examined in their study, Holm and Josendal observed that miscible displacement occurred at CO_2 densities ranging from 0.4 to 0.65 g/cm^3 depending on both the total amount of C_5-through-C_{30} hydrocarbons in the C_5^+ fraction of the reservoir oil and the distribution of hydrocarbons in this carbon number range.

2. Reservoir temperature is an important variable affecting MMP because of its effect on the pressure required to achieve the CO_2 density required for miscible displacement. A higher temperature results in a higher miscibility pressure requirement, other factors being equal.

3. MMP is inversely related to the total amount of C_5-through-C_{30} hydrocarbons present in the crude oil. The more of these hydrocarbons contained in the oil, the lower the miscibility pressure.

4. MMP is affected by the molecular weight distribution of the individual C_5-through-C_{30} hydrocarbons in the reservoir oil. Low-molecular-weight gasoline-range hydrocarbons are particularly effective in promoting miscibility and result in a lower miscibility pressure requirement, other factors being equal.

5. MMP also is affected but to a much lesser degree by the type of hydrocarbons present in the C_5-through-C_{30} fraction. For example, aromatics result in lower miscibility pressure.

6. Properties of the heavy fraction (i.e., $>C_{30}$ hydrocarbons) also influence MMP, although they are not as important as the total quantity of C_{30}^+ material.

7. Development of dynamic miscibility does not require the presence of C_2-through-C_4 hydrocarbons.

8. The presence of methane in the reservoir oil does not change the MMP appreciably.

Yellig and Metcalfe[5] also conducted experiments to determine the effect of oil composition and temperature on MMP. They varied the relative amount of a light fraction (C_1 + CO_2 + N_2), an intermediate fraction (C_{2-6}), and a heavy fraction (C_7^+) in the oil but maintained the same ratio of one component to another within each fraction with the exception of one oil. Like Holm and Josendal, they found only a negligible to a small effect of the light and intermediate-molecular-weight materials in the oil on MMP. Their published experiments did not test the effect of a variation in C_7^+ properties on MMP, but by correlating MMP vs. temperature only and subsequently predicting with this correlation the miscibility pressure for other oils that had been tested in a slim-tube miscibility pressure apparatus and that represented a significant variation of C_7^+ properties, they concluded that C_7^+ properties had a minor effect.

The Holm and Josendal and the Yellig and Metcalfe conclusions concerning the effect of C_5^+ or C_7^+ material on MMP are not in agreement. Moreover, Johnson and Pollin[6] reported observing an effect of oil composition on CO_2 miscibility pressure in their study, while Lee[7] apparently found little effect. From this brief discussion, it should be apparent that there is not yet a consensus concerning the quantitative effect of oil properties on miscibility pressure. However, there does seem to be sufficient evidence that oil composition, primarily the gasoline-and-heavier hydrocarbons, can affect miscibility pressure, even if the circumstances and magnitude of this effect are not defined entirely.

There have been several attempts to develop a correlation for estimating miscibility pressure.[3-11] The simplest

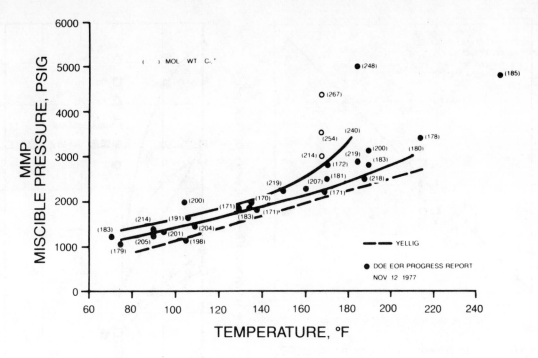

Fig. 8.3—Holm-Josendal dynamic miscibility displacement correlation for CO_2 (after Ref. 4).

method was published by Yellig and Metcalfe[5,12] and is shown in Fig. 8.1. A full-page version of this figure is given in Appendix E as Fig. E-4. In this correlation, reservoir temperature is the only parameter affecting miscibility pressure; oil composition is considered to have only a minor effect. Whenever Fig. 8.1 predicts miscibility pressure to occur below the oil bubble-point pressure, miscibility pressure should be set equal to the bubble-point pressure. The correlation was tested by Yellig and Metcalfe against 35 experimentally determined miscibility pressures, 26 of which were not used to develop the correlation. According to these authors, these oils had large variations in the relative amounts of light ($C_1 + N_2 + CO_2$), intermediate (C_2-through-C_6), and heavy (C_7^+) fractions and contained C_7^+ fractions with significantly different bulk properties and compositions. Temperatures ranged from 94 to 255°F. The correlation of Fig. 8.1 predicted miscibility pressures for all 35 oils within 725 psi and predicted 27 within 200 psi.

Holm and Josendal[3,4] and Cronquist[9] recommended correlations in which both oil composition and reservoir temperature are correlating parameters. The Holm and Josendal method, as extended by Mungan,[10] is shown in Fig. 8.2. A full-page version of this figure is also given in Appendix E, as Fig. E-5. Oil composition is characterized by the molecular weight of the pentanes-and-heavier fraction. According to this correlation, the effect of oil composition on MMP is not large at temperatures as low as 100°F, but the effect of oil composition becomes more pronounced as temperature increases above the 120-to-140°F range. This correlation projects an almost asymptotic increase in MMP with temperature for oils with C_5^+ molecular weight of about 240 or higher.

Holm and Josendal compared their correlation with the experimental miscibility pressure data shown in Fig. 8.3. Although many of the experimental data are in the range of temperature and C_5^+ molecular weight where the correlation does not predict a strong sensitivity to oil composition, several of the data points for high C_5^+ molecular weights and for high temperatures show the compositional effect trend predicted by the Holm and Josendal correlation. The correlation of Yellig and Metcalfe also is shown on this figure for comparison.

In the Cronquist correlation, oil composition is characterized by two parameters: molecular weight of the pentanes-and-heavier fraction and mole percent of methane and nitrogen. This correlation is given below in equation form.[9,*]

PMISC =

$$15.988(TEMP)^{0.744206 + 0.0011038(MWC5P) + 0.0015279(MPC1)},$$

$$\dots\dots\dots\dots\dots\dots\dots\dots\dots\dots\dots\dots\dots(8.1)$$

where

MWC5P = molecular weight of pentanes-and-heavier fraction,

MPC1 = mole percent of methane and nitrogen, and

TEMP = reservoir temperature, °F.

Fifty-eight experimental miscibility pressures contributed from various sources were used to develop the Cronquist correlation.[9] Stock-tank gravities of the test oils ranged from 23.7 to 44°API, reservoir temperatures ranged from 71 to 248°F, and experimental miscibility pressures ranged from 1,075 to 5,000 psi. The average error between experimental and predicted miscibility pressures was 310 psi; maximum error was 1,700 psi. Because the data used to develop the Cronquist correlation came from a variety of sources, the experimental methods used to determine the miscibility pressures were

*Cronquist, C.: personal communication, Gulf Universities Research Consortium (1978).

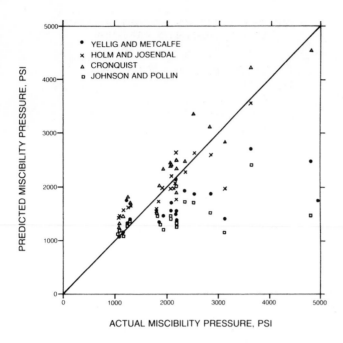

Fig. 8.4—Accuracy of screening guides for predicting CO_2 miscibility pressure.

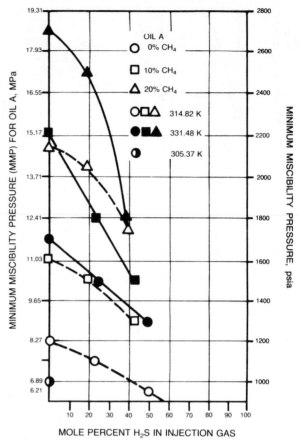

Fig. 8.5—Effect of temperature and impurities on acid-gas MMP (after Ref. 13).

not all the same; even different criteria for defining miscibility pressure were used. This undoubtedly contributed to the difficulty of obtaining a correlation able to predict all miscibility pressures with good accuracy.

Johnson and Pollin[6] correlated MMP vs. temperature and an oil characterization index. The characterization index was itself a function of oil density and effective molecular weight.

Although the various miscibility pressure correlations have some important differences, primarily with regard to the quantitative effect of the C_5^+ fraction of the oil, they also have a number of common features. All the correlations show temperature to be an extremely important parameter. Conversely, they all predict that light ends in the oil such as methane and nitrogen and intermediate-molecular-weight hydrocarbons in the oil, such as ethane, propane, and butane, have a small effect on CO_2 miscibility pressure.

Fig. 8.4 compares predictions of all four correlations against 19 experimental miscibility pressures obtained from displacements in a 20-ft-long, ¼-in.-diameter sandpacked tube. Experimental technique and criteria for defining miscibility pressure conformed to the recommendations in Sec. 2.8. These data were part of the data set used in developing the Cronquist correlation. The test oils were from various geographical locations: west Texas, midcontinent, Rocky Mountains, California, and Alaska. API gravities ranged from 24 to 39°, and reservoir temperatures ranged from 90 to 235°F. When the predicted miscibility pressure for each correlation was less than the oil's bubble-point pressure, it was set equal to the bubble-point pressure as recommended by Yellig and Metcalfe.

For these 19 data points, the average error in the Yellig and Metcalfe prediction was 580 psi with a maximum error of 2,370 psi. This method predicted too low a miscibility pressure for 16 of the 19 oils. The Holm

and Josendal predictions had an average error of 280 psi with a maximum error of 1,160 psi, while the Cronquist method gave an average error of 330 psi with a maximum error of 850 psi. The Holm and Josendal method predicted too high about as often as it predicted too low, whereas the Cronquist method predicted too high a miscibility pressure for 16 of the 19 oils. Average error of the Johnson and Pollin correlation for this data set was 790 psi with a maximum error of 3,400 psi. This method predicted miscibility pressures that were too low for 14 of the oils.

None of the correlations can be used with enough confidence for final design unless there is sufficient margin in the operating pressure to allow for the potential maximum error in the correlation. They should be used only for screening and preliminary design work. When oil composition is available, the Cronquist or Holm and Josendal correlations are preferred. They seem to be of comparable accuracy. The Cronquist method is useful also for conditions beyond the range of the Holm and Josendal chart. When oil composition is unavailable, rough estimates can be made from the Yellig and Metcalfe graph, but according to the comparisons of Fig. 8.4, there is potential for a substantial error.

All of the correlations discussed so far are for high-purity CO_2. Miscibility pressure can be affected substantially by various contaminants in the CO_2.[13,14] This is illustrated in Figs. 8.5 and 8.6 and in Table 8.1, which show the effect on miscibility pressure of im-

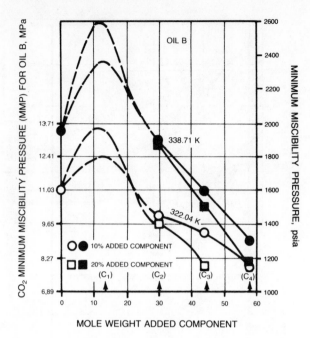

Fig. 8.6—Effect of temperature and mole weight of added component on CO_2 MMP (after Ref. 13).

TABLE 8.1—INFLUENCE OF CONTAMINANTS ON CO_2 MISCIBILITY PRESSURE

Contaminant	Contaminant in CO_2 (mol%)	Miscibility Pressure (psia)
West Texas oil at 109°F (previously unpublished data)		
methane	0	1,200
	10	1,800
	20	2,000
nitrogen	0	1,200
	10	3,300
	20	4,200
Rangely Oil at 160°F (from Ref. 14)		
methane	5	2,100
methane	5 ⎱	
nitrogen	10 ⎰	3,040

8.3 Properties of CO_2 and CO_2 Mixtures

Critical temperature for pure CO_2 is 88°F and critical pressure is 1,071 psia. Although in most miscible floods reservoir temperature will be above the critical value and the CO_2 technically is a gas, it still may be a relatively dense fluid with liquid-like density at the miscibility pressure and above. The densities of oil and CO_2 are similar at many reservoir conditions, and, depending on the particular temperature, pressure, and oil composition, CO_2 can be either less or more dense than reservoir oil. Figs. 8.7 and 8.8 show the compressibility factor and density of pure CO_2 vs. temperature and pressure.[15] Also, see Fig. E-9 in Appendix E for a full-page version of Fig. 8.7.

Compressibility factors for CO_2/methane mixtures are given in Ref. 16, and compressibility factors for a mixture of CO_2 and methane with small amounts of heavier hydrocarbons are found in Ref. 17. In this latter study, the experimental data were tested against values calculated by the corresponding-state method and the Benedict-Webb-Rubin equation of state. These comparisons showed that experimental data are needed for mixtures of CO_2 containing the heavier hydrocarbons, primarily in the critical region, until an improved generalized correlation is devised.

Fig. 8.9 shows the viscosity of pure CO_2 vs. temperature and pressure. A full-page version of this figure is also shown in Appendix E as Fig. E-10. Viscosity values for this figure were calculated from a correlation published by Kennedy and Thodos.[18] At temperatures of interest in miscible flooding, CO_2 viscosity is more gas-like than liquid-like. As a result, oil/CO_2 viscosity ratio will be unfavorable from the standpoint of viscous fingering.

CO_2 is partially soluble in water, and a portion of the CO_2 injected in a miscible flood will be dissolved, either by formation water or by water injected with the CO_2 for mobility ratio improvement. This reduces the volume of CO_2 available for miscible displacement of the oil. Fig. 8.10 gives the solubility of CO_2 in fresh water.[19] CO_2 solubility increases with pressure but decreases with in-

purities commonly found in CO_2 used for oilfield flooding.

Fig. 8.5 shows the effect of methane and hydrogen sulfide impurities in the CO_2 on miscibility pressure with a west Texas oil. Addition of methane to the CO_2 increases miscibility pressure. In this example, addition of 10 mol% methane increased the miscibility pressure at 105°F from about 1,200 psi to 1,600 psi, and addition of 20 mol% methane increased miscibility pressure to about 2,160 psi. Hydrogen sulfide, on the other hand, decreases the miscibility pressure requirement. Twenty mole percent hydrogen sulfide in the otherwise uncontaminated CO_2 lowered miscibility pressure by about 100 psi.

Fig. 8.6 shows the effect of intermediate-molecular-weight hydrocarbons in the CO_2 on miscibility pressure with another west Texas oil. Ethane and heavier hydrocarbons in this example lowered the miscibility pressure requirement, and the largest reduction was observed for the highest-molecular-weight impurity (C_4).

Table 8.1 compares the effect of nitrogen and methane when these materials contaminate CO_2 used for miscible flooding. The examples are for a third west Texas oil and for a Rocky Mountain oil. The data for the west Texas oil show that, as expected, miscibility pressure is higher with a given percentage of nitrogen contaminant than when the contaminant is methane. Whereas 10 mol% methane increased miscibility pressure by 600 psi in this example, the same percentage of nitrogen increased miscibility pressure by 2,100 psi. For the higher temperature and different oil composition of the Rocky Mountain oil, 10 mol% nitrogen increased miscibility pressure by less than half the amount observed for the west Texas oil.

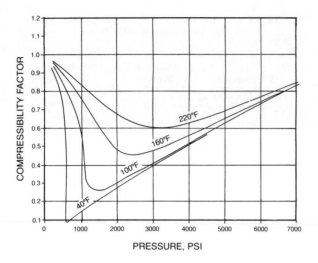

Fig. 8.7—Compressibility factors for CO_2 (after Ref. 15).

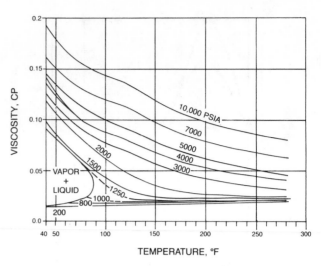

Fig. 8.9—Viscosity of CO_2 (after Ref. 18).

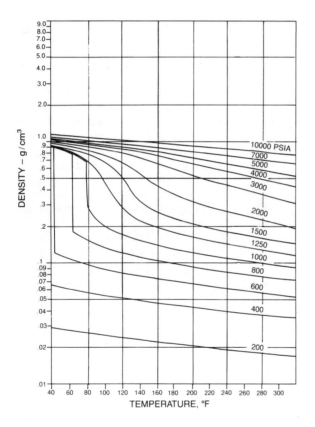

Fig. 8.8—CO_2 density vs. temperature at various pressures (after Ref. 15).

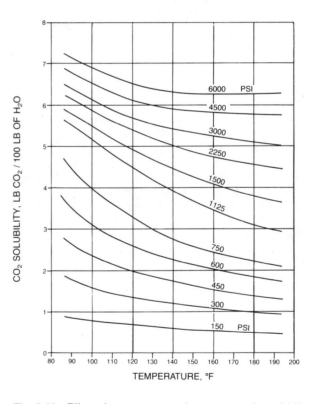

Fig. 8.10—Effect of temperature and pressure on the solubility of CO_2 in water (after Ref. 19).

creasing temperature. Solubility in brine decreases with increasing salinity of the brine, as shown in Fig. 8.11. [20]

Water also is soluble in CO_2. Before transporting CO_2 in a pipeline or reinjecting it in an oilfield flood, enough water must be removed by drying to prevent condensation and corrosion in lines. Fig. 8.12 shows solubility of water in CO_2 for a range of temperatures and pressures. [21]

Some phase behavior and thermodynamic properties of pure CO_2 are shown in Figs. 8.13 and 8.14. Fig. 8.13 shows temperature and pressure conditions necessary for the coexistence of two phases: vapor/liquid, vapor/solid (dry ice), or liquid/solid. The triple point, where all three phases coexist, occurs at $-70°F$ and 75 psi.

When water is present, CO_2 hydrates can form at appropriate temperatures and pressures. Fig. 8.13 also

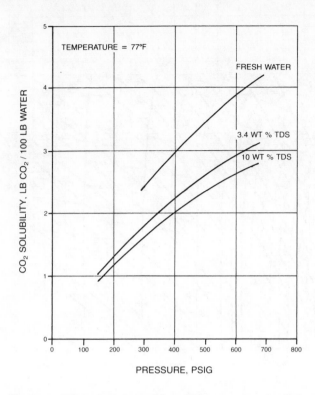

Fig. 8.11—Effect of salinity on CO_2 solubility in water (after Ref. 20).

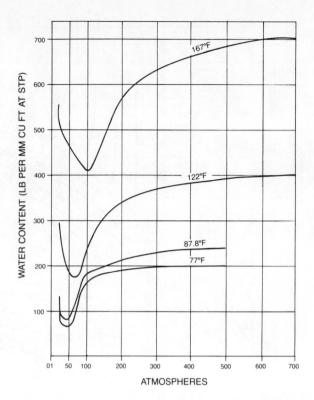

Fig. 8.12—Water vapor content of CO_2 at saturation (after Ref. 21).

shows these conditions.[22] CO_2 hydrates can occur at a temperature as high as 50°F if pressure is greater than 650 psia. Hydrate formation can be troublesome at valves and chokes where pressure is reduced suddenly and CO_2 cools because of expansion.

Fig. 8.14 is a thermodynamic diagram for pure CO_2 showing relationships of pressure, temperature, enthalpy, entropy, and specific volume.[23] This diagram is useful for evaluating conditions that might exist in flow lines, valves, chokes, or wells producing from CO_2 reservoirs. For example, suppose that the bottomhole flowing pressure in a well producing from a CO_2 reservoir is 950 psia and that the bottomhole flowing temperature is 100°F (Point 1, Fig. 8.14). For the assumption of adiabatic expansion of CO_2 up the well and through a surface choke (constant entropy), minimum allowable surface pressure is 600 psia for single-phase flow (Point 2). Below this pressure, liquid CO_2 will form.

Solution of CO_2 in oil swells the oil and reduces its viscosity. Simon and Graue[24] published correlations for estimating CO_2 solubility, oil swelling, and swollen oil viscosity for mixtures of CO_2 and dead oils. Other data on CO_2/oil solubility can be found in Refs. 25 through 30.

8.4 CO₂ Sources

Large volumes of CO_2 are required for oilfield flooding. A major flood in a 100-million-bbl field could require on the order of 50 to 100 billion cu ft of CO_2 over a period ranging from a few years to as long as 10 years or more. A small pilot test, on the other hand, might require a rate

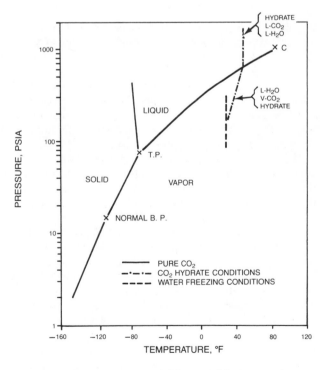

Fig. 8.13—Phase behavior of CO_2 and CO_2/water mixtures (after Refs. 22 and 84).

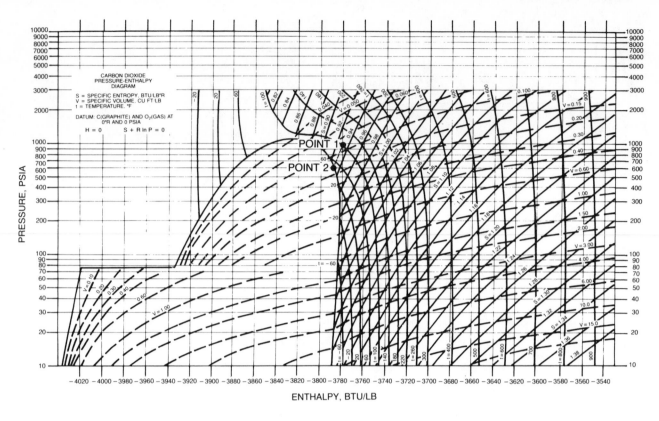

Fig. 8.14—CO_2 pressure/enthalpy diagram (after Ref. 23).

of only 0.5 to 1 million cu ft/D for a time period as short as 1 year. Potential sources of CO_2 include: natural CO_2 deposits; by-product from ammonia plants, other chemical plants, and oilfield acid gas separation facilities; stack gas from power plants; by-product from coal gasification and synthetic natural gas (SNG) plants; and flue gas from cement plants and limestone calcination facilities. CO_2 produced in the course of a flood can be reinjected, after purification, if necessary.

The cost of CO_2 and therefore the feasibility of a source for oilfield flooding depends on (1) the nature of the source, (2) its location relative to the oil field, and (3) the method used to transport the CO_2 to the oil field. Cost at the source can be quite different from one source to another. Natural deposits must be developed by drilling wells and installing producing facilities much as natural gas fields are developed. Exploration costs incurred to find the CO_2 reservoir must be included as well as the costs for field operation, royalty, and for whatever gas compression may be necessary. If power-plant stack gas is the source, the cost of CO_2 will include separation of CO_2 from the stack gas and compression from near atmospheric pressure to the desired delivery pressure. CO_2 from ammonia plants and coal gasification plants also is available at near atmospheric pressure, and compression costs will be incurred for these sources. The price negotiated with the CO_2 owner adds directly to the final delivered cost.

Transportation from source to oil field is usually a major component of CO_2 cost for oilfield flooding. Liquefaction at the source and transportation by refrigerated truck or tank car is feasible for small delivery rates on the order of 1 to 5 MMcf/D. This method of delivery ordinarily is used only for small pilot tests. The liquefied CO_2 is stored on site in insulated tanks and is pumped into the injection lines.[31-34] Vaporization occurs in the injection lines and wellbore.[31] In one pilot test, rail tank cars provided the short-term CO_2 injection-site storage.[32] In another project, located in a southern Louisiana marsh,[33] CO_2 was delivered first by refrigerated truck to an onshore storage site. From there it was transferred to insulated and refrigerated storage tanks located on barges for transport to the project site. The barges also served as the oilfield CO_2-storage facilities.

CO_2 for large, long-term projects is delivered much more economically by pipeline. Because of the relationship between flow rate and line diameter, there is an economy of scale in transmission by pipeline. For a fixed pressure drop, the required line diameter does not increase proportionally with an increase in flow capacity; and although there is some increase in the required wall thickness with pressure and although there are some costs such as surveys, ditching, engineering, etc. that are independent of capacity, the total cost for pipeline construction does not increase proportionally with line capacity.[35] Because of the economy of scale in pipeline transmission, the unit cost of CO_2 delivered to an oil field should decrease as the project size and injection rate increase, other factors being equal, and large projects may be required to economically utilize those CO_2 sources removed very far from the target oil field. For example, CO_2 from a source several hundred miles away may not be economically attractive for a project requiring several tens of millions of cubic feet per day,

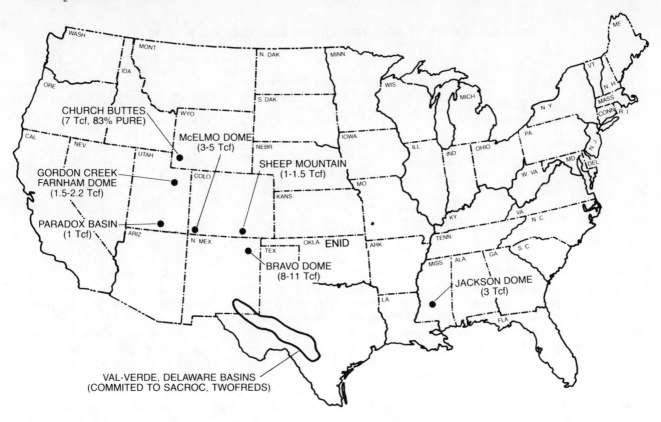

CHURCH BUTTES
(7 Tcf, 83% PURE)

McELMO DOME
(3-5 Tcf)

SHEEP MOUNTAIN
(1-1.5 Tcf)

GORDON CREEK
FARNHAM DOME
(1.5-2.2 Tcf)

PARADOX BASIN
(1 Tcf)

ENID

BRAVO DOME
(8-11 Tcf)

JACKSON DOME
(3 Tcf)

VAL-VERDE, DELAWARE BASINS
(COMMITED TO SACROC, TWOFREDS)

Fig. 8.15—Locations of several natural CO$_2$ deposits (after Ref. 35).

whereas this source may be attractive for a project requiring several hundred million cubic feet per day.

The Canyon Reef Carriers line supplying the SACROC project is an example of a large CO$_2$ pipeline. It consists of 40 miles of 12-in.-diameter pipe and 180 miles of 16-in.-diameter pipe, and has a capacity of 240 MMcf/D. Several papers have described design considerations for the SACROC line.[36-38] CO$_2$ was found to be transported most economically at pressures above the critical pressure so that two-phase flow could not occur. Dehydration was necessary to prevent liquid dropout and to minimize corrosion.

The pipeline for the Slaughter Estate project is an example of a small pipeline built to supply a pilot test with CO$_2$ from a nearby acid gas separation facility.[39] It is only 3 in. in diameter and 7 miles long and is constructed of Schedule 80, ASTM A106 steel because of high concentrations of hydrogen sulfide in the gas.

Further discussion of design considerations and costs for CO$_2$ pipelines can be found in Ref. 35. Operation of the pipeline above the supercritical pressure was found to be preferable generally in this study.

Natural Deposits

CO$_2$ occasionally occurs in reservoirs in nearly pure form or associated with nitrogen or hydrocarbon gases. Fig. 8.15 shows some of the areas where wells have produced nearly pure CO$_2$ or gas with a high concentration of CO$_2$.[8,35,40] The areas with potential for containing natural CO$_2$ deposits generally are concentrated in several western and southwestern states and central Mississippi.

In many of the natural gas fields of the Delaware and Val-Verde basins, CO$_2$ is a contaminant that is removed in treating plants before gas sales. The early CO$_2$ projects in the Kelly-Snyder (SACROC), Crossett, and Twofreds fields utilized much of the available CO$_2$ from this source. Wells in the other geographical areas apparently were drilled in search of natural gas or oil. At the time, CO$_2$ was not a commodity of interest and the potential reservoirs were not developed. Ultimate CO$_2$ reserves that might be contained in the areas indicated on Fig. 8.15 are speculative. Some areas are under development, while others are largely undeveloped.[35,40] One study estimates that the potential could be in the range of 23 to 30 Tcf.[35] Additional deposits could be discovered in the future.

If reserves should be large and well producing rates high, the cost for developing some of these known natural deposits could be favorable relative to the costs for developing impure and/or low-pressure sources such as power plants and coal gasification plants. As these known locations are developed, rising exploratory costs for finding additional CO$_2$ reservoirs will make this source more costly.

Ammonia Plants

CO$_2$ of approximately 98% purity is produced as a by-product of ammonia synthesis from natural gas. This high-quality CO$_2$ requires no further purification and can be used directly for oilfield flooding after compression, drying, and transmission. It is available at a few pounds per square inch gauge, saturated with water. A study by the Pullman Kellogg Co. for the U.S. DOE

TABLE 8.2—RESULTS OF SELECTED CO$_2$-MISCIBLE FLOOD FIELD TESTS

Project	Secondary or Tertiary Recovery	Viscosity Ratio	Slug Size (% HCPV)	Incremental Recovery (% OOIP)	Gross CO$_2$/Oil Ratio (Mcf/STB)	Net CO$_2$/Oil Ratio (Mcf/STB)
SACROC main flood	S	8	12 to 15	7*	6 to 7*	~4.6*
SACROC tertiary pilot test	T	8	10 to 18	3.5	12 to 20	5 to 14
Little Creek**	T	10	160**	18**	24	13.5
Willard-Wasson	T	20	20	8 to 12*	5 to 7*	3 to 4*
Twofreds	T	>25		>2†	26†	?
Mead-Strawn	S	50	25	17	?	?
Slaughter Estate	T	20	26	>16	<5	?
Levelland No. 736	T		>50	>6	<29	

*Estimated with numerical simulators utilizing test data.
**Numbers referenced to total HCPV in test area including pinchout volume.
†After approximately 0.3 HCPV of CO$_2$ injection; CO$_2$ injection continuing.

identified locations and CO$_2$ rates that are potentially available at ammonia plants in several major oil producing areas: Los Angeles basin, Utah, Colorado, Wyoming, Texas, New Mexico, and Louisiana.[35] Information on ammonia plant locations and capacities in other producing areas can be found in Ref. 41.

CO$_2$ is used in urea manufacture, and facilities for making urea sometimes are located at ammonia plants. In these cases, all or part of the CO$_2$ by-product from ammonia synthesis is not available for miscible flooding.

Individual ammonia plants provide only a very limited supply of CO$_2$ for miscible flooding. Supply rates for most plants are small, often less than 1 MMcf/D, but may occasionally be as high as 50 to 70 MMcf/D. If several plants are located in relatively close proximity, as occurs in the Donaldsonville-Geismer and Taft-Luling-Avondale areas of southern Louisiana,[35] there could be an opportunity to feed a common pipeline and to supply a larger rate. For individual plants with small rates, location relative to oil fields suitable for CO$_2$ flooding in general is a problem. However, a few of these plants may be valuable sources for pilot testing or for small-scale flooding of nearby oil fields.

SNG Manufacture

CO$_2$ is produced as a by-product in SNG manufacture. Currently, about 80 MMcf/D are being produced from two SNG facilities in Illinois utilizing naphtha and NGL as feedstocks,[42] but there is a large supply potential from future coal gasification plants. For some proposed coal gasification plant designs, as much as 300 MMcf/D of CO$_2$ could be generated as a by-product. This is about 1 Tcf over a 10-year period. The CO$_2$ is available at relatively high purity but at near atmospheric pressure in the sulfur recovery section of the plant. It will require dehydration and compression to pipeline transmission pressure. The number of plants that might actually be built, their locations, and the time frame for construction are uncertain at this time.

Power-Plant Stack Gas

Power-plant stack gas is the most widely distributed single source of CO$_2$. Depending on power-plant size and type of fuel burned, CO$_2$ from this source could be available at rates ranging from tens to hundreds of million cubic feet per day.[35,42] Locations of power plants and the approximate daily rates of CO$_2$ available

in the stack gas of each plant are given in Ref. 35 for the same oil-producing states listed previously for ammonia plant CO$_2$ sources. Information for other states is given in Refs. 43 and 44.

CO$_2$ is present in power-plant stack gas at low concentrations in the range of 6 to 16%, depending on whether the plant is fueled by gas, oil, or coal. Other constituents of the gas include nitrogen and oxygen in addition to fly ash and oxides of sulfur, if the plant is coal-fired. The stack gas is available at essentially atmospheric pressure. For oilfield flooding, the CO$_2$ must be purified, dehydrated, and transported to the field.

Methods for recovering CO$_2$ from flue gas include chemical or physical absorption with appropriate solvents and subsequent regeneration from the solvents, dry bed adsorption, and cryogenic separation. A study by the Pullman Kellogg Co. concluded chemical absorption with a monoethanolamine solution (MEA) probably would be the most cost effective of the various methods.[35,42] Problems potentially can arise because some of the constituents of flue gas such as sulfur oxides, even if present in small concentrations after bulk removal, can degrade the MEA solution. The potential severity of this problem as well as designs for reclaiming the MEA have not been established. Some technology development could be required. If MEA reclamation proves costly and if, in addition, bulk sulfur oxides removal would not otherwise be required and must be charged to the CO$_2$ separation plant, flue gas could be a costly CO$_2$ source.

Other Sources

Potential supply from other sources is small and most likely will be uneconomical unless found exceptionally close to a candidate oil field. By-product streams from hydrogen plants located at oil refineries are pure enough for oilfield use and need no further upgrading except for dehydration and compression. At acid gas separation plants, the CO$_2$ may be contaminated with H$_2$S. Otherwise, it will be highly concentrated. As with power-plant stack gas, flue gas from cement plants and limestone calcination contains CO$_2$ at a low concentration, and the CO$_2$ must be separated by the same methods mentioned for stack gas. CO$_2$ concentration also is low in the by-product stream from ethylene oxide and acrylonitrile plants and needs to be purified before use for miscible flooding.

Fig. 8.16—Locations of selected CO_2-miscible flooding field projects.

8.5 Field Experience

At least 33 field trials of this process have been undertaken as of April 1982, and with one exception, all were started in the 1970's or later. Many are still in progress.[45] Table 8.2 gives details of selected projects, and Fig. 8.16 shows locations of some of the projects to date. Most projects were small-scale pilot tests of less than 100 acres, although there have been at least three projects that were large enough in size to be considered commercial-size floods rather than pilot tests.[46-48] In some situations, CO_2 may improve oil recovery even when miscibility pressure cannot be attained. There have been at least three field tests where CO_2 was injected well below the miscibility pressure.[49,50,86]

Unlike the hydrocarbon miscible processes, the majority of CO_2 projects have tested tertiary recovery, reflecting the increasing need in the U.S. for tertiary recovery methods to prolong field life. The largest project, however, in the SACROC Unit of the Kelly-Snyder field is a secondary recovery flood.[47] CO_2 was injected continuously or in very large slugs in at least five projects, similar to continuous injection of the high-pressure gas solvent in vaporizing-gas drive flooding.[32,46,48,51,52,86] Moderate slug sizes were injected in the other projects, but in many of these tests no attempt was made to achieve miscible displacement at the trailing edge of the slug.[1,31,47,50,53-57] Instead, the CO_2 slug was driven immiscibly with water, leaving a residual CO_2 saturation in the reservoir. Alternate injection of water with the CO_2 slug to improve sweepout was tried in about half the field projects.[1,40,47,50,54-58] CO_2 misci-

ble flood field projects have been about equally divided between sandstone and carbonate reservoirs.

Extremely heterogeneous as well as reasonably homogeneous reservoirs have been included. Most of the floods have been in low-relief, essentially horizontal reservoirs, although at least three projects were carried out in high-relief reservoirs and were designed to be gravity-stable displacements.[33,52,86]

Gravities of the oils tested generally have been in the range of 30 to 50°API. Viscosities normally have been low, less than about 2 cp.

Short summaries of several projects are given in the following.

SACROC Unit, Kelly-Snyder Field[36-38,47,58,59]

Main Project. This secondary recovery flood operated by Chevron U.S.A. Inc. is by far the largest CO_2 miscible project. More than 33,000 acres of this 50,000-acre, 2.1-billion-bbl unit are being flooded with CO_2, and the remaining acreage is being waterflooded. Injection is into the 6,700-ft-deep Canyon Reef formation, an algal carbonate reef with massive carbonate buildup along a northeast-southwest trend and with thinner, gently sloping flanks. Formation thickness varies from an average of 800 ft on the crest to less than 50 ft on the flanks and averages 268 ft. The carbonate mass is not a reef in the classic sense, but rather a series of shelf limestones, an important distinction because of the difference in distribution of porosity in a reef and in shelf limestones. The reservoir is stratified into layers of porous and dense zones that are not continuous areally. Average permeability is 19 md and average porosity is 3.9%. The reservoir contains 42°API gravity oil of 0.35-cp viscosity.

From discovery in 1948 until 1953, production resulted primarily from solution gas drive. Early performance showed that ultimate primary recovery would be less than 20% OOIP. Water injection began in 1953 into centerline injectors located along the crest of the structure to flood outward from the crest toward the flanks of the reservoir. Calculations showed ultimate combined recovery would be less than 44%. Because of poor reservoir continuity between the crestal waterflood and the flank areas, pressure continued to decline in the flank areas, and in Jan. 1972 CO_2 injection began in the first of three project areas in downstructure portions of the field that had not responded to waterflooding.

The Phase 1, 2, and 3 CO_2 floods encompassed areas of approximately 11,000, 9,000, and 13,000 acres, respectively. The injection pattern was approximately a 160-acre inverted nine-spot with injection wells located off center of the pattern. Miscibility pressure was determined from slim-tube tests to be 1,600 psi. Viscosity ratio between oil and CO_2 was eight.

The CO_2 came from natural-gas treating plants in the Delaware basin of Texas where it was separated from the natural gas stream before being transported to SACROC in a 220-mile, 16-in.-diameter pipeline. Injection rates averaged about 150 MMcf/D.

In the Phase 1 area, a 3.7% HCPV slug of water was injected ahead of the CO_2 to repressure above the 1,600-psi level. Initially, water and CO_2 were injected in small, alternate slugs in a 0.47 res bbl ratio. Because of rapid CO_2 breakthrough, after about 1.6% HCPV CO_2 was injected, the injection ratio was increased to 1:1 and ultimately to 3:1. Total CO_2 injected into this area amounts to 14.3% HCPV. After total slug injection, CO_2 is being driven through the reservoir by continuous water injection.

CO_2 injection into the Phase 2 and 3 areas began in March 1973 and Nov. 1976. A CO_2 slug size of 12% HCPV was targeted for these areas. A small slug of water preceded CO_2 injection into Phase 2, but 22.1% HCPV water had been injected in Phase 3 before CO_2 injection began, and reservoir pressure was 2,696 psi at the time. Water/CO_2 injection ratios in these areas varied from 1:1 to 3:1. CO_2 breakthroughs also occurred in the Phase 2 and 3 areas after approximately 1 to 2% HCPV of CO_2 injection.

Although rapid CO_2 breakthrough was experienced in this flood, subsequent CO_2 production was not excessive. Alternate water/CO_2 injection and a well program to control CO_2 injection into selected zones were successful in moderating CO_2 production. Three plants were built to separate CO_2 from produced gas and to reinject it. The operator expects about 23% of the CO_2 to be produced ultimately. Reinjected CO_2 will account for about 17% of the total CO_2 injected.

Analysis of flood behavior shows that incremental recovery will amount to about 88 million STB or 6.7% OOIP in the pattern area. Gross CO_2 requirement is expected to be about 6.1 Mcf/STB incremental oil.

Tertiary Recovery Test.[53] A small tertiary-recovery pilot test was completed in a waterflooded area of the field. This test consisted of two adjacent 40-acre, normal five-spot patterns. Gross thickness of the reef in the pilot area was about 200 to 250 ft. For the pilot test, an attempt was made to isolate an interval approximately 80 ft thick that well tests showed produced 100% water. Because of the unconfined nature of the test and the attempt to run the test in an isolated section of the gross interval, there was considerable uncertainty concerning the amount of CO_2 that went into the test pattern and the loss of displaced oil by crossflow to adjacent intervals. Subsequent analysis with a streamtube model showed that CO_2 captured by the test patterns could have ranged from a low of 32% of the total injected to a high of 57%. If so, average CO_2 slug size for the two patterns was in the range of 10 to 18% HCPV. The CO_2 slugs were injected continuously, without alternate water injection, and then were displaced with water.

Oil recovery was only 3.4% OOIP in one of the five-spots and was 2.4% in the other. CO_2 production ranged from 32 to 58% of the cumulative CO_2 injected, depending on the volume of CO_2 that actually invaded the test patterns. CO_2 utilization for the test conditions was in the range of 12 to 20 Mcf gross CO_2 injected into the two patterns per incremental stock-tank barrel of oil recovered. The operator projected that a 30% HCPV slug in the best performing pattern would have recovered a maximum of 5.75% OOIP, with a CO_2 utilization of 15 Mcf/STB.

North Cross (Devonian) Unit, Crossett Field[48,58,60]

Shell Oil Co. operates this secondary recovery project in the 1,700-acre North Cross Unit, Crane County, TX. The flood is in the 5,400-ft-deep Devonian formation, a chalky, siliceous carbonate composed of microscopic tripolitic chert grains bound by limestone cement. The formation is characterized by high porosity, averaging 22%, and low, uniform permeability, averaging 5 md. Pay continuity between wells is very good. Reservoir oil is a 44°API-gravity, 0.42-cp viscosity crude.

The field was developed on regular 40-acre spacing and produced by solution gas drive until it was unitized in 1964. At this time, a program of partial pressure maintenance by updip residue gas injection began. Waterflooding was not attractive because of the low permeability to water, and in April 1972, CO_2 injection began through six wells downdip in an inverted nine-spot pattern arrangement. At that time, reservoir pressure was still generally above the 1,650-psi miscibility pressure, although pressure was below the crude's original bubble point, and a free gas saturation was present. CO_2 comes from the SACROC pipeline, and total injection rate has averaged 17 MMcf/D. Oil/CO_2 viscosity ratio is five.

The project is still in the early stages. Original plans called for injection of approximately 40% HCPV of pure CO_2 with reinjection of produced gases. No water is being injected with the CO_2 for mobility reduction, and there are no plans to inject water ultimately to drive the slug. By Nov. 1976, 15% HCPV CO_2 had been injected. Response to the flood was characterized by increased oil rate and declining GOR's in producers adjacent to CO_2 injectors. In one producer, oil rate increased from 40 to 300 BOPD and GOR decreased from 8,000 scf/bbl to 1,300 scf/bbl. Producing-well response has occurred on average after about 10% HCPV injection followed by CO_2 breakthrough on average after 16% HCPV injection. In early 1978, less than 2% of the total CO_2 injected had been produced, although monthly production was 8 to 10% of monthly injection. CO_2 injectivity has been less than originally anticipated at the start of the flood. The operator speculated that injectivity was reduced because of the CO_2/oil phase behavior and miscibility mechanism, which combined to leave a small saturation of heavy hydrocarbon in the reservoir.

Little Creek Field[32,58,61]

The Little Creek project was a tertiary recovery pilot test operated by Shell Oil Co. in Lincoln and Pike Counties, MS. Injection was into the 10,750-ft-deep lower Tuscaloosa sandstone, a combination structural/stratigraphic trap containing 102 million bbl OOIP. The reservoir is a series of sand bars deposited in an ancient river and has large-scale crossbeds of medium sand at the base that grade upward into small-scale laminations of very fine sand at the top. The formation has average porosity of 23%, average permeability of 33 md, and contained 39°API crude with a viscosity of 0.4 cp at the 248°F reservoir temperature. Formation volume factor was 1.32 RB/STB. Average waterflood residual oil saturation was 21%. The reservoir was produced by solution gas drive with limited water encroachment until 1962, when a peripheral line-drive waterflood was begun. When waterflooding was discontinued in 1970, cumulative recovery was almost 46% OOIP.

Testing of the CO_2 miscible process began in Feb. 1974. The test pattern was a semiconfined, 31-acre quarter nine-spot symmetry element. Three producers surrounded a single injector to the north and west, and the reservoir shale-out boundary confined fluids to the east and south. Backup water injectors also helped confine fluids to the test area. Reservoir thickness in the pilot test area varied from 10 to 30 ft. CO_2 was delivered by truck and rail cars from a commercial source. Viscosity ratio of the flood was about 10, and CO_2 miscibility pressure was 4,500 psig.

The pilot test area was repressured with water from 4,400 psig to 5,100 psig before almost 1.6 HCPV* of CO_2 was injected continuously, without alternate water slugs, at a rate of approximately 3 to 4 MMcf/D. Produced gas was reinjected. Finally, water was injected into both the CO_2 injector and a former producing well to drive any remaining banked-up oil into the other producers. A definite tertiary oil bank was produced in all pilot producers. CO_2 breakthrough occurred in all wells within 1 to 6 months after breakthrough of the tertiary oil bank, and most of the tertiary oil was produced after CO_2 breakthrough. There was evidence of preferential fluid movement in an east/west direction. When CO_2 injection was stopped, cumulative production amounted to 58% of the waterflood residual oil estimated for that portion of the test area bounded by the pilot injector and producers (or 42% of the waterflood residual oil if pinchout volume is included). At that time, 53% of the injected gas had been produced and recycled. Oil rate continued a rapid decline with subsequent water injection. Ultimate

*Based on total HCPV in test area, including pinchout volume.

production was projected to be as high as 65% of the waterflood residual oil in the area bounded by pilot injector and producers.

Mead-Strawn Field[31,62]

This Union Oil Co. project in Jones County, TX, was one of the early secondary-recovery pilot tests with CO_2. Injection was into the Strawn sand at 4,500 ft, a poorly developed lenticular sandstone of fine to very fine grain containing abundant, thinly interbedded shales and subordinate, thin limestone stringers. Solution gas furnished the primary drive mechanism. The test pattern was a slightly irregular, 33.5-acre normal five-spot and an adjacent area of about 9 acres drained by one producing well. The test area was surrounded by dry holes to the east, north, and west and by two water injection wells to the south and southwest that helped confine fluids within the test area. Average reservoir thickness in the test area was 10.3 ft. Average porosity was 11%, and average air permeability was 9 md. The reservoir contained a 41°API crude with a viscosity of 1.3 cp at original reservoir conditions.

Pressure in the test area had declined to 115 psig at the start of CO_2 flooding. After injecting water to raise the reservoir pressure, CO_2 injection began in Dec. 1964. Viscosity ratio between CO_2 and oil was approximately 50, and miscibility pressure was close to 1,800 psi. A 25% PV slug of CO_2 was followed by carbonated water and then by brine. Pressure at the flood front may have been below the miscibility pressure by several hundred pounds per square inch during part of the test.

The production wells affected by CO_2 flooding did not respond dramatically. However, production from these wells was characterized by higher oil flow rates that did not decline as rapidly as the rates of wells affected only by an adjacent waterflood. Although there was evidence of water channeling during repressuring, less than 5 to 10% of the injected CO_2 was produced during the displacement phase of the test. By comparing recovery from the pilot test producers with recovery from other wells that had been waterflooded only, the operator concluded that approximately 53 to 82% more oil was produced by the CO_2 flood. This amounted to an incremental recovery of roughly 6 to 11% PV. Two cores, one located at 60 ft and the other at 400 ft from the injector, were taken with the pressure-retaining core barrel. Average residual oil saturations of about 10 and 5% PV, respectively, were found in the CO_2-flooded zone. The operator also reported that about 1 bbl of oil was produced per reservoir barrel of CO_2 injected through subsequent pressure depletion.

Willard Unit, Wasson Field[54,63-66]

Phase 1. ARCO Oil & Gas Co. ran this test in Gaines County, TX, to evaluate CO_2 flooding in the San Andres carbonate formation. The San Andres formation at Wasson field is highly stratified with good porosity zones separated by dense intervals a few inches to a few feet thick. The dense, low-porosity zones act to restrict vertical movement of fluids. Both good porosity and dense zones are correlative for several well distances. The reservoir occurs at a depth of 5,200 ft, is 59 to 230 ft thick, and has an average porosity and permeability of

8.5% and 1.5 md. It contains a 32°API gravity oil that has a viscosity of 1.3 cp at reservoir conditions.

The Phase 1 test was about 425 acres in size. CO_2 was injected into eight wells that were on 20-acre well spacing, which was normal for the unit. The Phase 1 area had been partially waterflooded with about 22% HCPV of water injection before the CO_2-flood test began. Viscosity ratio in this flood was about 20. Miscibility pressure was 1,200 to 1,250 psi.

CO_2 injection began in Nov. 1972. The CO_2 source was acid gas effluent from the Wasson gas sweetening plant. This gas contained almost 4% H_2S. Small slugs of CO_2 and water were injected alternately in a ratio that varied from about 1:1 to 1.5:1 on a reservoir volume basis in an effort to improve mobility ratio. Because of start-up problems with injection wells and the CO_2 compressor, injection pressure exceeded fracturing pressure for several months early in the project life. This caused some early CO_2 breakthrough in wells both within and outside the test area that were located along the direction of preferred fracture orientation from the CO_2 injectors. However, subsequent CO_2 production was minimized by reducing injection pressure and by alternate injection of water. The project was shut down prematurely in Feb. 1975 because of an accident involving escape of the H_2S-containing gas. At that time, only 4.4% HCPV CO_2 had been injected (6.8% in one well) out of a planned 20% HCPV slug. A measure of incremental oil recovery was not obtained, but only 2.5% of the cumulative gas injected had been produced at this time despite the early extension of fractures in injection wells for a short period of time. The operator concluded from the CO_2 production behavior that there was no evidence of severe gravity segregation or areal sweep problems in this heterogeneous reservoir.

Minipilot Test. A small, three-well pilot test also was run to evaluate unit displacement behavior and CO_2 flow behavior in this stratified reservoir. Saturation changes occurring in the responding strata were measured by periodically logging a cased observation well located 100 ft from a CO_2 injector with compensated neutron and pulsed neutron logs. Pressure measurements and production tests were made in another well located 25 ft from the observation well.

At the start of the test in Aug. 1972, the reservoir in the vicinity of the three test wells was in a state of solution-gas-drive depletion. Water was injected until Dec. 1973 to waterflood the observation well thoroughly, at which time injection of small CO_2 slugs was alternated with injection of small slugs of water in a 1:1 reservoir volume ratio. After Feb. 1975, water was injected continuously, and in May 1976, cores were taken by the pressure core barrel technique at a location between the injector and observation wells.

Variable oil saturations were found in strata that had been swept by CO_2, ranging from about 3% PV to as high as 30% PV. From a preliminary analysis of test results using a miscible flood simulator, the operator projected that incremental oil recovery in a full-scale flood could be on the order of 12% of the initial oil in place in a typical pattern for injection of a 20% HCPV slug of CO_2 driven by water.

Twofreds [46]

HNG Fossil Fuels Co. operates this tertiary recovery project in Loving and Ward Counties, TX. CO_2 is being injected continuously through 14 wells into the 4,800-ft-deep Delaware sandstone. The reservoir contains approximately 4,400 productive acres with average pay thickness of 25 ft, 19% porosity, and 33-md permeability. OOIP is estimated to be 53 million bbl. Crude oil gravity is 36°API.

Injection of CO_2 from a separation plant in a nearby gas field began in the east 800 acres in Feb. 1974. At that time the average WOR was about 52. Oil production rose steadily from 27 B/D to about 520 B/D in early 1978. At that time CO_2 breakthroughs had occurred, but CO_2 production was still relatively minor. The project was still in an early stage when, after injection of roughly 0.26 HCPV CO_2, approximately 3% OOIP of incremental oil had been recovered.

Slaughter Estate Unit [1,58,67]

This tertiary recovery pilot test operated by Amoco Production Co. in Hockley County, TX, was conducted in the 5,000-ft-deep San Andres formation, a stratified carbonate of dense and permeable layers that appear to be continuous in the Slaughter Estate area. The formation in the test area has average porosity of 11.3%, average permeability of 5.8 md, and average net thickness of 77 ft. It contains a 32°API gravity oil of 2-cp viscosity at the 105°F reservoir temperature.

The pilot test consisted of two adjacent five-spot patterns located in an area of the field that had been produced by solution gas drive but never waterflooded. Each five-spot was 6 acres. From Nov. 1972 to Aug. 1976, the two patterns were waterflooded in preparation for the tertiary recovery test. Peak oil rate from the two center producers reached 400 BOPD and had declined to about 50 BOPD by the time the CO_2 injection began in Aug. 1976. The CO_2 stream was obtained from a nearby gasoline plant and contained 28% hydrogen sulfide. This CO_2/H_2S mixture was injected alternately with water in a 1:1 reservoir volume ratio until a total 26% HCPV slug had been injected by late 1979. The slug was driven with nitrogen, also injected alternately with water in a 1:1 ratio.

This has been one of the best-performing CO_2 flood tests. Peak tertiary oil rate reached 155 B/D, and incremental recovery has reached about 16% OOIP after roughly 0.8 HCPV of total fluid injection. The operator projects that ultimate incremental recovery will be no less than 20% OOIP.

CO_2 breakthrough occurred after about 10% HCPV of slug had been injected (approximately 20% HCPV total fluid). By the time incremental recovery had reached 16% OOIP, approximately 40% of the slug had been produced, and about 63% of the drive gas had been produced.

Other Field Tests

Amoco also has initiated small pilot tests in the Levelland and North Cowden fields in the San Andres and Greyburg formations of west Texas. [58] In each field the CO_2 process is being tested by two adjacent, normal five-spot patterns of about 12 acres in addition to smaller, single five-spot patterns of about 1.5 acres. After about 50% HCPV of CO$_2$ injection in the 1.5-acre Levelland No. 736 test, incremental recovery was about 6% OOIP. These tests are not expected to be completed until the early 1980's.

Shell Oil Co. conducted a logging and coring minipilot test in the Denver Unit of Wasson field, similar to the one run by ARCO in the Willard Unit of this field. Logging observation wells and post-test coring were used to determine the areal distribution of oil saturation left around a central injector after CO$_2$ was injected continuously without alternate water slugs. The test area was waterflooded before CO$_2$ injection.

In Oct. 1978, Shell also began a small tertiary recovery test in the Weeks Island field in south Louisiana.[58,68] Updip CO$_2$ injection was designed to accomplish a gravity-stabilized downdip displacement. This test appears to be an example of near-miscible displacement with CO$_2$.

CO$_2$ also was injected into a dipping sandstone reservoir in south Texas by Amoco Production Co. The CO$_2$ was injected at the gas/oil contact in the South Gillock field, followed by downdip oil displacement.[86] CO$_2$ injection has ended in this project, but results have not been published.

Texaco Inc. initiated another gravity-stable CO$_2$ flood in 1981 in the Bay St. Elaine field in southern Louisiana.[33] This field is located in the coastal marshes, and the CO$_2$ both is transported to the field by barge and is stored on-site on barges. It was necessary to add methane to the CO$_2$ to decrease density to the point where a gravity-stable displacement was feasible. A small amount of butane then was added to lower miscibility pressure to the desired level. Current plans call for using nitrogen as a drive fluid.

Three small tertiary recovery tests were run in the Granny's Creek,[55] Rock Creek,[56] and Griffithville[57] fields in West Virginia, and a small secondary recovery pilot test was run in the Hilly Upland[69] field, also in West Virginia.

8.6 Assessment of Process
General

Although field behavior to date has been encouraging, CO$_2$ miscible flooding projects have not yet produced sufficient data to evaluate this method's true potential. This situation should improve in the future as more and larger tests are completed.

Generalization of test results is made difficult by the wide variety of test conditions. Not only are floods being conducted in reservoirs of markedly different degrees of heterogeneity, but different slug sizes and even different versions of the process are being tested. Although optimal slug sizes, CO$_2$/water injection ratios, and preferred operating methods have not been established unambiguously from field testing to date, some important information and insights have been gained from the available data and from analysis of these data.

Like the hydrocarbon miscible processes, CO$_2$ miscible flooding is a method for both secondary and tertiary oil recovery. The Crossett project demonstrated that oil can be displaced and banked by CO$_2$ flooding in a reservoir containing a gas saturation at the start of CO$_2$ injec-

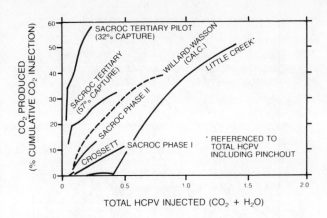

Fig. 8.17—CO$_2$ production in field tests (after Ref. 85).

tion.[48] In this field, pressure had been partially maintained by residue gas injection, and response at producers to the CO$_2$ injection was characterized by a decrease in GOR and an increase in oil producing rate. The ability of CO$_2$ to displace and to recover some of the residual oil left after waterflooding in the reservoir as well as in laboratory corefloods has been demonstrated by such tertiary recovery field tests as the Little Creek,[32] Slaughter Estate,[1] Levelland,[34,58] and SACROC[53] pilot tests and by the Twofreds project.[46]

CO$_2$ Breakthrough and Production

Most projects, both secondary recovery and tertiary recovery floods, have experienced early CO$_2$ breakthrough, usually after injection of 0.05 to 0.2 HCPV of total fluid (CO$_2$ or CO$_2$ plus water). This also is typical of behavior observed in field tests of miscible flooding with hydrocarbon solvents; but in most cases subsequent CO$_2$ production has not been excessive, and corrective measures such as alternate water injection,[1,47,54] zonal isolation,[47] and reducing the injection pressure[54] have been partially successful in moderating CO$_2$ production.

Fig. 8.17 compares the CO$_2$ breakthrough and cumulative CO$_2$ production behavior experienced in the SACROC, Crossett, and Little Creek floods. Also shown on this figure is the CO$_2$ production projected for a flood in the Willard Unit of Wasson field. The Willard calculation was made with a miscible flood simulator and utilized information obtained in the logging/coring pilot test run in this unit.[63,64] Relatively early CO$_2$ breakthrough occurred in all these tests. In the SACROC Phase 1 and 2 areas, breakthroughs in some patterns occurred after only a few percent HCPV total fluid had been injected and generally occurred before 0.1 HCPV of injection. Two curves are shown on this figure for the SACROC tertiary recovery pilot test, reflecting the uncertainty in estimating how much of the CO$_2$ slug actually invaded the test pattern. Breakthrough of CO$_2$ was exceptionally rapid in this test, occurring after injection of only a few percent of the HCPV of the double five-spot. Breakthrough occurred before 0.1 HCPV injection in the Crossett flood. Breakthrough from pattern sweepout had not been observed after 0.06 HCPV of combined CO$_2$ plus water injection in the Wasson-

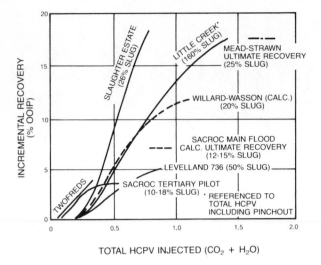

TOTAL HCPV INJECTED (CO_2 + H_2O)

Note: These curves are the author's estimates of project performance based on production data. They are not the official interpretations of the project operators.

Fig. 8.18—Incremental oil production in CO_2 field tests (after Ref. 85).

Willard Phase 1 area but was projected to occur at about 0.1 HCPV injection. At Little Creek, CO_2 reached one of the producers after 0.2 HCPV injection. This producer then was shut in until breakthrough occurred in the other two producers after approximately 0.4 HCPV injection. CO_2 breakthrough in the Slaughter Estate test[1] (not shown in Fig. 8.17) was observed after approximately 0.1 HCPV of CO_2 injection (approximately 0.2 HCPV of CO_2 plus water).

Ultimate CO_2 production depends on slug size as well as such factors as reservoir heterogeneity, mobility ratio, well pattern, and method of operation. Production of a significant fraction of the slug should be anticipated, perhaps in the range of 20 to 60%. Relatively minor production of CO_2 was observed only in the Mead-Strawn pilot test (less than 10% of the slug). Other factors being equal, more CO_2 will be produced the larger the slug size injected. Produced CO_2 can be compressed and reinjected, economics permitting. However, in some cases facilities may be needed to extract it from the total gas stream if the level of contamination would interfere with displacement efficiency.

Cumulative CO_2 production in Fig. 8.17 is expressed as percent of cumulative CO_2 injection. In the SACROC Phase 1 and 2 areas, cumulative production is projected to reach 23% of the 12 to 14% HCPV slug. A much higher level of production occurred in the SACROC tertiary recovery pilot, amounting to between 32 and 58% of the slug, depending on the fraction of injected CO_2 actually entering the pattern; but the increased CO_2 production was not necessarily caused by flooding in an area of high water saturation. The Canyon Reef formation at SACROC is extremely heterogeneous, and the tertiary recovery pattern may have been located in an area of greater-than-average stratification. (Some wells in the main-project secondary-recovery area also had very rapid breakthrough and excessive CO_2 production.) Although CO_2 production at Little Creek exceeded 50% of cumulative injection, a very large slug of approx-

imately 1.6 HCPV was injected. Projections showed that between 30 to 40% of a 20% HCPV slug would be produced during a full-scale flood at Willard-Wasson.[65]

Oil Recovery

Because waterflooding will be an alternative to CO_2 flooding in most cases, incremental rather than total recovery is an important measure of CO_2-flood effectiveness. Eq. 8.2 shows the relationship of factors that influence incremental recovery in a miscible flood.

$$\Delta N_p = V_p \cdot E_v \cdot E_c \cdot \left(\frac{S_{orw}}{B_{ow}} - \frac{S_{orm}}{B_{om}} \right), \quad \ldots \ldots \ldots (8.2)$$

where

ΔN_p = incremental oil recovery, STB,
E_v = miscible sweepout, fraction,
E_c = capture efficiency for incremental oil, fraction,
V_p = pore volume, res bbl,
S_{orw} = waterflood residual oil saturation, fraction,
S_{orm} = miscible flood residual oil saturation, fraction,
B_{ow} = oil formation volume factor for waterflood residual, RB/STB, and
B_{om} = oil formation volume factor for miscible flood residual, RB/STB.

Incremental recovery is affected by volumetric sweepout of the CO_2, by swept-zone residual oil saturations left both to waterflooding and CO_2 flooding, and by crossflow and resaturation, which reduce the fraction of displaced oil actually produced. It also is affected by the CO_2 slug size, effective mobility ratio of the flood, fluid used to drive the slug (gas or water), degree of heterogeneity, and well pattern, all of which affect sweepout and oil capture efficiency. Hydrocarbon liquid recovery, of course, is related directly to the FVF. Since the parameters in Eq. 8.2 vary from reservoir to reservoir, a range of outcomes should be expected from the CO_2-flood field tests.

There are only a few field tests to date from which ultimate incremental recovery can be inferred, either from direct measurement or from analysis of production data. Results of these projects are given in Table 8.2 and Fig. 8.18. Incremental recovery has ranged from a low of about 3.5% HCPV to a high of about 18% HCPV,* or, expressed in terms of PV rather than HCPV, it has ranged from 3 to 12% PV.

The Slaughter Estate tertiary recovery pilot test has been by far the best performer.[1,58,67] Incremental recovery in the 12-acre test was reported to be about 16.0% OOIP and increasing for only a 26% HCPV CO_2 slug.[67] A high incremental recovery, about 18% OOIP, also was reported for the Little Creek pilot test, although a very large slug size was injected amounting to about 160% HCPV.[32,51] In contrast, incremental recovery for the 80-acre SACROC tertiary recovery pilot test was only 3.5% OOIP for a 10 to 18% HCPV slug.[53] After 50% HCPV of CO_2 injection in the Levelland No. 736 tertiary recovery test, incremental recovery was only 6% OOIP.[58] The relatively low recovery for this latter proj-

*Incremental recovery at Little Creek was 18% if referenced to the total HCPV including pinchout volume, and 28% if referenced to the area bounded by pilot injector and producers.

ect for the large slug size is attributed by the project operator to an unexpectedly low oil saturation in the thoroughly waterflooded pilot location. The largest field trial to date, the SACROC main flood,[47] was a secondary recovery flood, and there was no direct measure of incremental oil. Simulator projections, after history matching performance, indicated an ultimate recovery of about 7% OOIP for slug sizes that varied between 12 and 15% HCPV.

The bulk of incremental oil production in the CO_2 flooding tests, as in the hydrocarbon miscible projects described in Chaps. 5 through 7, occurred after CO_2 breakthrough; the incremental oil was produced concurrently with CO_2. In the tertiary recovery floods, CO_2 breakthrough occurred shortly after or practically coincident with first production of the tertiary oil bank. This behavior undoubtedly is caused by a combination of viscous fingering, gravity segregation/override, channeling caused by stratification, and crossflow of the oil bank. The relative importance of these phenomena varied from project to project, but, as shown in Figs. 8.17 and 8.18, the general flood character of relatively early CO_2 breakthrough followed by a prolonged period of production of incremental oil along with CO_2 should be anticipated for the typical project.

Residual Oil Saturation

Oil saturation left behind the CO_2 front was determined in a few projects by coring with the pressure-retaining core barrel. This is a coring tool designed to capture the core and bring it to the surface at formation pressure, thereby preventing expulsion of core fluids by pressure depletion.[70] All these tests showed the oil saturation to be reduced below the residual saturation expected for waterflooding, although the magnitude of saturation reduction varied appreciably from one core to another and even between different strata in the same cores. In the Mead-Strawn sandstone, average saturations of 10 and 5% PV stock-tank oil (STO) were found at locations 60 and 400 ft away from a CO_2 injector.[31] A wide variation in oil saturation was found in the different strata swept by CO_2 in the stratified San Andres carbonate at Willard-Wasson. These saturations ranged from 3 to 30% PV STO at a location only 35 ft from a CO_2 injector.[64] Residual oil saturations of 2.5 to 4.9% STO were found in the dolomitized carbonate tested in the Little Knife pilot test.[71]

There are a number of potential causes of a residual oil saturation in miscible flooding that are discussed in Sec. 3.7. However, other than the saturations found in the few cores mentioned previously that were taken behind CO_2 flood fronts and results from a limited amount of CO_2 flooding in reservoir cores,[73] there is little other published information on the magnitude of CO_2 flood residual oil saturation to expect with different reservoir rocks and with different oils. When reservoirs are being evaluated for CO_2 flooding, testing for residual oil saturation seems to be warranted, either by laboratory flooding of reservoir cores (Sec. 3.7) or more directly by injecting CO_2 and coring behind the CO_2 front.

CO₂ Requirement

This monograph defines the gross CO_2 requirement as the ratio of gross CO_2 injected at the surface (makeup

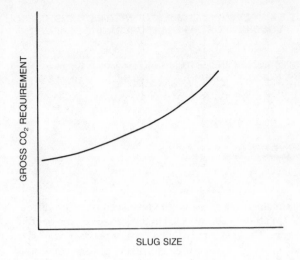

Fig. 8.19—Influence of slug size on the gross CO_2 requirement.

plus recycle) per stock tank barrel of incremental oil recovered. Eq. 8.3 shows the relationship of factors that influence the gross CO_2 requirement.

$$\frac{Q}{\Delta N_p} =$$

$$\frac{1}{B_{CO_2}} \left(\frac{1}{\frac{S_{orw}}{B_{ow}} - \frac{S_{orm}}{B_{om}}} \right) \left(\frac{\beta}{E_v} \right) \left(\frac{1}{E_c \cdot F} \right) , \ldots . (8.3)$$

where

Q = cumulative CO_2 injection, Mcf,
β = slug size actually entering the formation of interest, fraction PV,
F = fraction of injected fluid entering formation, and
B_{CO_2} = gas formation volume factor for CO_2, RB/Mcf.

The gross CO_2 requirement is affected by waterflood and CO_2 flood residual oil saturations, sweepout, capture efficiency, the gas formation volume factor, slug size, and the degree to which fluid at the injection well enters extraneous intervals. Gas formation volume factor is sensitive to reservoir pressure and temperature. More standard cubic feet of CO_2 are required to occupy a reservoir barrel of pore space as pressure increases and temperature decreases. Fig. 8.19 shows conceptually how the gross CO_2 requirement varies with slug size for a particular set of reservoir and fluid data. Although miscible sweepout increases with increasing slug size, it does not increase in proportion to the slug size. As a result, higher gross CO_2 requirements result for larger slug sizes. As the slug size decreases toward a value of zero, the gross CO_2 requirement tends to a value of $1/B_{CO_2}$ for these idealized conditions: (1) the slug is driven by water and 1 res bbl of CO_2 is left as a residual saturation for every reservoir barrel of oil that is displaced, (2) negligible CO_2 is dissolved in reservoir brine, (3) the oil formation volume factor, B_{ow}, is ap-

TABLE 8.3—EXAMPLE SIMULATOR PROJECTIONS OF CO$_2$ FLOODING FOR DIFFERENT OPERATING SCHEMES

Slug Size (% HCPV)	CO$_2$/Water Injection Ratio (RB/RB)	Oil Recovery (% S_{orw})	Gross CO$_2$ Requirement (Mcf/STB)
160	∞	55	24
20	∞	19	9
20	1:2*	27	6.5
20	1:2**	14	12

*No trapping assumed.
**Severe trapping assumed.

proximately unity, (4) all the displaced oil is produced, and (5) all the CO$_2$ injected at the surface enters the formation. The quantitative relationship between slug size and gross CO$_2$ requirement depends on conditions specific for each reservoir and is affected generally by the parameters shown in Eq. 8.3.

Table 8.2 shows the gross CO$_2$ requirements observed in several field tests. There was no direct measure of incremental recovery at Willard, Wasson, or in the SACROC main flood, and it was projected by the operators of these projects with a reservoir simulator using test data. The gross CO$_2$ requirements for the tests listed in Table 8.2 vary over a considerable range from about 5 to about 25 Mcf/STB. When reinjection of produced CO$_2$ is considered, the net CO$_2$ requirements are found to vary from about 3 to about 14 Mcf/STB. The high value for gross CO$_2$ requirement in the Little Creek test is attributed mainly to the large slug of CO$_2$ used in this small pilot test and to the relatively low value for waterflood residual oil saturation. Quantitative interpretation of the SACROC tertiary recovery test was clouded by uncertainty concerning the degree to which fluids were confined to the test interval and area.

Alternate CO$_2$/Water Injection

When mobile water interferes with the miscible displacement of oil, a higher CO$_2$ requirement will result for a given slug size in tertiary recovery floods or in floods where water injected to reduce CO$_2$ mobility invades the oil bank. When trapping of oil by mobile water is not significant, reservoir simulations show that alternate injection of water with CO$_2$ decreases the gross CO$_2$ requirement both in reservoirs where stratification is the dominant influence on flow behavior and in reservoirs where gravity is the dominant influence on flow behavior.[51,72] Table 8.1 shows results of simulations that support these remarks for a hypothetical reservoir in which gravity segregation was the dominant influence on flow behavior.

Even when oil trapping occurs, alternate water injection may still be a preferred method if the CO$_2$/water ratio can be adjusted so as not to create a large mobile water saturation in the oil bank. This conclusion assumes that in the absence of a mobile water saturation only a minor oil saturation is left behind the CO$_2$ front in CO$_2$-invaded regions. If appreciable oil is bypassed because of pore geometry and/or small-scale occlusions of low permeability, alternate injection of water might be harmful to subsequent extraction of a portion of this oil by diffusion (Sec. 3.7).

Alternate injection of water with the CO$_2$ slug has

been practiced in about half the field tests to date, but the improvement in process performance, if any, resulting from this method of operation is not well defined from field data. The operator of the Slaughter Estate project felt that alternate gas/water injection was "a substantially superior process over continuous gas injection" based on the recovery actually attained in that test and the recovery expected for comparable continuous gas injection projects.[1] Other operators have reported that the technique was beneficial in moderating the produced GOR after CO$_2$ breakthrough.[47,54] Mobility of the CO$_2$/water region also was measured directly in one field test and was found to be about as low as the mobility of drive water in the preceding waterflood.[74] However, there have been no tests where production performance for this method of operating has been compared directly in the same reservoir with the performance resulting from injection of an equivalent volume of CO$_2$ without water. The evidence to date supporting alternate water injection as a preferred method in CO$_2$ floods consists of the field test behavior cited, similar behavior reported for several condensing-gas drive floods (see Sec. 6.5, also Refs. 75 through 77), and simulation results such as those shown in Table 8.3.

Injectivity

Injectivity is an important consideration in any flood because of its effect on project life and number of injection wells that will be needed. Alternate CO$_2$/water injection reduces the relative permeability of both fluids, and a decrease in injectivity should be expected unless there are other, compensating phenomena occurring. The magnitude of injectivity reduction for each fluid depends on the relative permeability relationships for the rock in question and on the injection ratio.

Interactions between CO$_2$ and reservoir rock or between CO$_2$ and reservoir brine also can affect injectivity. Calcium carbonate is attacked by the carbonic acid formed by solution of CO$_2$ in formation or injected water and subsequently converted to the bicarbonate form by the following reaction.

$$CaCO_3 + CO_2 + H_2O \rightarrow Ca^{++} + 2(HCO_3)^- . \quad . . . (8.4)$$

Dissolution of carbonate from the walls of flow channels or dissolution of carbonate cementing material may increase permeability.[53,78] There also is a possibility, as yet unsupported by published data, that in some instances the reaction may cause a reduction in permeability by releasing particles that migrate and subsequently plug flow channels.[78]

Dissolved CO$_2$ also decreases the solubility of calcium sulfate. Solution of CO$_2$ in brines saturated with calcium sulfate, as might be found in reservoirs containing gypsum or anhydrite, will precipitate some of the calcium sulfate. Potentially, this precipitation could reduce permeability and injectivity.

When CO$_2$ and some reservoir fluids mix, as many as four phases — two liquids, a gas, and a solid — are in equilibrium (see Sec. 2.7). A potential consequence of this complicated phase behavior is that a heavy liquid phase or a solid hydrocarbon precipitate may reduce CO$_2$ permeability out into the reservoir.[48]

Lower-than-anticipated CO$_2$ injectivity has been observed in a number of field trials.[1,48] This seems to be particularly true in the low-temperature carbonate reservoirs of west Texas. Lower-than-anticipated water injectivity also has been observed when CO$_2$ and water were injected alternately.[1] However, injectivity in the Willard-Wasson test, even with alternate injection of CO$_2$ and water, did not deteriorate appreciably after fillup.[54]

Operating Problems and Solutions

Operating problems have been more severe than in waterflooding. Corrosion, leaks, and scaling occur in addition to occasional precipitation of a heavy hydrocarbon from the crude. Several excellent papers have been published concerning operating practices in CO$_2$ floods.[32-34,38,39,47,48,54,79-81] One paper treats the additional problems arising from high concentrations of hydrogen sulfide in the CO$_2$.[39]

In projects where CO$_2$ is to be injected continuously, without water, corrosion should be minimum in the injection system if the CO$_2$ is dehydrated before injection. Corrosive conditions will exist, however, in injection wells when water and CO$_2$ are injected alternately for mobility control and in producing wells where water and CO$_2$ are produced concurrently.

Dehydration is the measure normally taken to control corrosion in the CO$_2$ oilfield distribution system. Other than this, special precautions such as internally coating the pipe or using special steels do not appear to be common operating practice, except when high concentrations of hydrogen sulfide are present. For this latter situation, one operator reported using stringent precautions to obtain satisfactory welds in addition to using a pipe with sufficiently low Rockwell-C hardness (<22) to resist sulfide stress cracking.[39]

Use of separate CO$_2$ and water distribution lines is standard operating procedure where these fluids are injected alternately. As an added precaution in some projects where alternate CO$_2$/water injection was practiced, the water line to an injection well was disconnected and capped with a blind flange when the well was on CO$_2$ injection, and the same procedure was followed for the CO$_2$ connecting line when the well was on water injection.[39,54]

When water is injected alternately with CO$_2$, the procedure followed by some operators has been to protect meter runs, wellheads, wellhead valves, tubing, and packers with a combination of plastic or metallic coatings and stainless steel parts.[34,38,39,54] At the beginning of the SACROC project,[38] plastic-coated meter runs corroded severely at any point of coating damage, as did coated gate valves. Nickel-plated valves also had severe corrosion where the nickel coating was damaged. Check-valve disk action against valve bodies resulted in almost complete failure of all valves. Later, meter runs were constructed using 316 SS pipe and fittings, 316 SS gate valves with 316 SS or ceramic gates, and 316 SS check valves. Some aluminum bronze gate valves with ceramic gates also were used successfully.

Injection wellheads at SACROC originally were equipped with 410 SS wellheads and valves, but both wellheads and valves were subject to severe pitting-type corrosion that occurred primarily under deposits from suspended matter in the injection water.[38] Although measures such as plastic coating the wellheads and valve bodies and replacement of valve gate and seat material with 316 SS prolonged the useful life of many wellheads, the operator planned a replacement program using all 316 SS wellheads. Injection wellhead components other than valves at Levelland,[34] another project where alternate CO$_2$/water injection was practiced, are constructed of carbon steel that is coated internally with plastic. The wellhead valves are of two types: 316 SS and 410 SS designed to meet NACE 2 construction standards.[34] The Levelland operator reported this design to be working satisfactorily.

Injection well tubing at both SACROC[38] and Levelland[34] was protected by plastic coating. Plastic-coated packer performance at SACROC was good, with minimal problems resulting from mechanical and handling damage. Proper handling to prevent mechanical damage is an important factor in achieving good performance of plastic-coated tubing.

Injection wells at Crossett,[81] Little Creek,[32] and Bay St. Elaine[33] had conventional wellheads with no special provisions for corrosion protection. No water was injected with the CO$_2$ in these projects, and the CO$_2$ was essentially dry and noncorrosive. However, a stainless steel tubing hanger and lower master valve were used at Little Creek, and the injection well at Bay St. Elaine was equipped both with a stainless steel gravel-pack setting and a gravel-pack packer with a stainless steel bore.

Tubing-casing leaks have been a problem in several projects.[38,54] Several measures have been used to prevent high-pressure gas leaks and to prevent access of fluids to the engaged threads of tubing couplings. Modified couplings with special Teflon® seals were used at Crossett[81] and Little Creek.[32] Polyphenylene sulfide was used as a coupling coating material at SACROC.[38]

The combination of CO$_2$ and brine production causes a corrosive environment in producing wells and surface facilities. Corrosion has been combatted with a combination of inhibitors, coatings, and stainless steel. At SACROC,[38] downhole batch treatment of corrosion inhibitors reduced rod pump failures, while inhibitors, thin film-epoxy modified resin coatings on tubing, and glass-filled epoxy coating on submersible pump and motor parts satisfactorily protected the tubing and submersible pumps. At Little Creek,[32] stainless steel tubing hangers and wellhead controls were installed in addition to plastic-coated tubing. Although these measures plus monthly batch-type tubing inhibitor treatments drastically curtailed producing well corrosion, they did not control it completely, and corrosion still accounted for some significant producing well problems. The operator at Little Creek speculated that continuous inhibition of the tubing by circulation of an inhibitor-solvent solution might have been a more satisfactory treatment.

Producing wells in the Bay St. Elaine project[33] were equipped with stainless steel gravel-pack settings, packers with stainless steel bores, Monel™-coated gas lift valves, and stainless steel wellhead surface control equipment. These wells were completed with uncoated tubing, and the operator planned to batch treat them periodically with a corrosion inhibitor.

Liberation of CO$_2$ from solution in the brine in producing wells forces the equilibrium of Eq. 8.4 to the left side, precipitating calcium carbonate scale from the supersaturated solution. Calcium carbonate scale formation is a relatively slow process and application of phosphonate and phosphate ester scale-inhibitor treatments downhole and at surface facilities proved satisfactory in preventing scale formation problems at SACROC.[38]

Other producing-well problems have included freezing and asphalt-paraffin plugging.[32] As CO_2 expands through a surface choke, sufficient heat may be absorbed to cause freezing at, and downstream from, the choke. Wellhead heaters can be used to control this problem.[32,33] Cooling caused by the expanding CO_2 also may aggravate plugging in the well tubulars, chokes, and surface production lines and equipment by asphalt-like heavy hydrocarbons and paraffin.

8.7 Guidelines for Identifying Reservoirs With CO_2 Miscible Flooding Potential

Rough guidelines for screening reservoirs to identify potential CO_2 flooding prospects have been given in several publications.[8,82,83] These screening guidelines are very broad and are intended only to help identify candidate reservoirs that might warrant a more thorough evaluation to assess their CO_2 miscible flooding suitability. These guidelines are discussed in this section.

For a reservoir to be a CO_2-miscible flooding candidate, miscibility pressure must be attainable over a significant volume of the reservoir. Miscibility pressure for CO_2 often is significantly lower than the pressure required for miscibility with natural gas, flue gas, or nitrogen, which gives CO_2 a major advantage compared with these gases. Opportunities for miscible flooding with natural gas, flue gas, or nitrogen are limited because the high pressure required for dynamic miscibility is unattainable in many reservoirs. However, this often is not the case with CO_2, and CO_2 miscibility can be attained at shallower depths for a much wider spectrum of oils than is possible with the other gases. Even so, CO_2 miscibility will not be attainable in all reservoirs. Generally, reservoirs shallower than about 2,500 ft will not be candidates because at this shallow depth even a relatively low miscibility pressure usually cannot be attained without fracturing the reservoir. Miscibility pressure increases with decreasing API oil gravity, and reservoirs containing oils with gravities lower than about 27°API generally will not be CO_2-miscible flood candidates because of a high miscibility pressure requirement. For screening purposes miscibility pressure can be estimated by the methods discussed in Sec. 8.2. For serious design, miscibility pressure should be determined experimentally by a technique similar to the one recommended in Sec. 2.8.

Oil viscosity and reservoir heterogeneity also determine the suitability of a reservoir for CO_2 flooding. Because CO_2 has a low viscosity, the viscosity ratio with reservoir oils invariably will be unfavorable; and, because of this, the mobility ratio of the displacement will be unfavorable unless the CO_2 relative permeability is sufficiently reduced by alternate water injection, semisolid or heavy-liquid precipitation, or other factors

to keep the mobility ratio favorable. An unfavorable mobility ratio adversely affects sweepout and can hasten CO_2 slug destruction in the gas-driven slug process by viscous fingering. For these reasons, reservoirs containing oils of relatively high viscosity are not suitable candidates for CO_2 miscible flooding. There is no single viscosity cutoff above which process performance becomes precipitously worse. Rather, performance in terms of oil recovered per cubic foot of CO_2 injected decreases steadily with increasing oil viscosity. The upper limit for oil viscosity is determined by the economics specific to the reservoir being evaluated, but 10 to 12 cp has been suggested in several publications as a rough screening guide.[8,82,83]

As in hydrocarbon-miscible flooding, severe reservoir heterogeneity causing excessive production of CO_2 is to be avoided. Although some CO_2 production is to be expected even in the best-performing floods and although compression and reinjection of produced CO_2 may be economically sound in specific projects, severe channeling caused by extreme stratification or fracturing can reduce the ratio of oil recovered per gross cubic foot of CO_2 injected to an uneconomical value, and reservoirs with these characteristics should be avoided. Waterflood history, geology, logs, and well transient tests can be indications of reservoir heterogeneity.

As with the hydrocarbon-miscible processes, economic factors determine the minimum oil saturation that is acceptable for CO_2 flooding. However, as a rough guideline, oil saturation should not be less than about 20% PV in those portions of the reservoir that will be swept miscibly.

References

1. Rowe, H.G., York, S.D., and Ader, J.C.: "Slaughter Estate Unit Tertiary Pilot Performance," paper SPE 9796 presented at the 1981 SPE Symposium on Enhanced Oil Recovery, Tulsa, April 5–8.
2. Doscher, T. et al.: Carbon Dioxide for the Recovery of Crude Oil — A Literature Search to June 30, 1979, U. of Southern California, U.S. DOE Contract No. ET-78-C-05-5785 (1978).
3. Holm, L.W. and Josendal, V.A.: "Mechanisms of Oil Displacement by Carbon Dioxide," J. Pet. Tech. (Dec. 1974) 1427–36; Trans., AIME, 257.
4. Holm, L.W. and Josendal, V.A.: "Effect of Oil Composition on Miscible-Type Displacement by Carbon Dioxide," Soc. Pet. Eng. J. (Feb. 1982) 87–98.
5. Yellig, W.F. and Metcalfe, R.S.: "Determination and Prediction of CO_2 Minimum Miscibility Pressures," J. Pet. Tech. (Jan. 1980) 160–65.
6. Johnson, J.P. and Pollin, J.S.: "Measurement and Correlation of CO_2 Miscibility Pressures," paper SPE 9790 presented at the 1981 SPE Symposium on Enhanced Oil Recovery, Tulsa, April 5–8.
7. Lee, J.L.: "Effectiveness of Carbon Dioxide Displacement Under Miscible and Immiscible Conditions," Research Report RR 40, Petroleum Recovery Inst., Calgary (March 1979).
8. Enhanced Oil Recovery — An Analysis of the Potential for Enhanced Oil Recovery from Known Fields in the United States — 1976 to 2000, Natl. Petroleum Council, Washington, DC (Dec. 1976).
9. Cronquist, C.: "Carbon Dioxide Dynamic Miscibility With Light Reservoir Oils," Proc., Fourth Annual U.S. DOE Symposium, Tulsa, Aug. 28–30, 1978, Vol. 1b-Oil.
10. Mungan, N.: "Carbon Dioxide Flooding — Fundamentals," J. Cdn. Pet. Tech. (Jan.–March 1981) 87–92.
11. Dumyuskin, I.I. and Namiot, A.Y.: "Mixing Conditions of Oil with Carbon Dioxide," Neft. Khozyaistvo (March 1978) 59–61.

12. Holm, L.W. and Josendal, V.A.: "Discussion of Determination and Prediction of CO$_2$ Minimum Miscibility Pressure," *J. Pet. Tech.* (May 1980) 870–71.

13. Metcalfe, R.S.: "Effects of 'Impurities' on Minimum Miscibility Pressures and Minimum Enrichment Levels for CO$_2$ and Rich Gas Displacements," paper SPE 9230 presented at the 1980 SPE Annual Technical Conference and Exhibition, Dallas, Sept. 21–24.

14. Graue, D.J. and Zana, E.: "Study of Possible CO$_2$ Flood in the Rangely Field, Colorado," *J. Pet. Tech.* (July 1981) 1312–18.

15. Sage, B.H. and Lacey, W.N.: *Some Properties of the Lighter Hydrocarbons, Hydrogen Sulfide, and Carbon Dioxide,* Monograph on API Research Project 37, API, Dallas (1955).

16. Reamer, H.H. *et al.*: "Methane-Carbon Dioxide System in the Gaseous Region," *Ind. Eng. Chem.* (1944) 36.

17. Simon, R. *et al.*: "Compressibility Factors for CO$_2$-Methane Mixtures," *J. Pet. Tech.* (Jan. 1977) 81–85.

18. Kennedy, J.T. and Thodos, G.: "The Transport Properties of Carbon Dioxide," *AIChE J.* (Dec. 1961) **7**, 625.

19. Dodds, W.S., Stutzman, L.F., and Sollami, B.J.: "Carbon Dioxide Solubility in Water," *I&EC Chem. and Eng. Data Series* (1956) **1**.

20. Stewart, P.B. and Munjal, P.: "Solubility of Carbon Dioxide in Pure Water, Synthetic Sea Water, and Synthetic Sea Water Concentrates at −5°C to 25°C and 10 to 45-Atm Pressure," *J. Chem. and Eng. Data 15* (1970) No. 1, 67.

21. Weibe, R. and Gaddy, V.L.: "Vapor Phase Composition of Carbon Dioxide-Water Mixtures at Various Temperatures and at Pressure to 700 Atmospheres," *J. Am. Chem. Soc.* (1941) **63**, 475–77.

22. Unruh, C.H. and Katz, D.L.: "Gas Hydrates of Carbon Dioxide-Methane Mixtures," *Trans.*, AIME (1949) 83–86.

23. Canjar, L.N. *et al.*: "Thermodynamic Properties of Nonhydrocarbons," *Hydrocarbon Processing, Petroleum Refiner* (Jan. 1966) 135.

24. Simon, R. and Graue, D.J.: "Generalized Correlations for Predicting Solubility, Swelling, and Viscosity Behavior of CO$_2$-Crude Oil Systems," *J. Pet. Tech.* (Jan. 1965) 102–06.

25. Welker, J.R. and Dunlop, D.D.: "Physical Properties of Carbonated Oils," *J. Pet. Tech.* (Aug. 1963) 873–76.

26. *Bibliography of Vapor-Liquid Equilibrium Data for Hydrocarbon Systems,* API, Div. of Refining (1963) Table 18, 32.

27. *Engineering Data Book,* Natural Gas Processors Suppliers Assn., Tulsa (1957).

28. Poettman, F.H. and Katz, D.L.: "CO$_2$ in a Natural Gas Condensate System," *Ind. and Eng. Chem.* (1946) **38**, 530.

29. Poettman, F.H.: "Vaporization Characteristics of CO$_2$ in a Natural Gas-Crude Oil System," *Trans.*, AIME (1951) **192**, 141–44.

30. Jacoby, R.H. and Rzasa, M.J.: "Equilibrium Vaporization Ratios for Nitrogen, Methane, Carbon Dioxide, Ethane and Hydrogen Sulphide in Absorber Oil-Natural Gas and Crude Oil-Natural Gas Systems," *Trans.*, AIME (1952) **195**, 99–110.

31. Holm, L.W. and O'Brien, L.J.: "Carbon Dioxide Test at the Mead-Strawn Field," *J. Pet. Tech.* (April 1971) 431–42.

32. Hansen, P.W.: "A CO$_2$ Tertiary Recovery Pilot, Little Creek Field, Mississippi," paper SPE 6747 presented at the 1977 SPE Annual Technical Conference and Exhibition, Denver, Oct. 9–12.

33. Palmer, F.S., Nute, A.J., and Peterson, R.L.: "Implementation of a Gravity Stable, Miscible CO$_2$ Flood in the 8000-Foot Sand, Bay St. Elaine Field," paper SPE 10160 presented at the 1981 SPE Annual Technical Conference and Exhibition, San Antonio, Oct. 5–7.

34. Macon, R.S.: "Design and Operation of the Levelland Unit CO$_2$ Injection Facility," paper SPE 8410 presented at the 1979 SPE Annual Technical Conference and Exhibition, Las Vegas, Sept. 23–26.

35. Hare, M. *et al.*: *Sources and Delivery of Carbon Dioxide for Enhanced Oil Recovery,* Report FE-2515-24 prepared by Pullman Kellogg Co., Houston, for the U.S. DOE, Contract EX-76-C-01-2515 (Dec. 1978).

36. Smith, R.L.: "Sacroc Initiates Landmark CO$_2$ Injection Project," *Pet Eng.* (Dec. 1971) 43–47.

37. West, J.M. and Kroh, C.G.: "Carbon Dioxide Compressed to Supercritical Values," *Pet. Eng.* (Dec. 1971) 63–66.

38. Newton, L.E. Jr. and McClay, R.A.: "Corrosion and Operation Problems, CO$_2$ Project, Sacroc Unit," paper SPE 6391 presented at the 1977 SPE Permian Basin Oil and Gas Recovery Conference, Midland, March 10–11.

39. Adams, G.H. and Rowe, H.G.: "Slaughter Estate Unit CO$_2$ Pilot — Surface and Downhole Equipment Construction and Operation in the Presence of Hydrogen Sulfide Gas," *J. Pet. Tech.* (June 1981) 1065–74.

40. Stalkup, F.I.: "Carbon Dioxide Miscible Flooding: Past, Present, and Outlook for the Future," *J. Pet. Tech.* (Aug. 1978) 1102–12.

41. North American Production Capacity Data, Circular Z-57, Natl. Fertilizer Development Center, Tennessee Valley Authority, Mussell Shoals, AL (Jan. 1975).

42. Rump, M.W., Hare, M., and Porter, R.E.: *The Supply of Carbon Dioxide for Enhanced Oil Recovery,* report to the U.S. DOE by Pullman Kellogg Co., Houston, Contract No. EX-76-C-01-2515 (Dec. 1977).

43. *Steam Electric Plant Factors,* Natl. Coal Assn., Washington, DC (1978).

44. *Statistics of Privately-Owned Electric Utilities in the United States,* Federal Power Commission, Washington, DC (1972).

45. "Steam Dominates Enhanced Oil Recovery," *Oil and Gas J.* (April 5, 1982) 139.

46. Thrash, J.C.: "Twofreds Field — Tertiary Oil Recovery Project," paper SPE 8382 presented at the 1979 SPE Annual Technical Conference and Exhibition, Las Vegas, Sept. 23–26.

47. Kane, A.V.: "Performance Review of a Large Scale CO$_2$-Wag Project SACROC Unit-Kelly Snyder Field," *J. Pet. Tech.* (Feb. 1979) 217–31.

48. Pontious, S.B. and Tham, M.J.: "North Cross (Devonian) Unit CO$_2$ Flood — Review of Flood Performance and Numerical Simulation Model," *J. Pet. Tech.* (Dec. 1978) 1706–16.

49. Stright, D.H. Jr. *et al.*: "Carbon Dioxide Injection Into Undersaturated Viscous Reservoirs — Simulation and Field Application," *J. Pet. Tech.* (Oct. 1977) 1248–58.

50. Reid, T.B. and Robinson, H.J.: "Lick Creek Meakin Sand Unit Immiscible CO$_2$-Waterflood Project," paper SPE 9795 presented at the 1981 SPE/DOE Symposium on Enhanced Oil Recovery, Tulsa, April 5–8.

51. Youngren, G.K. and Charlson, G.S.: "History Match Analysis of the Little Creek CO$_2$ Pilot Test," *J. Pet. Tech.* (Nov. 1980) 2042–52.

52. Perry, G.E. and Kidwell, C.M.: "Weeks Island 'S' Sand Reservoir B Gravity Stable Miscible CO$_2$ Displacement," paper presented at Fifth Annual U.S. DOE Symposium on Enhanced Oil and Gas Recovery and Improved Drilling Recovery, Aug. 22–24, 1979.

53. Graue, P.J. and Blevins, T.R.: "SACROC Tertiary CO$_2$ Pilot Project," paper SPE 7090 presented at the 1978 SPE Symposium on Improved Methods for Oil Recovery, Tulsa, April 16–19.

54. Johnston, J.W.: "A Review of the Willard (San Andres) Unit CO$_2$ Injection Project," paper SPE 6388 presented at the 1977 SPE Permian Basin Oil and Gas Recovery Conference, Midland, TX, March 10–11.

55. Conner, W.D.: "Granny's Creek CO$_2$ Injection Project, Clay County, West Virginia," *Proc.*, U.S. DOE Symposium on Enhanced Oil and Gas Recovery and Improved Drilling Methods, Tulsa, Aug. 30–Sept. 1, 1977.

56. San Filippo, G.P. and Guckert, L.G.S.: "Development of a Pilot Carbon Dioxide Flood in the Rock Creek-Big Injun Field, Roane Co., West Virginia," *Proc.*, U.S. DOE Symposium on Enhanced Oil and Gas Recovery and Improved Drilling Methods, Tulsa, Aug. 30–Sept. 1, 1977.

57. Beeler, P.F.: "West Virginia CO$_2$ Oil Recovery Project Interim Report," *Proc.*, U.S. DOE Symposium on Enhanced Oil and Gas Recovery and Improved Drilling Methods, Tulsa, Aug. 30–Sept. 1, 1977.

58. *Improved Oil-Recovery Field Reports,* SPE, Dallas (1975–1982).

59. DiCharry, R.M., Perryman, T.L., and Ronquille, J.D.: "Evaluations and Design of a CO$_2$ Miscible Flood Project — SACROC Unit, Kelly-Snyder Field," *J. Pet. Tech.* (Nov. 1973) 1309–18; *Trans.*, AIME, **215**.

60. Henderson, L.E.: "The Use of Numerical Simulation to Design a Carbon Dioxide Miscible Displacement Project," *J. Pet. Tech.* (Dec. 1974) 1327–34.

61. "Shell Starts Miscible CO$_2$ Pilot in Mississippi's Little Creek Field," *Oil and Gas J.* (Sept. 2, 1974) 52–53.

62. Holm, L.W.: "Status of CO$_2$ and Hydrocarbon Miscible Oil Recovery Methods," *J. Pet. Tech.* (Jan. 1976) 76–84.

63. Charlson, G.S., Bilhartz, H.L., and Stalkup, F.I.: "Use of Time-Lapse Logging Techniques in Evaluating the Willard Unit CO$_2$ Flood Mini-Test," paper 7049 presented at the 1978 SPE Sym-

posium on Improved Methods for Oil Recovery, Tulsa, April 16–19.

64. Bilhartz, H.L. and Charlson, G.S.: "Coring for In-Situ Saturations in the Willard Unit CO_2 Flood Mini-Test," paper SPE 7050 presented at the 1978 SPE Symposium on Improved Methods for Oil Recovery, Tulsa, April 16–18.

65. Bilhartz, H.L. et al.: "A Method for Projecting Full-Scale Performance of CO_2 Flooding in the Willard Unit," paper SPE 7051 presented at the 1978 SPE Symposium on Improved Methods for Oil Recovery, Tulsa, April 16–19.

66. Bilhartz, H.L.: "Case History: A Pressure Core Hole," paper SPE 6389 presented at the 1977 SPE Permian Basin Oil and Gas Recovery Conference, Midland, TX, March 10–11.

67. Ader, J.C. and Stein, M.H.: "Slaughter Estate Unit CO_2 Pilot Reservoir Description via a Black Oil Model Waterflood History Match," paper SPE 10727 presented at the 1982 SPE/DOE Symposium on Enhanced Oil Recovery, Tulsa, April 4–7.

68. Perry, G.E.: "Weeks Island 'S' Sand Reservoir B Gravity Stable Miscible CO_2 Displacement, Iberia Parish, Louisiana," Proc., U.S. DOE Symposium on Enhanced Oil and Gas Recovery and Improved Drilling Methods, Tulsa, Aug. 30–Sept. 1, 1977.

69. Watts, R.J. et al.: "A Single CO_2 Injection Well Minitest in a Low-Permeability Eastern Carbonate Reservoir — A Preliminary Report," paper SPE 9430 presented at the 1980 SPE Annual Technical Conference and Exhibition, Dallas, Sept. 21–24.

70. Hagedorn, A.R. and Blackwell, R.J.: "Summary of Experience with Pressure Coring," paper SPE 3692 presented at the 1972 SPE Annual Meeting, San Antonio, Oct. 8–11.

71. Desch, J.B. et al.: "Enhanced Oil Recovery by CO_2 Miscible Displacement in the Little Knife Field, Billings County, North Dakota," paper SPE 10696 presented at the 1982 SPE/DOE Symposium on Enhanced Oil Recovery, Tulsa, April 4–7.

72. Warner, H.R.: "An Evaluation of Miscible CO_2 Flooding in Waterflooded Sandstone Reservoirs," J. Pet. Tech. (Oct. 1977) 1339–47.

73. Spence, A.P. and Watkins, R.W.: "The Effect of Microscopic Core Heterogeneity on Miscible Flood Residual Oil Saturation," paper SPE 9229 presented at the 1980 SPE Annual Technical Conference and Exhibition, Dallas, Sept. 21–24.

74. Henry, R.L. et al.: "Utilization of Composition Observation Wells in a West Texas CO_2 Pilot Flood," paper SPE 9786 presented at the 1981 SPE/DOE Symposium on Enhanced Oil Recovery, Tulsa, April 5–8.

75. Holloway, H.D. and Fitch, R.A.: "Performance of a Miscible Flood with Alternate Gas-Water Displacements," J. Pet. Tech. (April 1964) 372–76.

76. Brannan, G. and Whitington, H.M. Jr.: "Enriched Gas Miscible Flooding — A Case History of the Levelland Unit Secondary Miscible Project," J. Pet. Tech. (Aug. 1977) 919–24.

77. Griffith, J.D. and Horne, A.L.: "South Swan Hills Solvent Flood," Proc., Ninth World Pet. Cong., Tokyo (1975) 4, 269–78.

78. Ross, G.D. et al.: "The Dissolution Effects of CO_2-Brine Systems on the Permeability of U.K. and North Sea Calcareous Sandstones," paper SPE 10685 presented at the 1982 SPE/DOE Symposium on Enhanced Oil Recovery, Tulsa, April 4–7.

79. Graham, B.D. et al.: "Design and Implementation of a Levelland Unit CO_2 Tertiary Pilot," paper SPE 8831 presented at the 1980 SPE/DOE Symposium on Enhanced Oil Recovery, Tulsa, April 20–23.

80. Crockett, D.H.: "A Profile Control Program Utilized in the SACROC Unit CO_2 Injection Program," Proc., 22nd Annual Southwestern Petroleum Short Course, Texas Tech. U., Lubbock (1975) 159–64.

81. Frey, R.P.: "Operating Practices in the North Cross CO_2 Flood," Proc., 22nd Annual Southwestern Petroleum Short Course, Texas Tech. U., Lubbock (1975) 165–68.

82. The Potential and Economics of Enhanced Oil Recovery, Report prepared by Lewin and Assocs. Inc. for U.S. DOE (April 1976).

83. Enhanced Oil Recovery Potential in the United States, Office of Technology Assessment, U.S. Congress, Washington, DC (Jan. 1978).

84. Robinson, D.B. and Mehta, B.R.: "Hydrates in the Propane-Carbon Dioxide-Water System," J. Cdn. Pet. Tech. (Jan.–March 1971) 33.

85. Stalkup, F.I.: "Status of Miscible Displacement," paper SPE 9992 presented at the 1982 SPE International Petroleum Exhibition and Symposium, Beijing, China, March 18–26.

86. McRee, B.C.: "CO_2: How It Works, Where It Works," Pet. Eng. Intl. (Nov. 1977) 52–63.

Chapter 9
Conclusions

9.1 The State of the Art

Miscible flooding with hydrocarbon solvents, flue gas, nitrogen, and CO_2 can be applied today in selected reservoirs without the need for additional technology development. A base technology exists today that is ready for application, although numerous technical unknowns still remain that contribute to uncertainty in project design optimization and performance forecasting, and these uncertainties must be taken into account in project-risk evaluation.

Two unresolved issues concerning the optimization of design variables for CO_2 flooding, and also for the other miscible processes to some extent, are especially worthy of mention: (1) Should water be injected alternately with the CO_2 and, if so, at what ratio? (2) Should water or gas that is miscible with the CO_2 be used to drive the CO_2 slug?

Considering viscosity ratio alone, an unfavorable mobility ratio would be expected for a CO_2-miscible flood, which could result in an inefficient sweepout. Alternate water injection is a technique to improve the sweepout of a low-viscosity injection fluid such as CO_2 by reducing its relative permeability. However, the complex CO_2/oil phase behavior observed at relatively low reservoir temperatures raises the possibility that precipitation of liquid and/or solid phases in the CO_2/oil transition zone might reduce CO_2 mobility below the level anticipated from normal viscosity and relative permeability considerations. As discussed in Chap. 8, CO_2 injectivity was lower than expected in many of the field tests, but whether this was caused by a near-wellbore phenomenon or by a reduced CO_2 mobility throughout the swept region has not been established conclusively. It is important to know the effective CO_2 mobility. Unnecessary injection of water or overinjection, if some water should be required, not only increases the complexity and expense of field operation but also decreases injectivity unnecessarily and lengthens project life. Also, in some situations water overinjection can trap some of the oil and render it inaccessible to CO_2.

In selecting a drive fluid, a choice must be made whether to maintain a miscible displacement at the trailing edge of the slug. Water as the drive fluid immiscibly displaces the CO_2 slug and dissipates the slug by leaving it as a residual phase.* Gases such as natural gas, flue gas, and nitrogen will displace the CO_2 slug miscibly above a minimum pressure. The potential advantage of this latter method compared with water drive is that the slug might be utilized more efficiently—i.e., the slug might be propelled intact farther into the reservoir by the gas rather than being left continually as a residual phase, and a given slug size might achieve a greater miscible sweepout and greater oil recovery compared with water drive. However, viscous fingering and mixing of the oil/solvent/drive gas by dispersion are mechanisms that act to destroy the integrity of gas-driven solvent slugs and, therefore, limit the effectiveness of this method. In addition, the mobility ratio with gas as the drive fluid will be more unfavorable than with water, which acts to reduce the sweepout of a gas drive compared with a water drive, although alternate drive gas/water injection will moderate this disadvantage. The overall issue to be resolved here is the magnitude of improvement in oil recovery for a given slug size attained by miscible gas drive compared with water drive (if there is any improvement) and whether the extra recovery justifies the added expense of the drive gas.

Oil recovery and producing-rate projections generally still have a large range of uncertainty despite extensive past laboratory research, field testing, and reservoir simulator development. Pilot testing may be advisable to define performance more accurately before proceeding with large-scale floods for projects where the economic risk would be unacceptable.

More detailed and comprehensive reservoir engineering, cost determination and control, and project monitoring are necessary for the typical miscible flood project than for waterflooding. There is a correspondingly greater requirement for trained manpower.

9.2 Comparison of Processes

Figs. 9.1 and 9.2 compare incremental recovery and slug effectiveness for selected first-contact miscible, condensing-gas drive, and CO_2-miscible floods where there are sufficient data to make the comparisons. Slug

---

*Some CO_2 also is dissolved in the residual oil left to CO_2 displacement. Also, some of the trapped or residual CO_2 is subsequently dissolved by and slowly produced with the drive water, although this does not prevent slug dissipation.

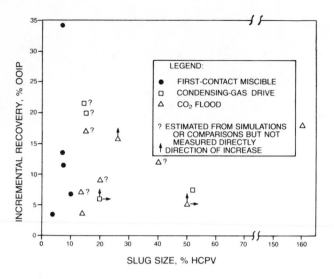

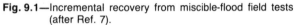

Fig. 9.1—Incremental recovery from miscible-flood field tests (after Ref. 7).

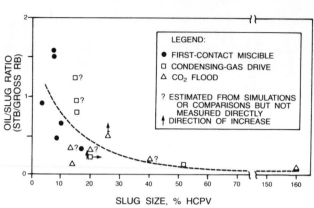

Fig. 9.2—Slug effectiveness in miscible-flood field tests (after Ref. 7).

effectiveness is defined here to be the incremental oil recovered per reservoir barrel of total solvent slug injected (makeup plus recycle). Conceptually, for a given process in a given reservoir, incremental recovery should increase and slug effectiveness defined in this manner eventually should decrease with increasing slug size, and these concepts roughly are supported by Figs. 9.1 and 9.2.

It should be kept in mind that incremental recovery and slug effectiveness depend on numerous other variables besides slug size, such as reservoir heterogeneity, temperature and pressure, crude oil composition, and flow regime, to mention just a few. These conditions varied widely between the tests shown in these figures, which should introduce a substantial data-point scatter, even for projects of a given process type, when results are plotted as a function of slug size only. Comparisons are clouded further because there is little overlap between slug sizes used in the first-contact miscible floods and the other processes. Within the data scatter there is no clearcut evidence that on average a given process is performing appreciably better or worse technically than the others for a given slug size. This is not to say that one process may not actually perform better than the others for given reservoir conditions, but no overall trend is evident in the composite field data.

The data in Fig. 9.2 suggest that 15 to 20% HCPV condensing-gas-drive slugs are performing as effectively as first-contact miscible slugs. Admittedly the data to support this conclusion are extremely limited, but if true, the LPG is being utilized more effectively in the rich-gas slugs since they contain only 30 to 50% LPG.

9.3 Outlook for Miscible Flooding

Future miscible flooding in the U.S. with hydrocarbon solvents most likely will be limited because of cost and possible short supply of these solvents. However, supply and cost in some of the other major oil-producing countries as well as conservation policies in these countries could be favorable for projects with hydrocarbon

solvents. Because of the high miscibility pressure requirement, flue gas and nitrogen injection also should have relatively limited application to deeper, higher pressure reservoirs where miscibility pressure is attainable. CO_2 flooding has the most potential for future miscible flood projects in the U.S. The magnitude of this potential depends heavily on the availability and cost of future CO_2 supplies.

Fig. 9.3 shows some past projections of producing rate and ultimate recovery potential in the U.S. for CO_2-miscible flooding.[1-5] Ultimate recovery projections were made by the Natl. Petroleum Council (NPC), the Energy Research and Development Admin. (ERDA), the Congressional Office of Technology Assessment (OTA), and the U.S. DOE. With the exception of OTA, these studies found ultimate recovery could be on the order of 5 to 7 billion barrels. The OTA estimated ultimate recovery as high as 12 billion barrels.

Projections of producing rate vary markedly, particularly in the 1990-to-1995 time frame, when the projections range from several hundred thousand barrels per day to well over a million barrels per day. The rate of future oil production from CO_2 miscible flooding will be affected by a number of factors that cannot be foreseen precisely. These include (1) magnitude of CO_2 supply and rate at which future supplies can be developed and delivered to oil fields, (2) cost of delivered CO_2, (3) volume of purchased CO_2 that on average will be required to recover a stock-tank barrel of incremental oil, and (4) amount of pilot testing that individual operators might require before trying a semiproved process.* Various assumptions were made in the different studies to account for these factors, which to a large degree have caused the differences in projected recovery rates. The studies all show, however, that CO_2-miscible flooding could make a significant contribution to future U.S. oil supply.

---

*The rate of future oil production from CO_2 miscible flooding also will be affected by oil price. The projections of Fig. 9.3 are for an oil price of approximately $20/bbl in 1976 U.S. dollars.

9.4 Areas for Further Study and Development

Reservoir Description

Because of the substantial injection fluid cost in miscible flooding, calculated project economics sometimes may range from favorable to unfavorable within the range of uncertainty in performance projections. Sweepout, overall recovery, and producing rates for potential field projects are calculated predominantly with reservoir simulators, but important reservoir data needed for these calculations often are known inaccurately or not available at all. Significant improvement in the reservoir description of the spatial distributions of rock properties and fluid saturations probably would have greater impact in reducing the overall uncertainty in miscible-flood performance projections than advances in any other area of miscible-flood technology.

Greater emphasis should be put on more effective application and integration of the information potentially available from reservoir geological analysis, geophysical techniques, and modern logs with the information traditionally available from core analysis, well testing, and conventional logs to define reservoir heterogeneity more accurately. Although advances already have been made in determining meaningful waterflood residual-oil saturations and the residual oil left in solvent-swept zones, further advances are needed here as well.

Reservoir Simulators

The advantages and shortcomings of current miscible-flood reservoir simulators were discussed in Chap. 4. Improved simulators are needed both to utilize fully the more detailed information potentially available from improved reservoir description and to account more accurately for some displacement phenomena. In particular, improved capability is needed for simulating viscous fingering and for simulating dispersion and the effect it has on solvent slug dilution, moderation of viscosity and density contrasts, and mixing of transition zone fluids into multiphase regions. In addition, improved methods are needed to account during simulation for the factors that result in a miscible-flood residual-oil saturation and the factors that govern solvent mobility (especially for CO_2). Much progress has been achieved already in incorporating realistic phase-behavior predictions into miscible flood simulations, but further improvement is warranted to increase the size and to reduce the cost of compositional simulations and to account more accurately for the effects of phase behavior on miscibility, multiphase flow in the solvent/oil transition zone, and liquid/solid precipitation. The flexibility to account for some or all of these phenomena simultaneously during a field study is needed. Yet this must be done for computing costs that are economically feasible.

Phase Behavior and Miscibility

Although there has been some good recent work to understand the complexities of CO_2/reservoir fluid phase behavior, more work is needed to understand the mechanism of dynamic miscibility and to substantiate the relationship between MMP and phase behavior at low temperatures where liquid/liquid extraction appears

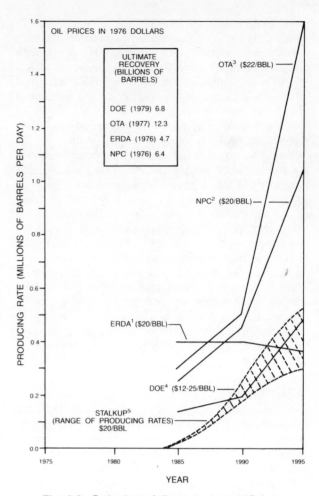

Fig. 9.3—Projections of oil recovery in the U.S. from CO_2-miscible flooding.

important. As this understanding advances, improved phase-behavior prediction methods may be needed for more realistic representation of CO_2/reservoir fluid phase behavior.

There is a need to adopt industry standards for the experimental measurement of dynamic MMP. The variety of experimental methods and interpretation techniques used by different investigators in the past undoubtedly has resulted in publication of some erroneous data and also has caused considerable uncertainty in comparing and correlating data from different sources as well as communication difficulty for all involved. The method discussed by Yellig and Metcalfe[6] and recommended in this monograph or a modification of this method, if warranted by subsequent information, could serve as such a standard.

Mobility

The importance of having an improved understanding of the factors affecting CO_2 mobility was emphasized in Sec. 9.1. More laboratory research is needed to understand the combined effects of phase behavior, rock properties, water saturation, and saturation hysteresis on solvent mobility in general but for CO_2 in particular because of the application potential anticipated for the CO_2-miscible method. This is one of the most important of all areas for additional mechanistic research. Additional understanding is needed also of the degree to which CO_2/brine/rock interactions can affect mobility through solution of rock constituents or through precipitation of brine constituents. As the factors affecting CO_2 mobility are better understood, improved methods will be needed for estimating CO_2 mobility in specific reservoirs of interest.

A number of gaps remain in our ability to relate mobility and mobility ratio to sweepout. As pointed out in Sec. 3.1, experimental data and theoretical analyses are lacking to define the mobility best characterizing movement of the solvent front when mobile water is present. Possible choices are (1) average total mobility of the solvent/water region, (2) total mobility at the backside of the solvent/water shock front, or (3) solvent mobility alone at either the backside of the solvent front or at the average solvent saturation. A similar dilemma prevails for the oil bank when mobile water is present. There is also a lack of experimental data and theoretical analysis for characterizing the effective mobility ratio in multifront displacements, such as in tertiary recovery or in miscible slug flooding. Improved methods are needed for estimating effective mobility ratio and sweepout in all these situations.

Sweepout

Poor sweepout caused by an unfavorable mobility ratio has been one of the most significant problems in past miscible flooding. Various methods to improve sweepout have been researched in the past, and solvent/water injection is used frequently in field applications, but there is still substantial need for improved methods of sweepout control. Research in this area should be continued. Work is needed still to define and to quantify the benefits vs. the disadvantages of solvent/water injection and to define more unequivocally the circumstances under which the method can be used to best advantage. This is especially true for tertiary recovery, where some controversy remains concerning the advisability of water injection. Greater attention should be given to utilizing and improving mechanical and chemical methods for injection and producing-well profile control. Foams have shown some promise in the laboratory for mobility reduction, and field experiments are needed to evaluate their commercial potential.

There are very few published volumetric sweepout data. Data particularly are limited or lacking for slug processes, tertiary recovery, and dynamic miscible displacement. Additional scaled-model data are needed to improve our understanding of sweepout and displacement behavior under these conditions and to serve as test cases for evaluating the quality of reservoir simulator projections.

The mixing-parameter method is a generalization of the Koval analytical technique, and it is used commonly for approximating the influence of viscous fingering on sweepout and oil recovery in coarsely-gridded reservoir simulations. However, there are only limited data to aid in mixing-parameter estimation for specific reservoir ap-

plications. More attention needs to be given to this problem, especially for applications involving reservoir heterogeneity, tertiary recovery, and solvent slugs. The limits of the method need to be defined better, and there is a need for overall performance data from well-defined experiments—both laboratory models and field trials—from which representative values of the mixing parameter can be estimated.

Unit Displacement Efficiency

Mechanisms that could result in oil being bypassed by the solvent front were discussed in Sec. 3.7. The amount of bypassed oil that might result from these mechanisms currently is related poorly to reservoir rock properties and brine saturation. Although some of the originally bypassed oil may be recovered subsequently by diffusion or by continued sweepout of low-permeability occlusions, methods for estimating the amount and rate of this recovery in reservoir floods are in early stages of development and poorly tested, and there are only limited published data against which theories and prediction methods can be tested adequately.

Dispersion

Dispersion plays an important role in miscible displacement. It is responsible for dilution of solvent slugs and loss of miscibility. Moderation of viscosity and density contrasts by dispersive mixing can affect sweepout in some situations where gravity tonguing is a factor. In dynamic miscible displacements, the amount of miscible flood residual oil is affected by the degree of dispersion in the miscible transition zone. Past research has been concerned with dispersion caused by molecular diffusion and microscopic convective mixing. Additional research is needed to understand and to quantify the macroscopic dispersion caused by reservoir heterogeneities that are too large to be observed in cores but are too small to be correlated between wells and included explicitly in the reservoir description.

References

1. *Research and Development in Enhanced Oil Recovery, Part I,* Lewin and Inc. for U.S. ERDA, Washington, DC (Nov. 1976).
2. *Enhanced Oil Recovery—An Analysis of the Potential for Enhanced Oil Recovery from Known Fields in the United States—1976 to 2000,* Natl. Petroleum Council, Washington, DC (Dec. 1976).
3. *Enhanced Oil Recovery Potential in the United States,* Office of Technology Assessment, U.S. Congress, Washington, DC (Jan. 1978).
4. *Projections of Enhanced Oil Recovery, 1985–1995,* TR/ES/79-30, DOE/EIA-0183/11, U.S. DOE, Energy Information Agency (Sept. 1979).
5. Stalkup, F.I.: "Carbon Dioxide Injection for Improved Oil Recovery," *Exploration and Economics of the Petroleum Industry,* Southwestern Legal Foundation, Matthew Bender, San Francisco (1981) **19,** 285.
6. Yellig, W.F. and Metcalfe, R.S.: "Determination and Prediction of CO_2 Minimum Miscibility Pressures," *J. Pet. Tech.* (Jan. 1980) 160–68.
7. Stalkup, F.I.: "Status of Miscible Displacement," *J. Pet. Tech.* (April 1982) 815–26.

Appendix A
Example Desk Top Calculation for Tertiary Recovery by Continuous Solvent Injection

Assume that flow is Region II type with a single gravity tongue.

$$\frac{M_s}{M_{ob}} = 31.$$

$$\frac{M_s}{M_{wf}} = 24.$$

Assume that the mobility ratio for solvent and oil bank sweepout is 25 and that

$$S_{orw} = 0.27, \ S_{orm} = 0.1,$$

$$S_{ob} = 0.65, \ S_{wr} = 0.2,$$

$$S_{wt} = 0.35.$$

Then

$$\text{HCPVI} = \frac{D_{vs}(1 - S_{wt} - S_{orm})}{(1 - S_{wr})} = 0.69 D_{vs},$$

and

$$D_{vob} = D_{vs} + D_{vob} = D_{vs} + \frac{(S_{orw} - S_{orm})}{(S_{ob} - S_{orw})} D_{vs}$$

$$= 1.45 D_{vs}.$$

Assume that the displacement follows the $M=25$ sweepout curve of Fig. 4.20.

1. Solvent breakthrough at $D_{vs} = 0.19$, and HCPVI $= 0.69(0.19) = 0.13$. Oil bank breakthrough at $D_{vs} = 0.19/1.45 = 0.13$, and HCPVI $= 0.69(0.13) = 0.09$.

2. Assume that $D_{vs} = 0.19$, $E_s = 0.19$. Then $D_{vob} = 1.45(0.19) = 0.275$, $E_{ob} = 0.23$. Recovery* (%OOIP) is expressed by

$$\frac{[E_s(S_{orw} - S_{orm}) - (E_{ob} - E_s)(S_{ob} - S_{orw})]100}{S_{oi}}$$

$$= \frac{0.19(17) - (0.04)(38)}{0.8} = 2.14,$$

and

$$\text{HCPVI} = 0.19(0.69) = 0.13.$$

3. Assume that $D_{vs} = 0.5$, $E_s = 0.315$. Then $D_{vob} = 1.45(0.5) = 0.725$, $E_{ob} = 0.385$. Recovery (%OOIP) is

$$\frac{0.315(17) - 0.05(38)}{0.8} = 3.38,$$

and

$$\text{HCPVI} = 0.69(0.5) = 0.345.$$

4. Assume that $D_{vs} = 0.8$, $E_s = 0.41$. Then $D_{vob} = 1.45(0.8) = 1.16$, $E_{ob} = 0.495$. Then, recovery (%OOIP) is

$$\frac{0.41(17) - 0.085(38)}{0.8} = 4.68,$$

and

$$\text{HCPVI} = 0.69(0.8) = 0.55.$$

*Assumes that FVF is the same for oil at discovery (S_{oi}), oil at start of miscible flood (S_{orw}, S_{ob}), and miscible-flood residual oil (S_{orm}).

Appendix B
Example Desk Top Calculation for the Little Creek CO_2-Flood Pilot Test

This test is described in Sec. 8.5 and Refs. 1 and 2, and the data are as follows.

$$T = 248°F,$$

$$P \cong 5,600 \text{ psi},$$

$$\mu_{CO_2} \cong 0.055 \text{ cp},$$

$$\rho_{CO_2} = 0.64 \text{ g/cm}^3,$$

$$\mu_o \cong 0.4 \text{ cp},$$

$$\mu_w \cong 0.25 \text{ cp},$$

$$\rho_w \cong 0.96 \text{ g/cm}^3,$$

$$k = 75 \text{ md},$$

$$h = 20 \text{ ft},$$

$$i \cong 3.5 \text{ MMcf/D} \cong 1,830 \text{ RB/D},$$

$$S_{wr} = 0.56, \text{ and}$$

$$S_{orw} = 0.21.$$

The actual test pattern is shown in Fig. B-1. Since the only data for gravity-tongue dominated sweepout are for a five-spot (Fig. 3.30 or Fig. 4.20) or for a linear cross section (Figs. 3.25 through 3.29), assume that sweepout behavior for the pilot area can be approximated roughly by five-spot behavior.

1. Estimate mobility ratios and saturations.

a. Estimate tertiary oil bank saturations from the oil/water fractional flow curve (not shown) using the Welge tangent method:

$$S_{ob} \cong 0.36$$

and

$$S_{wt} \cong 0.63.$$

b. Estimate mobility ratios from relative permeability data (not shown):

$$\frac{M_{CO_2}}{M_{ob}} \cong 8$$

and

$$\frac{M_{CO_2}}{M_{wf}} \cong 15.$$

2. Estimate CO_2 dissolved in water and miscible-flood residual oil. Assume that $S_{orm} \cong 0.05$ (reservoir conditions), CO_2 solubility in water $\cong 0.097$ RB/RB water, and CO_2 solubility in residual oil $\cong 0.34$ RB/RB oil. In the CO_2-swept zone, 1 res bbl of total pore space contains 0.32 res bbl of CO_2-displaceable volume, 0.05 res bbl of miscible-flood residual oil, and 0.63 res bbl of water. Therefore, we must inject

$$\frac{0.32 + 0.05(0.34) + 0.63(0.097)}{0.32} = 1.24$$

displaceable volumes of CO_2 to have one displaceable volume of free CO_2 in the reservoir.

3. Estimate the viscous/gravity force ratio for the idealized equivalent five-spot quadrant shown by the dashed lines in Fig. B-1.

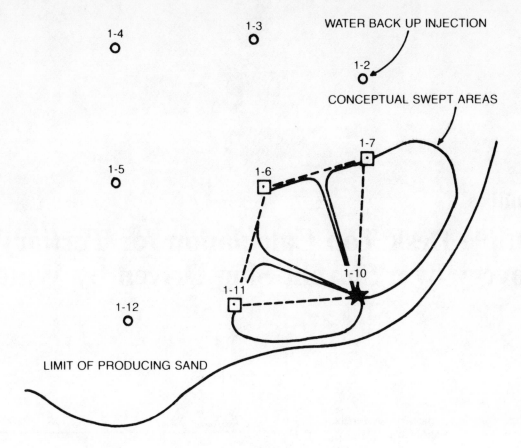

1-4

1-3

WATER BACK UP INJECTION

1-2

CONCEPTUAL SWEPT AREAS

1-7

1-5

1-6

1-11

1-10

1-12

LIMIT OF PRODUCING SAND

Fig. B-1—Little Creek field CO_2-pilot-test area.

a. Assume that approximately two-thirds of the total injection goes into the volume bounded by the dashed line.

b. Use the formula of Fig. 3.30 for viscous/gravity ratio:

$$R_{v/g} = 512(4) \left(\frac{2}{3} \right) \frac{i\mu_o}{k\Delta\rho h^2}.$$

The first constant, 512, applies if one-fourth of the total injection goes into the quadrant; 512(4) is the appropriate equation coefficient if all total injection goes into the quadrant; and 512(4)(2/3) applies if two-thirds of the total injection enters the quadrant.

If two-thirds of the total injection at Little Creek went into the volume bounded by the dashed lines in Fig. B-1 (assumed), the viscous/gravity ratio is given by

$$1,367 \cdot \frac{(1,830)(0.4)}{(75)(0.32)(20)^2} = 104.$$

According to Figs. 3.30 and 3.12, sweepout should be dominated largely by gravity tonguing, although flow could be in the beginning of the Region III transition zone.

4. Follow procedure outlined in Appendix A to estimate oil recovery vs. displaceable volumes of free CO_2 injected. Use the $M = 15$ sweepout curve of Fig. 4.20.

References

1. Hansen, P.W.: "A CO_2 Tertiary Recovery Pilot, Little Creek Field, Mississippi," paper SPE 6474 presented at the 1977 SPE Annual Technical Conference and Exhibition, Denver, Oct. 9–12.
2. Youngren, G.K. and Charlson, G.S.: "History Match Analysis of the Little Creek CO_2 Pilot Test," *J. Pet. Tech.* (Nov. 1980) 2042–52.

Appendix C

Example Desk Top Calculation for Tertiary Recovery by a Solvent Slug Driven by Water

Data for this example are the same as for the example given in Appendix A. Additional data are $S_{srw} = 0.17$, solvent slug size = 25% HCPV.

Calculations for Solvent-Injection Phase

$$HCPVI = \frac{D_{vs}(1 - S_{wt} - S_{orm})}{(1 - S_{wr})}$$

$$= 0.69 D_{vs}.$$

$$D_{vob} = D_{vs} + \Delta D_{vob} = D_{vs} + \frac{(S_{orw} - S_{orm})}{(S_{ob} - S_{orw})} D_{vs}$$

$$= 1.45 D_{vs}.$$

During this phase of the displacement, solvent and oil bank sweepout are assumed to follow the $M = 25$ curve, Figs. 4.20 and C-1.

1. Calculate solvent and oil bank breakthroughs. Solvent breakthrough at $D_{vs} = 0.19$ is given by

$$HCPVI = 0.69(0.19) = 0.13$$

and oil bank breakthrough at $D_{vs} = 0.19/1.45 = 0.13$ is given by

$$HCVPI = 0.69(0.13) = 0.09.$$

2. Assume that $D_{vs} = 0.19$, $E_s = 0.19$. Then $D_{vob} = 1.45(0.19) = 0.275$, and $E_{ob} = 0.23$. Recovery* (%OOIP) is given by

$$\frac{[E_s(S_{orw} - S_{orm}) - (E_{ob} - E_s)(S_{ob} - S_{orw})]100}{S_{oi}}$$

$$= \frac{0.19(17) - (0.04)(38)}{0.8} = 2.14$$

and

$$HCPVI = 0.19(0.69) = 0.13.$$

3. At completion of slug injection, $D_{vs} = 0.25/0.69 = 0.36$, $E_s = 0.26$, $D_{vob} = 1.45(0.36) = 0.52$, and $E_{ob} = 0.315$. Recovery (%OOIP) is given by

$$\frac{0.26(17) - 0.055(38)}{0.8} = 2.9.$$

Calculations for Displacement Period Between End of Solvent-Slug Injection and Time at Which the Last of the Solvent Slug is Left as a Residual Saturation to Drive Water

Assume that sweepout of solvent and oil bank still follows the $M = 25$ curve. Then pseudodisplaceable volumes injected for estimating solvent sweepout is given by

$$D_{vs} = D'_{vs} + \Delta D_{vs} = 0.36 + \frac{D_{vw}(1 - S_{orm} - S_{wt} - S_{srw})}{(1 - S_{orm} - S_{wt})}$$

$$= \overset{.}{0}.36 + 0.69 D_{vw},$$

and

$$HCPVI = 0.69 D'_{vs} + \frac{D_{vw}(1 - S_{orm} - S_{srw} - S_{wt})}{S_{oi}}$$

$$= 0.25 + 0.48 D_{vw}.$$

1. At $D_{vw} = 0.2$, $E_w = 0.2$, $D_{vs} = 0.36 + 0.2(0.69) = 0.5$, $E_s = 0.31$, $D_{vob} = 1.45(0.5) = 0.73$, and $E_{ob} = 0.385$. Recovery (%OOIP) is given by

$$\frac{0.31(17) - (0.075)(38)}{0.8} = 3.03,$$

and

$$HCPVI = 0.25 + 0.48(0.2) = 0.35.$$

2. At $D_{vw} = 0.35$, $E_w = 0.35$, $D_{vs} = 0.36 + 0.69(0.35) = 0.6$, $E_s = 0.35$, $D_{vob} = 0.87$, and $E_{ob} = 0.43$. Recovery is 3.64%OOIP and HCPVI is 0.42.

At this value of D_{vw} the sweepouts of water and solvent are the same, and all solvent is left as a residual saturation.

Calculations for Displacement Period Where Drive Water Displaces Remaining Tertiary Oil Bank

$$\frac{M_{dw}}{M_{ob}} = 1.5$$

Assume that sweepout of the water and remaining tertiary oil bank eventually follows an $M = 1.5$ curve. There are no data available, so the sweepout curve for the rest of the calculation is drawn in using judgment and the $M = 1.5$ single-front curve as a guide (see Fig. C-1).

$$D_{vob} = D'_{vob} + \Delta D_{vw} = 0.87 + \Delta D_{vw}$$

and

$$D_{vw} = D'_{vw} + \Delta D_{vw} = 0.35 + \Delta D_{vw}.$$

Also,

$$HCPVI = 0.42 + 0.48 \times \Delta D_{vw},$$

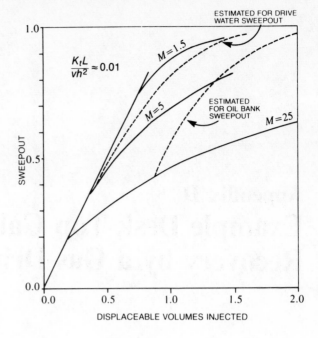

Fig. C-1—Vertical sweepout in a linear cross section for Region II flow.

and recovery* (%OOIP) is given by

$$\frac{[E'_s(S_{orw} - S_{orm}) - (E_{ob} - E_w)(S_{ob} - S_{orw})]100}{S_{oi}},$$

where E'_s is solvent sweepout when the last of the solvent slug is left as a residual saturation to the water drive.

1. At $\Delta D_{vw} = 0.3$, $D_{vw} = 0.35 + 0.3 = 0.65$, $E_w = 0.595$, $D_{vob} = 0.87 + 0.3 = 1.17$, and $E_{ob} = 0.65$. Recovery (%OOIP) is 4.83 and HCPVI is $0.42 + 0.48(0.3) = 0.56$.

2. At $\Delta D_{vw} = 0.6$, $D_{vw} = 0.95$, $E_w = 0.79$, $D_{vob} = 1.47$, and $E_{ob} = 0.835$. Recovery (%OOIP) is 5.3 and HCPVI is 0.72.

3. At $\Delta D_{vw} = 1.0$, $D_{vw} = 1.35$, $E_w = 0.93$, $D_{vob} = 1.87$, and $E_{ob} = 0.95$. Recovery (%OOIP) is 6.49 and HCPVI is 0.90. Ultimate recovery (%OOIP) is 7.4.

*Assumes that FVF is the same for oil at discovery (S_{oi}), oil at start of miscible flood (S_{orw}, S_{ob}), and miscible-flood residual oil (S_{orm}).

Appendix D
Example Desk Top Calculation for Tertiary Recovery by a Gas-Driven Slug

Data are as follows.

$$S_{orw} = 0.25,$$
$$S_{orm} = 0.03,$$
$$S_{ob} = 0.65,$$
$$S_{org} = 0.30,$$
$$S_{wr} = 0.2,$$
$$S_{wt} = 0.35,$$

$$\frac{M_s}{M_{ob}} = 17,$$

$$\frac{M_s}{M_{wf}} = 25,$$

$$\frac{M_{dg}}{M_{ob}} = 67,$$

$$\frac{M_{dg}}{M_{wf}} = 100, \text{ and}$$

$$B_o = 1.0.$$

For small slugs, assume that oil bank, solvent, and drive-gas sweepout are characterized predominantly by the drive-gas/waterflood-region mobility ratio. Assume that a ratio of approximately 85 characterizes the flood (Fig. D-1). Slug size is 5% HCPV.

Displacement Until Miscibility is Lost

Total displaceable volumes of slug plus drive gas injected is given by

$$D_{vt} = \frac{Q_s + Q_{dg}}{V_p(1 - S_{orm} - S_{wt})},$$

where Q_s is volume of solvent injected and Q_{dg} is volume of drive gas injected.

Pseudodisplaceable volumes injected for calculating sweepout of leading edge of oil bank is given by

$$D_{vob} = D_{vt} + D_{vt} \frac{(S_{orw} - S_{orm})}{(S_{ob} - S_{orw})} = 1.55 D_{vt},$$

and

$$\text{HCPVI} = \frac{D_{vt}(1 - S_{wt} - S_{orm})}{(1 - S_{wr})} = 0.78 D_{vt}.$$

From Fig. D-1, ultimate miscible sweepout, E_m, is 0.19. From Fig. D-2, solvent breakthrough occurs at $D_{vt} \cong 0.12$, and oil breakthrough occurs at $D_{vt} \cong 0.12/1.55 = 0.08$.

1. At $D_{vt} = 0.12$, $E_t = 0.12$, $D_{vob} = 0.19$, and $E_{ob} = 0.18$. Recovery* (%OOIP) is given by

$$\frac{[E_t(S_{orw} - S_{orm}) - (E_{ob} - E_t)(S_{ob} - S_{orw})]100}{S_{oi}}$$

$$= \frac{[0.12(22) - (0.06)(40)]}{0.8} = 0.3,$$

where E_t is the total sweepout of solvent plus drive gas, and HCPVI is 0.09.

2. At $D_{vt} = 0.21$, $E_t = 0.19$, $D_{vob} = 0.33$, and $E_{ob} = 0.28$. Recovery (%OOIP) is 0.73 and HCPVI is 0.16.

---

*Assumes that FVF is the same for oil at discovery (S_o), oil at start of miscible flood (S_{orw}, S_{ob}), miscible-flood residual oil (S_{orm}), and residual oil to immiscible gas drive (S_{org}).

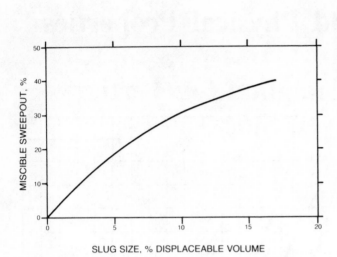

Fig. D-1—Miscible sweepout in viscous-finger-dominated displacements as estimated from the data of Lacey *et al*.

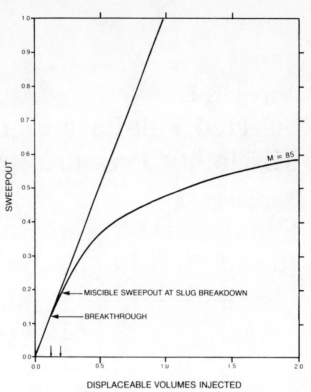

Fig. D-2—Sweepout of a single front for viscous-finger-dominated flow in a five-spot pattern.

Recovering Remaining Tertiary Oil Bank by Injecting Drive Gas

$$D_{vob} = D'_{vob} + \Delta D_{vob},$$

where D'_{vob} is pseudodisplaceable volumes injected when miscibility is lost and

$$\Delta D_{vob} = \Delta D_{vt} \frac{(S_{ob} - S_{org})}{(S_{ob} - S_{orw})} = \frac{35}{40} \Delta D_{vt} = 0.88 \Delta D_{vt},$$

where ΔD_{vt} is displaceable volumes of drive gas injected after miscibility is lost. Also,

$$HCPVI = \frac{\Delta D_{vt}(1 - S_{wt} - S_{org})}{(1 - S_{wr})} = 0.44 \Delta D_{vt}.$$

1. At $\Delta D_{vt} = 0.29$, $D_{vt} = 0.29 + 0.21 = 0.5$, $E_t = 0.37$, $\Delta D_{vob} = 0.29(0.88) = 0.26$, $D_{vob} = 0.26 + 0.33 = 0.59$, and $E_{ob} = 0.40$. Recovery* (%OOIP) is given by

*Assumes that FVF is the same for oil at discovery (S_{oi}), oil at start of miscible flood (S_{orw}, S_{ob}), miscible-flood residual oil (S_{orm}), and residual oil to immiscible gas drive (S_{org}).

$$\{[E_m(S_{orw} - S_{orm}) - (E_t - E_m)(S_{org} - S_{orw})$$

$$- (E_{ob} - E_t)(S_{ob} - S_{orw})]100\} \div (1 - S_{wr})$$

$$= \frac{0.19(22) - (0.18)(5) - (0.03)(40)}{0.8} = 2.6$$

and

$$HCPVI = 0.16 + 0.44(0.29) = 0.29.$$

2. At $\Delta D_{vt} = 0.5$, $D_{vt} = 0.21 + 0.5 = 0.71$, $E_t = 0.425$, $\Delta D_{vob} = 0.44$, $D_{vob} = 0.33 + 0.44 = 0.77$, and $E_{ob} = 0.44$. Recovery (%OOIP) is 3 and HCPVI is 0.38.

3. If $\Delta D_{vt} = 0.9$, $D_{vt} = 0.21 + 0.9 = 1.11$, $E_t = 0.49$, $\Delta D_{vob} = 0.88(0.9) = 0.79$, $D_{vob} = 0.33 + 0.79 = 1.12$, and $E_{ob} = 1.11$. The immiscible gas front has just overtaken the leading edge of the tertiary oil bank and ultimate recovery (%OOIP) is 3.3 and HCPVI is $0.16 + 0.44(0.9) = 0.56$.

Reference

1. Lacey, J.W., Faris, J.E., and Brinkman, F.H.: "Effect of Bank Size on Oil Recovery of the High Pressure Gas-Driven LPG-Bank Process," *J. Pet. Tech.* (Aug. 1961) 806–12; *Trans.*, AIME, **222**.

Appendix E
Selected Full-Page Figures for Estimating Miscibility Pressure and Physical Properties

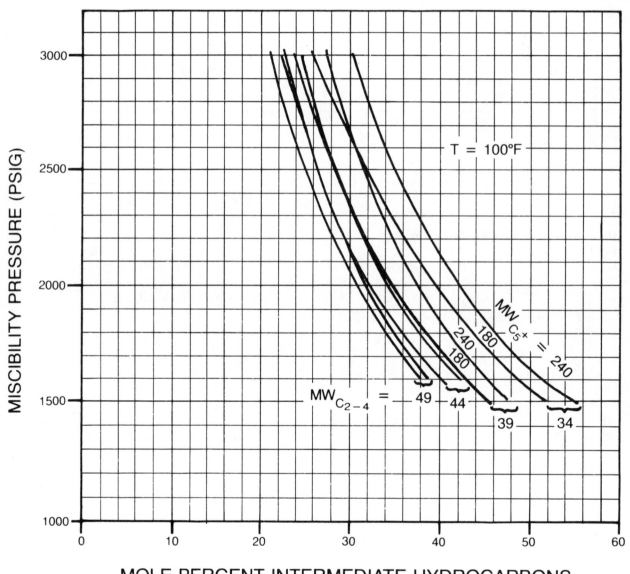

Fig. E-1—Condensing-gas-drive miscibility-pressure correlation, $T = 100°F$.

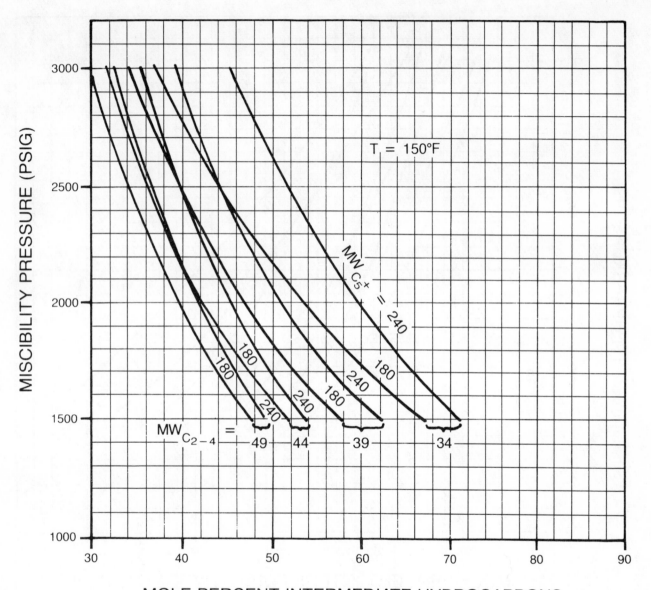

Fig. E-2—Condensing-gas-drive miscibility-pressure correlation, $T = 150°F$.

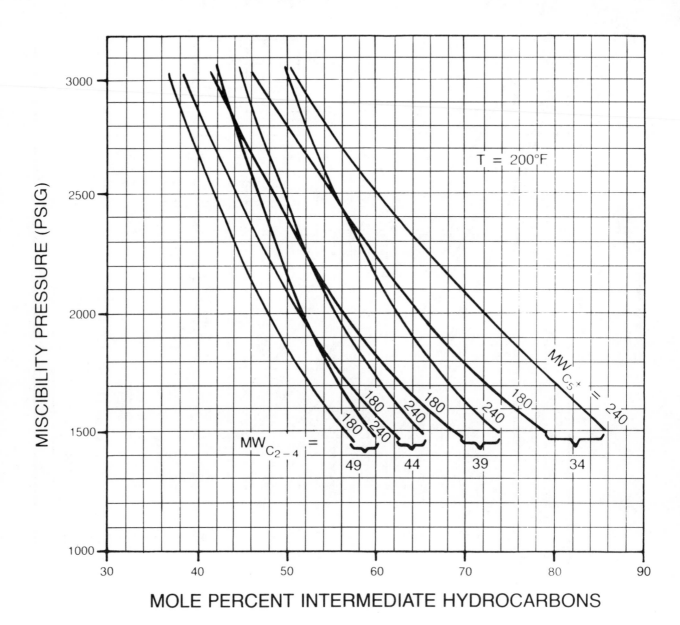

Fig. E-3—Condensing-gas-drive miscibility-pressure correlation, $T = 200°F$

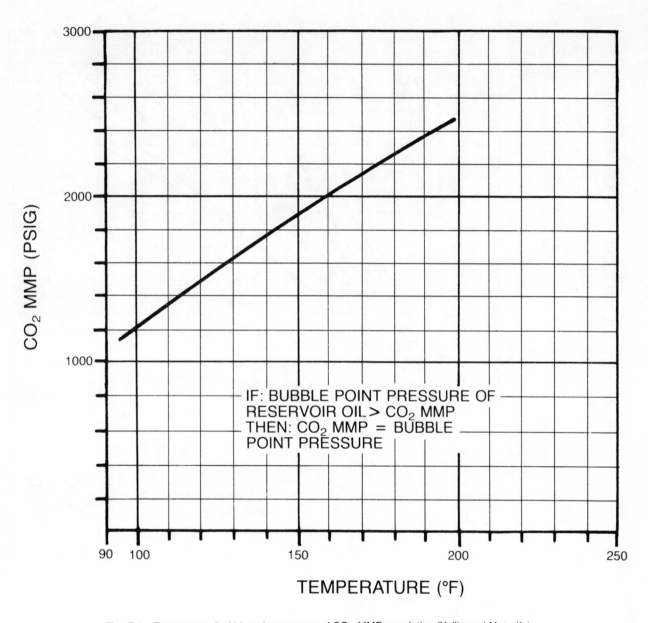

Fig. E-4—Temperature/bubble-point pressure of CO_2 MMP correlation (Yellig and Metcalfe).

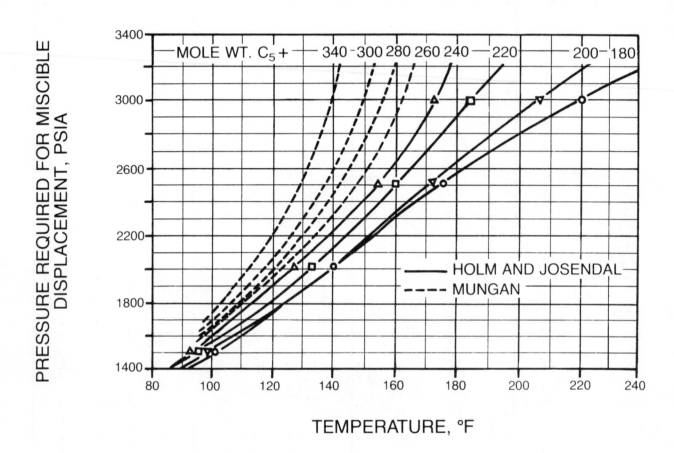

Fig. E-5—Pressure required for miscible displacement in CO_2 flooding (Holm and Josendal).

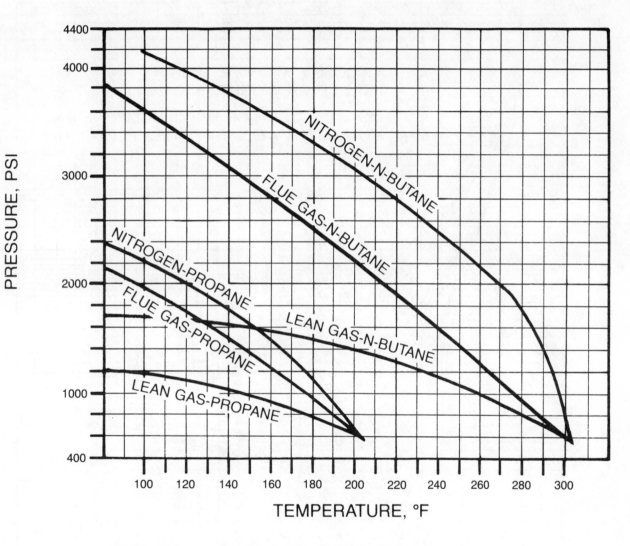

Fig. E-6—Minimum pressure for miscibility, miscible slug/driving-gas combinations.

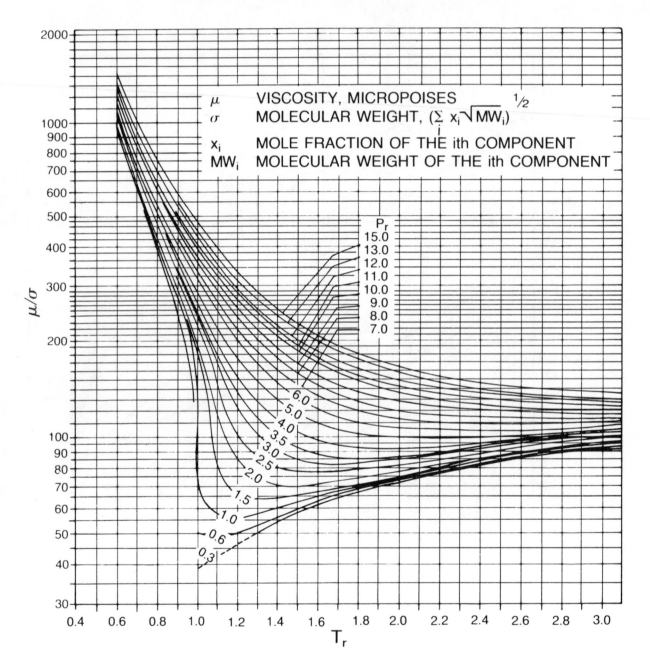

Fig. E-7—Viscosity of hydrocarbon systems as a function of reduced pressure.

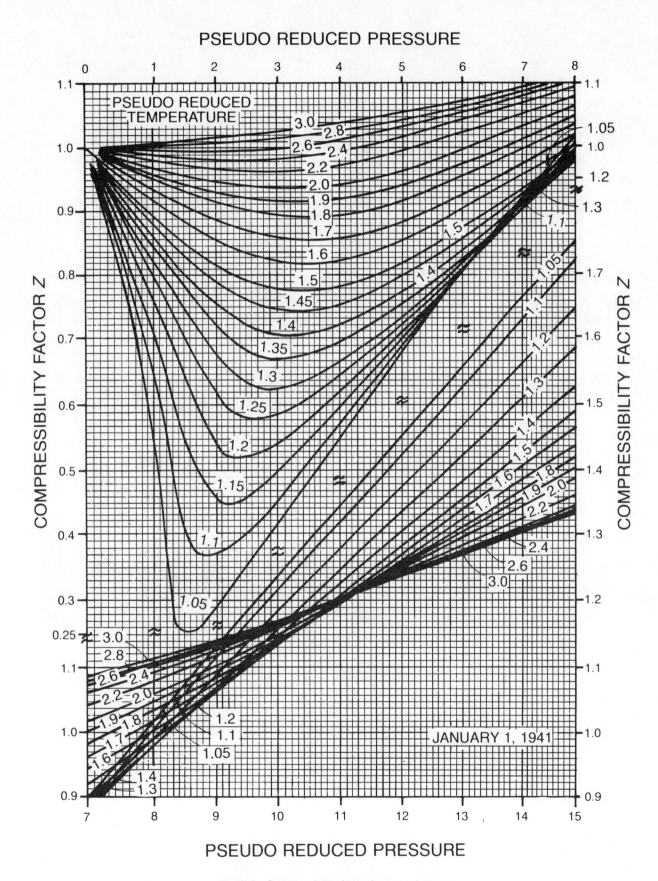

Fig. E-8—Compressibility factor for natural gases.

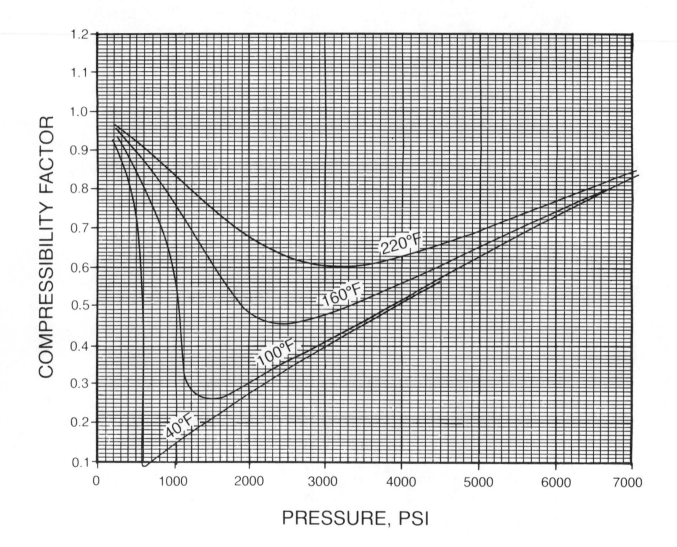

Fig. E-9—Compressibility factors for CO_2.

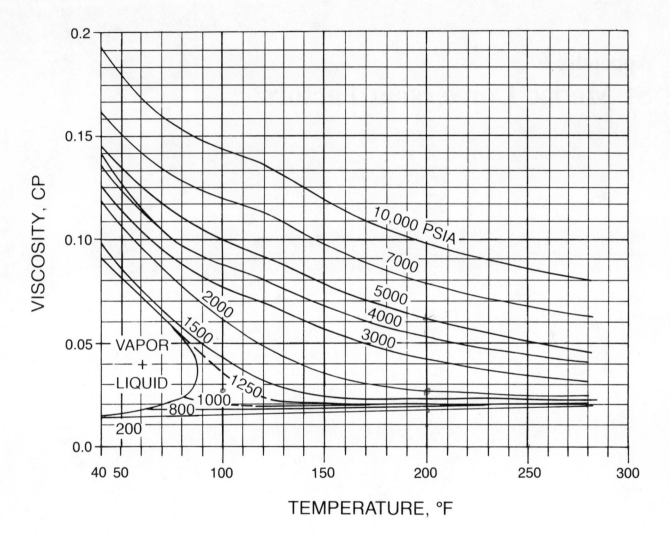

Fig. E-10—Viscosity of CO_2.

Appendix F
SI Metric Conversion Factors

$$
\begin{array}{rll}
\text{acre} \times 4.046\ 873 & \text{E}+03 & = \text{m}^2 \\
\text{acre-ft} \times 1.233\ 489 & \text{E}+03 & = \text{m}^3 \\
\text{°API} \quad 141.5/(131.5+\text{°API}) & & = \text{g/cm}^3 \\
\text{atm} \times 1.013\ 250^* & \text{E}+05 & = \text{Pa} \\
\text{bbl} \times 1.589\ 873 & \text{E}-01 & = \text{m}^3 \\
\text{cp} \times 1.0^* & \text{E}-03 & = \text{Pa·s} \\
\text{cu in.} \times 1.638\ 706 & \text{E}+01 & = \text{cm}^3 \\
\text{cu ft} \times 2.831\ 685 & \text{E}-02 & = \text{m}^3 \\
\text{degree} \times 1.745\ 329 & \text{E}-02 & = \text{rad} \\
\text{°F} \quad (\text{°F}-32)/1.8 & & = \text{°C} \\
\text{ft} \times 3.048^* & \text{E}-01 & = \text{m} \\
\text{in.} \times 2.54^* & \text{E}+00 & = \text{cm} \\
\text{lbf} \times 4.448\ 222 & \text{E}+00 & = \text{N} \\
\text{lbm} \times 4.535\ 924 & \text{E}-01 & = \text{kg} \\
\text{mile} \times 1.609\ 344^* & \text{E}+00 & = \text{km} \\
\text{psi} \times 6.894\ 757 & \text{E}+00 & = \text{kPa} \\
\text{°R} \times \tfrac{5}{9} & & = \text{K} \\
\text{scf} \times 2.863\ 640 & \text{E}-02 & = \text{std m}^3 \\
\text{sq in.} \times 6.451\ 6^* & \text{E}+00 & = \text{cm}^2 \\
\text{sq ft} \times 9.290\ 304^* & \text{E}-02 & = \text{m}^2 \\
\text{yard} \times 9.144^* & \text{E}-01 & = \text{m}
\end{array}
$$

*Conversion factor is exact.

Nomenclature

a = dimensionless mass transfer coefficient, $K^1 L/v$

A = area

B = formation volume factor, volume at reservoir conditions divided by volume at surface conditions

B_{om} = formation volume factor of residual oil to miscible flood

B_{ow} = formation volume factor of residual oil to waterflood

C = concentration

d_p = particle diameter

D = effective molecular diffusion coefficient

D_o = molecular diffusion coefficient

D_v = displaceable volumes injected

E = effective viscosity ratio in Koval equation

E_A = areal sweepout, fraction

E_c = capture efficiency, fraction

E_m = miscible sweepout

E_I = vertical sweepout, fraction

E_v = volumetric sweepout, fraction

f = fraction of hydrocarbon phase that is mobile

f_{se} = solvent fractional flow in effluent

f_w = water fractional flow

F = formation electrical resistivity factor

g = gravitational acceleration

G = quantity of material that has diffused across a plane

h = thickness

H = Koval heterogeneity factor

i = injection rate per well

k = absolute permeability

k_H = horizontal permeability

k_o = effective permeability to oil

k_{ro} = relative permeability to oil

k_{roi} = relative permeability to oil at initial conditions

$k_{rs,wr}$ = solvent relative permeability at irreducible water saturation

$k_{rw,sr}$ = water relative permeability at residual solvent saturation

k_v = vertical permeability

K = EH (Koval method) or K-value or equilibrium ratio of vapor and liquid mole fractions, y/x

K^1 = mass transfer coefficient

K_e = effective dispersion coefficient that approximates mixing caused both from true dispersion and from dispersion of components between flowing and bypassed phases

K_ℓ = longitudinal dispersion coefficient

K_t = transverse dispersion coefficient

L = length

M = mobility ratio, displacing fluid/displaced fluid

$\bar{M}$ = average mobility ratio

MW_i = molecular weight of component i

N = number of equilibrium stages

N_p = cumulative recovery

N_{pv} = oil recovery in pore volumes

p = pressure

p_c = critical pressure

p_R = reduced pressure

P_c = capillary pressure

Q = cumulative injection

q_t = total injection rate

r_m = mean radius of displacement

R_{so} = solubility of solvent in residual oil

R_{sw} = solubility of solvent in water

$R_{v/g}$ = ratio of viscous to gravity forces

S = saturation, fraction at reservoir conditions

S_{ob} = oil saturation in oil bank, fraction at reservoir conditions

S_{oi} = initial oil saturation at discovery, fraction at reservoir conditions

S_{org} = residual oil saturation to immiscible gas drive, fraction at reservoir conditions

S_{orm} = residual oil saturation to miscible flooding, fraction at reservoir conditions

S_{orw} = residual oil saturation to waterflood, fraction at reservoir conditions

S_{sr} = residual solvent saturation to waterflood, fraction at reservoir conditions

S_{wi} = initial water saturation, fraction at reservoir conditions

S_{wr} = residual or irreducible water saturation, fraction at reservoir conditions

S_{wt} = water saturation at the trailing edge of the oil bank, fraction at reservoir conditions

$\bar{S}_o$ = oil saturation at stock-tank conditions, fraction

t = time
T = temperature
T_c = critical temperature
T_R = reduced temperature
u = volumetric or Darcy velocity
u_c = critical rate, Darcy velocity
u_{st} = stable rate, Darcy velocity
v = interstitial velocity
V = molal volume
V_D = displaceable volume
V_{pD} = dimensionless pore volumes
V_{pDi} = dimensionless pore volumes injected
W = width
x = mole fraction in liquid phase
y = mole fraction in vapor phase
x,y,z = directions in a rectangular coordinate system
α = angle
β = slug size, fraction of PV
γ = vL/K_ℓ
δ = distance from 50% concentration front

λ_c = critical finger width
μ = viscosity
μ_{ge} = effective gas viscosity
μ_{oe} = effective oil viscosity
ρ = density
σ = inhomogeneity factor or interfacial tension
ϕ = porosity, fraction
ω = mixing parameter

Subscripts
BT = breakthrough
g = gas
i = component i or initial conditions
o = oil
ob = oil bank
ow = oil/water
s = solvent
sw = solvent/water
w = water
wf = waterflood region

Bibliography

A

Acs, G., Doleschall, S. and Farkas, E.: "General Purpose Compositional Model," paper SPE 10515 presented at the Sixth SPE Symposium on Reservoir Simulation, New Orleans, Jan. 31–Feb. 3, 1982.

Adams, G.H. and Rowe, H.G.: "Slaughter Estate Unit CO_2 Pilot—Surface and Downhole Equipment Construction and Operation in the Presence of Hydrogen Sulfide Gas," *J. Pet. Tech.* (June 1981) 1065–74.

Ader, J.C. and Stein, M.H.: "Slaughter Estate Unit CO_2 Pilot Reservoir Description via a Black Oil Model Waterflood History Match," paper SPE 10727 presented at the SPE/DOE Third Joint Symposium on Enhanced Oil Recovery, Tulsa, April 4–7, 1982.

Agan, J.B. and Fernandes, R.J.: "Performance Prediction of a Miscible-Slug Process in a Highly Stratified Reservoir," *J. Pet. Tech.* (Jan. 1962) 81–86.

Alani, G.H. and Kennedy, H.T.: "Volumes of Liquid Hydrocarbons at High Temperatures and Pressures," *Trans.*, AIME (1960) **219**, 288–92.

Albright, J.N. and Pearson, C.F.: "Location of Hydraulic Fractures Using Microseismic Techniques," paper SPE 9509 presented at the 1980 SPE 55th Annual Technical Conference and Exhibition, Dallas, Sept. 21–24, 1980.

Aris, R. and Amundson, N.R.: "Some Remarks on Longitudinal Mixing or Diffusion in Fixed Beds," *AIChE J.* (1957) **3**, 280.

Aronofsky, J.S.: "Mobility Ratio—Its Influence on Flood Patterns During Water Encroachment," *Trans.*, AIME (1952) **195**, 15–24.

Ausburn, B.E., Nath, A.K., and Wittick, T.R.: "Modern Seismic Methods—An Aid to the Petroleum Engineer," *J. Pet. Tech.* (Nov. 1978) 1519–30.

Aziz, K. and Settari, A.: *Petroleum Reservoir Simulation*, Applied Science Publishers Ltd., London (1979).

B

Bain, J.S., Nordberg, M.O., and Hamilton, T.M.: "3D Seismic Applications in the Interpretation of Dunlin Field, U.K. North Sea," *J. Pet. Tech.* (March 1981) 407–12.

Baker, L.E.: "Effects of Dispersion and Dead-End Pore Volume in Miscible Flooding," *Soc. Pet. Eng. J.* (June 1977) 219–27.

Balch, A.H. *et al.*: "The Use of Vertical Seismic Profiles and Surface Seismic Profiles to Investigate the Distribution of Aquifers in the Madison Group and Red River Formation, Powder River Basin, Wyoming-Montana," paper SPE 9312 presented at the 1980 SPE Annual Technical Conference and Exhibition, Dallas, Sept. 21–24.

Barstow, W.F. and Watt, G.W.: "Fifteen Years of Progress in Catalytic Treating of Exhaust Gas," paper SPE 5347 presented at the 1975 SPE Rocky Mountain Regional Meeting, Denver, April 7–9.

Baugh, E.G.: "Performance of Seeligson Zone 20-B Enriched Gas-Drive Project," *J. Pet. Tech.* (March 1960) 29–33.

Beeler, P.F.: "West Virginia CO_2 Oil Recovery Project Interim Report," *Proc.*, 1977 ERDA Symposium on Enhanced Oil, Gas Recovery, and Improved Drilling Methods, Tulsa.

Benedict, M., Webb, G.B., and Rubin, L.C.: "An Empirical Equation for Thermodynamic Properties of Light Hydrocarbons and Their Mixtures—Constants for Twelve Hydrocarbons," *Chem. Eng. Prog.* (1951) **47**, 419.

Benham, A.L., Dowden, W.E., and Kunzman, W.J.: "Miscible Fluid Displacement—Prediction of Miscibility," *Trans.*, AIME (1960) **219**, 229–37; *Miscible Displacement*, Reprint Series, SPE, Dallas (1965) **8**, 123–31.

Benham, A.L. and Olson, R.W.: "A Model Study of Viscous Fingering," *Soc. Pet. Eng. J.* (June 1963) 138–44; *Trans.*, AIME, **228**.

Bennett, G.S.: "A Study of the Foam-Drive Process for Removal of Brine from a Consolidated Sandstone Core," MS thesis, Pennsylvania State U., University Park (1963).

Bentsen, R.G. and Nielson, R.F.: "A Study of Plane Radial Miscible Displacement in a Consolidated Porous Medium," *Soc. Pet. Eng. J.* (March 1965) 1–5; *Trans.*, AIME, **234**.

Bernard, G.G.: "Effect of Foam on Recovery of Oil by Gas Drive," *Prod. Monthly* (1963) **27**, No. 1, 18–21.

Bernard, G.G.: U.S. Patent No. 3,330,351 (1967).

Bernard, G.G. and Holm, L.W.: "Effect of Foam on Permeability of Porous Media to Gas," *Soc. Pet. Eng. J.* (Sept. 1964) 267–74.

Bernard, G.G. and Holm, L.W.: "Use of Surfactant to Reduce CO_2 Mobility in Oil Displacement," *Soc. Pet. Eng. J.* (Aug. 1980) 281–92.

Bernard, G.G., Holm, L.W., and Jacobs, W.L.: "Effect of Foam on Trapped Gas Saturation and on Permeability of Porous Media to Water," *Soc. Pet. Eng. J.* (Dec. 1965) 295–300.

Besserer, G.J., Serra, J.W., and Best, D.D.: "An Efficient Phase Behavior Package for Use in Compositional Reservoir Simulation Studies," paper SPE 8288 presented at the 1979 SPE Annual Technical Conference and Exhibition, Las Vegas, Sept. 23–26.

Bibliography of Vapor-Liquid Equilibrium Data for Hydrocarbon Systems, API, Div. of Refining (Bibliography No. 1, 1963) Table 18, 32.

Bicher, L.B. Jr. and Katz, D.L.: "Viscosity of Natural Gases," *Trans.*, AIME (1944) **155**, 246–52.

Bilhartz, H.L.: "Case History: A Pressure Core Hole," paper SPE 6389 presented at the 1977 SPE Permian Basin Oil and Gas Recovery Conference, Midland, TX, March 10–11.

Bilhartz, H.L. and Charlson, G.S.: "Coring for In-Situ Saturations in the Willard Unit CO_2 Flood Mini-Test," paper SPE 7050 presented at the 1978 SPE Symposium on Improved Methods for Oil Recovery, Tulsa, April 16–18.

Bilhartz, H.L. Jr. and Charlson, G.S.: "Field Polymer Stability Studies," paper SPE 5551 presented at the 1975 SPE Annual Technical Conference and Exhibition, Dallas, Sept. 28–Oct. 1.

Bilhartz, H.L. et al.: "A Method for Projecting Full-Scale Performance of CO_2 Flooding in the Willard Unit," paper SPE 7051 presented at the 1978 SPE Symposium on Improved Methods for Oil Recovery," Tulsa, April 16–18.

Blackwell, R.J.: "Laboratory Studies of Microscopic Dispersion Phenomena," Soc. Pet. Eng. J. (March 1962) 1–8; Trans., AIME, 225; Miscible Displacement, Reprint Series, SPE, Dallas (1965) 8, 69–76.

Blackwell, R.J., Rayne, J.R., and Terry, W.M.: "Factors Influencing the Efficiency of Miscible Displacement," Trans., AIME (1959) 216, 1–8; Miscible Displacement, Reprint Series, SPE, Dallas (1965) 8, 197–204.

Blackwell, R.J. et al.: "Recovery of Oil by Displacement with Water-Solvent Mixtures," Trans., AIME (1960) 21, 293–300; Miscible Displacement, Reprint Series, SPE, Dallas (1965) 8, 103–10.

Blanton, J.R., McCaskill, N., and Herbeck, E.F.: "Performance of a Propane Slug Pilot in a Watered-Out Sand—South Ward Field," J. Pet. Tech. (Oct. 1970) 1209–14.

Bleakley, W.B.: "Journal Survey Shows Recovery Projects Up," Oil and Gas J. (March 25, 1974) 69–78.

Block, W.E. and Donovan, R.W.: "An Economically Successful Miscible Phase Displacement Project," J. Pet. Tech. (Jan. 1961) 35–40.

Bond, D.G. and Bernard, G.G.: "Rheology of Foams in Porous Media," paper presented at the 1966 SPE/AIChE Symposium, AIChE 58th Annual Meeting, Dallas, Feb. 7–10.

Bond, D.C. and Holbrook, O.C.: U.S. Patent No. 2,866,507 (1958).

Bossler, R.B. and Crawford, P.B.: "Miscible-Phase Floods May Precipitate Asphalt," Oil and Gas J. (1959) 57, No. 9, 137.

Botset, H.G.: "The Electrolytic Model and Its Application to the Study of Recovery Problems," Trans., AIME (1946) 165, 15–25.

Bouck, L.S., Hearn, C.L., and Dohy, G.: "Performance of a Miscible Flood in the Bear Lake Cardium Unit, Pembina Field, Alberta, Canada," J. Pet. Tech. (June 1975) 672–78.

Bournazel, C. and Jeanson, B.: "Fast Water-Coning Evaluation Method," paper SPE 3628 presented at the 1971 SPE Annual Meeting, New Orleans, Oct. 3–6.

Brannan, G. and Whittington, H.M. Jr.: "Enriched Gas Miscible Flooding—A Case History of the Levelland Unit Secondary Miscible Project," J. Pet. Tech. (Aug. 1977) 919–24.

Brigham, W.E.: "Mixing Equations in Various Geometries," paper SPE 4585 presented at the 1973 SPE Annual Meeting, Las Vegas, Sept. 30–Oct. 3.

Brigham, W.E.: "Mixing Equations in Short Laboratory Cores," Soc. Pet. Eng. J. (Feb. 1974) 91–99; Trans., AIME, 257.

Brigham, W.E., Reed, P.W., and Dew, J.N.: "Experiments on Mixing During Miscible Displacement in Porous Media," Soc. Pet. Eng. J. (March 1961) 1–8; Trans., AIME, 222.

Brinkley, T.W.: "Sunray's LPG Pilot Test is Paying Off," World Oil (Aug. 1, 1960) 85–93.

Brooks, R.E.: "Miscible Drive Field Application—In the Bisti Lower Gallup Pool," J. Pet. Tech. (May 1958) 22–24.

Brooks, R.E.: "Miscible Displacement in Bisti Pool, San Juan Co., New Mexico," paper 1037G presented at the 1958 AIME Mid-Continent Section Petroleum Conference on Production and Reservoir Engineering, Tulsa, March 20–21.

Brown, G.G. et al.: Natural Gasoline and the Volatile Hydrocarbons, Natural Gas Assn. of America, Tulsa (1948).

Brownscombe, E.R., McNeese, C.R., and Miller, C.C.: "Improved Oil Recovery by Miscible Displacement," Proc., Sixth World Pet. Cong. (1963) 2, 563–75.

Buckley, S.E. and Leverett, M.C.: "Mechanisms of Fluid Displacement in Sands," Trans., AIME (1942) 146, 107–16.

Burt, R.A. Jr.: "High Pressure Miscible Gas Displacement Project, Bridger Lake Unit, Summit County," paper SPE 3487 presented at the 1971 SPE Annual Meeting, New Orleans, Oct. 3–6.

Burton, M.B. Jr. and Crawford, P.B.: "Application of the Gelatin Model for Studying Mobility Ratio Effects," Trans., AIME (1956) 207, 333–37.

C

Camy, J.P. and Emanual, A.S.: "Effect of Grid Size in the Compositional Simulation of CO_2 Injection," paper SPE 6894 presented at the 1977 SPE Annual Technical Conference and Exhibition, Denver, Oct. 9–12.

Canjar, L.N. et al. "Thermodynamic Properties of Nonhydrocarbons," Hydrocarbon Processing, Petroleum Refiner (Jan. 1966) 135.

Caraway, G.E. and Lowrey, L.L.: "Generating Flue Gas for Injection Releases Sales Gas," Oil and Gas J. (July 28, 1975) 126–32.

Cardenas, R.L. et al.: "Laboratory Design of a Gravity-Stable, Miscible CO_2 Process," paper SPE 10270 presented at the 1981 SPE Annual Technical Conference and Exhibition, San Antonio, Oct. 5–7; accepted for publication in J. Pet. Tech.

Caudle, B.H. and Dyes, A.B.: "Improving Miscible Displacement by Gas-Water Injection," Trans., AIME (1958) 213, 281–84; Miscible Displacement, Reprint Series, SPE, Dallas (1965) 8, 111–14.

Chambers, F.T.: "Tertiary Oil Recovery Combination Water-Gas Miscible Flood—Hibberd Pool," Prod. Monthly (Jan. 1968) 17.

Chandrasekhar, S.: Hydrodynamic and Hydromagnetic Stability, Oxford Clarendon Press, London (1961).

Chappelear, J.E. and Hirasaki, G.J.: "A Model of Oil-Water Coning for 2-D Areal Reservoir Simulation," Trans., AIME (1976) 261, 65–72.

Charlson, G.S., Bilhartz, H.L., and Stalkup, F.I.: "Use of Time-Lapse Logging Techniques in Evaluating the Willard Unit CO_2 Flood Mini-Test," paper SPE 7049 presented at the 1978 SPE Symposium on Improved Methods for Oil Recovery, Tulsa, April 16–18.

Chase, C.A. and Todd, M.R.: "Numerical Simulation of CO_2 Flood Performance," paper SPE 10514 presented at the 1982 SPE Symposium on Reservoir Simulation, New Orleans, Jan. 31–Feb. 6.

Chaudhari, N.M.: "A Numerical Solution with Second Order Accuracy for Compressible Stable Miscible Flow," Soc. Pet. Eng. J. (April 1973) 84–92.

Chaudhari, N.M.: "An Improved Numerical Technique for Solving Multidimensional Miscible Displacement Equations," Soc. Pet. Eng. J. (Sept. 1971) 277–84.

Christian, L.D. et al.: "Planning a Tertiary Oil Recovery Project for Jay-Little Escambia Creek Fields Unit," J. Pet. Tech. (Aug. 1981) 1534–44.

Chuoke, R.L., Van Meurs, P., Van der Poel, C.: "The Instability of Slow, Immiscible, Viscous Liquid-Liquid Displacements in Permeability Media," Trans., AIME (1959) 216, 188–94.

Clancy, J.P., Kroll, D.E., and Gilchrist, R.E.: "How Nitrogen is Produced and Used for Enhanced Recovery," *World Oil* (Oct. 1981) 233–42.

Claridge, E.L.: "A Method for the Design of Graded Viscosity Banks," *Soc. Pet. Eng. J.* (Oct. 1978) 315–24.

Claridge, E.L.: "A Trapping Hele-Shaw Model for Miscible-Immiscible Flooding Studies," *Soc. Pet. Eng. J.* (Oct. 1973) 255–60.

Claridge, E.L.: "CO_2 Flooding Strategy in a Communicating Layered Reservoir," *J. Pet. Tech.* (Dec. 1982) 2746–56.

Claridge, E.L.: "Prediction of Recovery in Unstable Miscible Flooding," *Soc. Pet. Eng. J.* (April 1972) 143–55.

Coan, S.W.: "Profitable Miscible Flood in a Watered-Out Reservoir," *Oil and Gas J.* (Dec. 12, 1966) 111.

Coats, K.H.: "An Equation of State Compositional Model," *Soc. Pet. Eng. J.* (Dec. 1981) 687–98.

Coats, K.H.: "Reservoir Simulation: State of the Art," *J. Pet. Tech.* (Aug. 1982) 1633–42.

Coats, K.H., George, W.D., and Marcum, B.E.: "Three Dimensional Simulation of Steamflooding," *Soc. Pet. Eng. J.* (Dec. 1974) 573–92; *Trans.*, AIME, **257**.

Coats, K.H. and Smith, B.D.: "Dead-End Pore Volume and Dispersion in Porous Media," *Soc. Pet. Eng. J.* (March 1964) 73–84; *Trans.*, AIME, **231**.

Coats, K.H. *et al.*: "Simulation of Three-Dimensional, Two-Phase Flow in Oil and Gas Reservoirs," *Soc. Pet. Eng. J.* (Dec. 1967) 377–88; *Trans.*, AIME, **240**.

Collins, R.E.: *Flow of Fluids Through Porous Media*, Reinhold Publishing Co., New York City (1961) 201.

Collona, J., Iffly, R., and Millet, J.L.: "Water Coning in Underground Gas Reservoirs," *Rev. L'Inst. Franc. Pétrole* (1969) **24**, No. 1, 121–44.

Cone, C.: "Case History of the University Block 9 (Wolfcamp) Field—A Gas-Water Injection Secondary Recovery Project," *J. Pet. Tech.* (Dec. 1970) 1485–91.

Connally, C.A. Jr.: "Tertiary Miscible Flood in Phegly Unit, Washington County, Colorado," paper SPE 3775 presented at the 1972 SPE Improved Oil Recovery Symposium, Tulsa, April 16–19.

Conner, W.D.: "Granny's Creek CO_2 Injection Project, Clay Co., West Virginia," *Proc.*, ERDA Symposium on Enhanced Oil and Gas Recovery, and Improved Drilling Methods, Tulsa, Aug. 30–Sept. 1, 1977.

Cordiner, F.S. and Popp, M.P.: "Application of Miscible Phase Displacement Techniques to a Water-Drive Reservoir, Johnson Field, Cheyenne Co., Nebraska," *Drill. and Prod. Prac.*, API (1962) 144.

Corteville, J., Van Quy, N., and Simandoux, P.: "A Numerical and Experimental Study of Miscible or Immiscible Fluid Flow in Porous Media with Interphase Mass Transfer," paper SPE 3481 presented at the 1971 SPE Annual Meeting, New Orleans, Oct. 3–6.

Cottin, R.H.: "Optimization of the Completion and Exploitation of a Well Under Two Phase Coning," *Rev. L'Inst. Franc. Pétrole* (1971) **26**, 171–97.

Craig, D.R. and Bray, J.A.: "Gas Injection and Miscible Flooding," *Proc.*, Eighth World Pet. Cong. (1971) **3**, 275–85.

Craig, F.F. Jr.: *The Reservoir Engineering Aspects of Waterflooding*, Monograph Series, SPE, Dallas (1971) **3**.

Craig, F.F. Jr.: "A Current Appraisal of Field Miscible Slug Projects," *J. Pet. Tech.* (May 1970) 529–36.

Craig, F.F. Jr.: "Laboratory Model Study of Single Five-Spot and Single Injection-Well Pilot Waterflooding," *Trans.*, AIME (1965) **234**, 1454–60.

Craig, F.F. Jr., and Lummus, J.L.: U.S. Patent No. 3,185,634 (1965).

Craig, F.F. Jr. and Owens, W.W.: "Miscible Slug Flooding—A Review," *J. Pet. Tech.* (April 1960) 11–15.

Craig, F.F. Jr. *et al.*: "A Laboratory Study of Gravity Segregation in Frontal Drives," *Trans.*, AIME (1957) **210**, 275–82.

Craig, F.F. Jr. *et al.*: "Optimized Recovery Through Cooperative Geology and Reservoir Engineering," *J. Pet. Tech.* (July 1977) 755–60.

Crane, F.E. and Gardner, G.H.F.: "Measurements of Transverse Dispersion in Granular Media," *J. Chem. Eng. Data* (1961) **6**, 283.

Crane, F.E., Kendall, H.A., and Gardner, G.H.F.: "Some Experiments of the Flow of Miscible Fluids of Unequal Density Through Porous Media," *Soc. Pet. Eng. J.* (Dec. 1963) 277–80; *Trans.*, AIME, **288**.

Crank, J.: *The Mathematics of Diffusion*, second edition, Oxford U. Press, New York City (1975).

Crockett, D.H.: "A Profile Control Program Utilized in the SACROC Unit CO_2 Injection Program," *Proc.*, 22nd Annual Southwestern Petroleum Short Course, Texas Tech. U., Lubbock (1975) 159–64.

Croes, G.A., Geertsma, J., and Swartz, N.: "Theory of Dimensionally Scaled Models of Petroleum Reservoirs," *Trans.*, AIME (1956) **207**, 118–22.

Cronquist, C.: "Carbon Dioxide Dynamic Miscibility With Light Reservoir Oils," *Proc.*, Fourth Annual U.S. DOE Symposium on Enhanced Oil and Gas Recovery and Improved Drilling Methods, Tulsa, Aug. 28–30, 1978.

Culham, W.E., Farouq Ali, S.M., and Stahl, C.D.: "Experimental and Numerical Simulation of Two-Phase Flow with Interphase Mass Transfer in One and Two Dimensions," *Soc. Pet. Eng. J.* (Sept. 1969) 323–37.

D

Darlow, B.L., Ewing, R.E., and Wheeler, M.F.: "Mixed Finite Element Methods for Miscible Displacement Problems in Porous Media," paper SPE 10501 presented at the 1982 SPE Symposium on Reservoir Simulation, New Orleans, Jan. 31–Feb. 3.

Deans, H.A. and Majoros, S.: *The Single-Well Chemical Tracer Method for Measuring Residual Oil Saturation*, Final Report for U.S. DOE, Contract No. DE-AS19-79BC20006 performed at Rice U. (Oct. 1980).

Deans, H.A. and Shallenberger, L.K.: "Single Well Chemical Tracer Method to Measure Connate Water Saturation," paper SPE 4755 presented at the 1974 SPE Improved Oil Recovery Symposium, Tulsa, April 22–24.

DeLaney, R.P. and Fish, R.M.: "Judy Creek CO_2 Flood Performance Predictions," paper 80-31-23 presented at the 1980 Annual Technical Meeting of the Petroleum Society of CIM, Calgary, May 25–28.

Deming, J.R.: "Fundamental Properties of Foams and Their Effects on the Efficiency of the Foam Drive Process," MS thesis, Pennsylvania State U., University Park (1964).

Deppe, J.C.: "Injection Rate—The Effect of Mobility Ratio, Area Swept and Pattern," *Soc. Pet. Eng. J.* (June 1961) 81–91.

Des Brisay, C.L. and Daniel, E.L.: "Supplemental Recovery Development of the Intisar 'A' and 'D' Reef Fields, Libyan Arab Republic," *J. Pet. Tech.* (July 1972) 785–90.

Des Brisay, C.L., Gray, J.W., and Spivak, A.: "Miscible Flood Performance of the Intisar 'D' Field, Libyan Arab Republic," *J. Pet. Tech.* (Aug. 1975) 935–43.

Des Brisay, C.L. *et al.*: "Review of Miscible Flood Performance, Intisar 'D' Field, Socialist People's Libyan Arab Jamahiriya," *J. Pet. Tech.* (Aug. 1982) 1651–60.

Desch, J.B. *et al.*: "Enhanced Oil Recovery by CO_2 Miscible Displacement in the Little Knife Field, Billings County, North Dakota," paper SPE 10696 presented at the 1982 SPE/DOE Symposium on Enhanced Oil Recovery," Tulsa, April 4–7.

DiCharry, R.M., Perryman, T.L., and Ronquille, J.D.: "Evaluations and Design of a CO_2 Miscible Flood Project—SACROC Unit, Kelly-Snyder Field," *J. Pet. Tech.* (Nov. 1973) 1309–18; *Trans.*, AIME, **215**.

Dickey, P.A.: *Petroleum Development Geology*, The Petroleum Publishing Co., Tulsa (1979).

Dietz, D.N.: *Proc.*, Acad. Sci. Amst. B. (1953) **56**, 83.

Dietz, D.N.: "A Theoretical Approach to the Problem of Encroaching and By-Passing Edge Water," Akademi Van Wetershappen, Amsterdam, *Proc., Ser. B. Physical Science* (1953) **56**, 83.

Dodds, W.S., Stutzman, L.F., and Sollami, B.J.: "Carbon Dioxide Solubility in Water," *I&EC Chem. and Eng. Data Series*, 1 (1956).

Doepel, G.W. and Sibley, W.P.: "Miscible Displacement—A Multilayer Technique for Predicting Reservoir Performance," *J. Pet. Tech.* (Jan. 1962) 73–80.

Dolan, J.P. *et al.*: "Liquid, Gas, and Dense Fluid Viscosity of n-Butane," *J. Chem. and Eng. Data* (July 1963) **8**, 396–99.

Donahoe, C.W. and Buchanan, R.D.: "Economic Evaluation of Cycling Gas-Condensate Reservoirs with Nitrogen," *J. Pet. Tech.* (Feb. 1981) 263–70.

Dougherty, E.L.: "Mathematical Models of an Unstable Miscible Displacement," *Soc. Pet. Eng. J.* (June 1963) 155–163; *Miscible Processes*, Reprint Series, SPE, Dallas (1965) **8**, 151–59.

Drennon, M.D., Kelm, C.H., and Whittington, H.M.: "A Method for Appraising the Feasibility of Field-Scale CO_2 Miscible Flooding," paper SPE 9323 presented at the 1980 SPE Annual Technical Conference and Exhibition, Dallas, Sept. 21–24.

Dumoré, J.M.: "Stability Considerations in Downward Miscible Displacement," *Soc. Pet. Eng. J.* (Dec. 1964) 356–62; *Trans.*, AIME, **231**.

Dumyuskin, I.I. and Namiot, A.Y.: "Mixing Conditions of Oils with Carbon Dioxide," *Neft. Khozyaistvo* (March 1978) 59–61.

Dyes, A.B. *et al.*: "Alternate Injection of HPG and Water—A Two-Well Pilot," paper SPE 4082 presented at the 1972 SPE Annual Meeting, San Antonio, Oct. 8–11.

Dyes, A.B., Caudle, B.H., and Erikson, R.A.: "Oil Production After Breakthrough—As Influenced by Mobility Ratio," *Trans.*, AIME (1954) **201**, 81–86.

Dykstra, H. and Parsons, R.L.: "The Prediction of Oil Recovery by Waterflooding," *Secondary Recovery of Oil in the United States*, second edition, API, New York (1950) 160.

E

Eckles, W.W., Prihoda, C., and Holden, W.W.: "Unique Enhanced Oil and Gas Recovery for Very High-Pressure Wilcox Sands Uses Cryogenic Nitrogen and Methane Mixture," *J. Pet. Tech.* (June 1981) 971–84.

Eilerts, C.K., Charlson, H.A., and Mullens, N.B.: "Effect of Added Nitrogen on Compressibility of Natural Gas," *World Oil*, Production Section (July 1948) 144–60.

Elkins, L.F. and Skov, A.M.: "Some Field Observations of Heterogeneity of Reservoir Rocks and Its Effects on Oil Displacement Efficiency," paper SPE 282 presented at the 1962 SPE Production Research Symposium, Tulsa, April 12–13.

Elliot, J.K. and Johnson, H.C.: "Control of Injection Gas Composition in Enriched Gas-Drive Projects," *Trans.*, AIME (1959) **216**, 398–99.

Engineering Data Book, ninth edition, Gas Processors Suppliers Assn., Tulsa (1972); fifth revision (1981).

Engineering Data Book, NGSMA (1957).

Enhanced Oil Recovery—An Analysis of the Potential for Enhanced Oil Recovery from Known Fields in the United States—1976 to 2000, Natl. Petroleum Council, Washington, DC (Dec. 1976).

Enhanced Oil Recovery Potential in the United States, Office of Technology Assessment, U.S. Congress, Washington, DC (Jan. 1978).

F

Fahien, R.W. and Smith, J.M.: "Mass Transfer in Packed Beds," *AIChE J.* (1955) **1**, 28.

Farr, J.B.: "High Resolution Seismic as a Reservoir Analysis Technique," paper SPE 7440 presented at the 1978 SPE Annual Technical Conference and Exhibition, Houston, Oct. 1–4.

Faulkner, B.L.: "Reservoir Engineering Design of a Tertiary Miscible Gas Drive Pilot Project," paper SPE 5539 presented at the 1975 SPE Annual Meeting, Dallas, Sept. 28–Oct. 1.

Fay, C.H. and Prats, M.: "The Application of Numerical Methods to Cycling and Flooding Problems," *Proc.*, Third World Pet. Cong. (1951) **2**, 555–63.

Fitch, R.A. and Griffith, J.D.: "Experimental and Calculated Performance of Miscible Floods in Stratified Reservoirs," *J. Pet. Tech.* (Nov. 1964) 1289–98.

Fitzgerald, J.G. and Nielsen, R.F.: "Oil Recovery from a Consolidated Core by Light Hydrocarbons in the Presence of High Water Saturations," *Mineral Industries Experimental Station Circular No. 66*, Pennsylvania State U., University Park (1964) 171.

Francis, A.W.: "Ternary Systems of Liquid Carbon Dioxide," *J. Phys. Chem.* **58** (1954) 1099–1114.

Frey, R.P.: "Operating Practices in the North Cross CO_2 Flood," *Proc.*, 22nd Annual Southwestern Petroleum Short Course, Texas Tech. U., Lubbock, 159–64 (1975).

Fried, A.N.: "The Foam Drive Process for Increasing the Recovery of Oil," RI 5866, USBM (1961).

Fussell, D.D. and Yanosik, J.L.: "An Iterative Sequence for Phase Equilibrium Calculations Incorporating the Redlich-Kwong Equation of State," *Soc. Pet. Eng. J.* (June 1978) 173–82.

Fussell, L.T.: "A Technique for Calculating Multiphase Equilibria," *Soc. Pet. Eng. J.* (Aug. 1979) 203–08.

Fussell, L.T. and Fussell, D.D.: "An Iterative Technique for Compositional Reservoir Models," *Soc. Pet. Eng. J.* (Aug. 1979) 211-20.

G

Gardner, A.O. Jr., Peaceman, D.W., and Pozzi, A.L. Jr.: "Numerical Calculation of Multidimensional Miscible Displacement by the Method of Characteristics," *Soc. Pet. Eng. J.* (March 1964) 26-36; *Trans.*, AIME, **231**; *Miscible Processes*, Reprint Series, SPE, Dallas (1965) **8**, 161-71.

Gardner, G.H.F., Downie, J., and Kendall, H.A.: "Gravity Segregation of Miscible Fluids in Linear Models," *Soc. Pet. Eng. J.* (June 1962) 95-104.

Gardner, J.W., Orr, F.M., and Patel, P.D.: "The Effect of Phase Behavior on CO_2 Flood Displacement Efficiency," *J. Pet. Tech.* (Nov. 1981) 2067-81.

Gardner, J.W. and Ypma, J.G.J.: "An Investigation of Phase Behavior—Macroscopic Bypassing Interaction in CO_2 Flooding," paper SPE 10686 presented at the 1982 SPE/DOE Symposium on Enhanced Oil Recovery, Tulsa, April 4-7.

Gernet, J.M. and Brigham, W.E.: "Meadow Creek Unit Lakota 'B' Combination Water-Miscible Flood," *J. Pet. Tech.* (Sept. 1964) 993-97.

Gillund, G.N. and Patel, C.: "Depletion Studies of Two Contrasting D-2 Reefs," paper 80-31-37 presented at the 1980 Annual Technical Meeting of the Petroleum Soc. of CIM, Calgary, May 25-28.

Giraud, A. *et al.*: "A Laboratory Investigation Confirms the Relative Inefficiency of True Miscible Drives and Outlines New Concepts for Maximizing Oil Recovery by Gas Injection," paper SPE 3486 presented at the 1971 SPE Annual Meeting, New Orleans, Oct. 3-6.

Glasser, S.R.: "History and Evaluation of an Experimental Miscible Flood in the Rio Bravo Field," *Prod. Monthly* (Jan. 1964) 17-20.

Glasstone, S.: *Textbook of Physical Chemistry*, second edition, Van Nostrum, New York City (1946) 791.

Goddard, R.R.: "Fluid Dispersion and Distribution in Porous Media Using the Frequency Response Method With a Radioactive Tracer," *Soc. Pet. Eng. J.* (June 1966) 143-52; *Trans.*, AIME, **237**.

Gonzales, M.H. and Lee, A.L.: "Graphical Viscosity Correlation for Hydrocarbons," *AIChE J.* (March 1968) 242-43.

Goss, M.J.: "Determination of Dispersion and Diffusion of Miscible Liquids in Porous Media by a Frequency Response Method," paper SPE 3525 presented at the 1971 SPE Annual Meeting, New Orleans, Oct. 3-6.

Graham, B.D. *et al.*: "Design and Implementation of a Levelland Unit CO_2 Tertiary Pilot," paper SPE 8831 presented at the 1980 SPE/DOE Symposium on Enhanced Oil Recovery, Tulsa, April 20-23.

Graue, D.J. and Blevins, T.R.: "SACROC Tertiary CO_2 Pilot Project," paper SPE 7090 presented at the 1978 SPE Symposium on Improved Oil Recovery Methods, Tulsa, April 16-19.

Graue, D.J. and Zana, E.: "Study of a Possible CO_2 Flood in the Rangely Field, Colorado," *J. Pet. Tech.* (July 1981) 1312-18.

Greenkorn, R.A., Johnson, C.R., and Haring, R.E.: "Miscible Displacement in a Controlled Natural System," *J. Pet. Tech.* (Nov. 1965) 1329-35; *Trans.*, AIME, **234**.

Griffith, J.D., Baiton, N., and Steffesen, R.J.: "Ante Creek—A Miscible Flood Using Separate Gas and Water Injection," *J. Pet. Tech.* (Oct. 1970) 1232-36.

Griffith, J.D. and Cyca, L.G.: "Performance of South Swan Hills Miscible Flood," *J. Pet. Tech.* (July 1981) 1319-26.

Griffith, J.D. and Horne, A.L.: "South Swan Hills Solvent Flood," *Proc.*, Ninth World Pet. Cong. (1975) **4**, 269-78.

Grinstead, W.C. Jr. and Barfield, E.C.: "Late Developments in the Seeligson Zone 20-B-07 Enriched Gas-Drive Project," *J. Pet. Tech.* (March 1962) 261-65.

Gupta, V.S.: "Measurement and Prediction of Phase Equilibria in Liquid Carbon Dioxide-Hydrocarbon Ternary Systems," PhD dissertation, Pennsylvania State U. (1969).

H

Haag, J.W.: "Analysis and Design of a Deep Reservoir, High Volume Nitrogen Injection Project in the R-1 Sand, Lake Barre Field," paper SPE 10159 presented at the 1981 SPE Annual Technical Conference and Exhibition, San Antonio, Oct. 5-7.

Habermann, B.: "The Efficiencies of Miscible Displacement as a Function of Mobility Ratio," *Trans.*, AIME (1960) **219**, 264-72; *Miscible Processes*, Reprint Series, SPE, Dallas (1965) **8**, 206-14.

Hagedorn, A.R. and Blackwell, R.J.: "Summary of Experience with Pressure Coring," paper SPE 3692 presented at the 1972 SPE Annual Meeting, San Antonio, Oct. 8-11.

Hall, H.N. and Geffen, T.M.: "A Laboratory Study of Solvent Flooding," *Trans.*, AIME (1957) **210**, 48-57; *Miscible Processes*, Reprint Series, SPE, Dallas (1965) **8**, 133-42.

Handy, L.L.: "An Evaluation of Diffusion Effects in Miscible Displacement," *Trans.*, AIME (1959) **216**, 61-63.

Haney, R.E.D. and Bliss, H.: "Compressibilities of Nitrogen-Carbon Dioxide Mixtures," *Ind. Eng. Chem.* (Nov. 1944) 985-89.

Hansen, P.W.: "A CO_2 Tertiary Recovery Pilot, Little Creek Field, Mississippi," paper SPE 6747 presented at the 1977 SPE Annual Technical Conference and Exhibition, Denver, Oct. 9-12.

Hardy, J.H. and Robertson, N.: "Miscible Displacement by High-Pressure Gas at Block 31," *Pet. Eng.* (Nov. 1975) 24-28.

Hare, M. *et al.*: "Sources and Delivery of Carbon Dioxide for Enhanced Oil Recovery," Report FE-2515-24 prepared by Pullman Kellogg Co. for the U.S. DOE, Contract No. EX-76-C-01-2515 (Dec. 1978).

Harris, D.G. and Hewitt, C.H.: "Synergism in Reservoir Management—The Geologic Perspective," *J. Pet. Tech.* (July 1977) 761-70.

Hartsock, J.H. and Slobod, R.L.: "The Effect of Mobility Ratio and Vertical Fractures on the Sweep Efficiency of a Five-Spot," *Prod. Monthly* (Sept. 1961) **26**, 2-7.

Harvey, M.T., Shelton, J.L. and Kelm, C.H.: "Field Injectivity Experiences With Miscible Recovery Projects Using Alternate Rich Gas and Water Injection," *J. Pet. Tech.* (Sept. 1977) 1051-55.

Hassinger, R.C. and von Rosenberg, D.V.: "A Mathematical and Experimental Examination of Transverse Dispersion Coefficients," *Soc. Pet. Eng. J.* (June 1968) 195-204; *Trans.*, AIME, **243**.

Hawthorne, R.G.: "Two-Phase Flow in Two-Dimensional Systems—Effects of Rate, Viscosity, and Density on Fluid Displacement in Porous Media," *Trans.*, AIME (1960) **219**, 81-87.

Helfferich, F.G.: "Theory of Multicomponent, Multiphase Displacement in Porous Media," *Soc. Pet. Eng. J.* (Feb. 1981) 51-62.

Henderson, L.E.: "The Use of Numerical Simulation to Design a Carbon Dioxide Miscible Displacement Project," *J. Pet. Tech.* (Dec. 1974) 1327-34.

Henry, R.L. *et al.*: "Utilization of Composition Observation Wells in a West Texas CO_2 Pilot Flood," paper SPE 9786 presented at the 1981 SPE/DOE Symposium on Enhanced Oil Recovery, Tulsa, April 5-8.

Henry, R.L. and Metcalfe, R.S.: "Multiple Phase Generation During CO_2 Flooding," paper SPE 8812 presented at the 1980 SPE/DOE Symposium on Enhanced Oil Recovery, Tulsa, April 20-23; accepted for publication in *Soc. Pet. Eng. J.*.

Herbeck, E.R. and Blanton, J.R.: "Ten Years of Miscible Displacement in Block 31 Field," *J. Pet. Tech.* (June 1961) 543-49.

Herbeck, E.F., Heintz, R.C., and Hastings, J.R.: *Fundamentals of Tertiary Oil Recovery*, Energy Communications, Dallas (1977).

Higgins, R.V. and Leighton, A.J.: "A Computer Method to Calculate Two-Phase Flow in Any Irregularly Bounded Porous Medium," *Trans.*, AIME (1962) **225**, 679-83.

Higgins, R.V. and Leighton, A.J.: "Computer Prediction of Water Drive of Oil and Gas Mixtures Through Irregularly Bounded Porous Medium," *Trans.*, AIME (1962) **225**, 679-83.

Hill, S.: "Genie Chemique," *Chem. Eng. Sci.* (1952) **1**, No. 6, 246.

Holloway, H.D. and Fitch, R.A.: "Performance of a Miscible Flood with Alternate Gas-Water Displacements," *J. Pet. Tech.* (April 1964) 372-76.

Holm, L.W.: "Foam Injection Test in the Higgins Field, Illinois," *J. Pet. Tech.* (Dec. 1970) 1499-1506.

Holm, L.W.: "The Mechanism of Gas and Liquid Flow Through Porous Media in the Presence of Foam," *Soc. Pet. Eng. J.* (Dec. 1968) 359-69.

Holm, L.W.: "Propane-Gas-Water Miscible Floods in Watered-Out Areas of the Adena Field, Colorado," *J. Pet. Tech.* (Oct. 1972) 1264-70.

Holm, L.W.: "Status of CO_2 and Hydrocarbon Miscible Oil Recovery Methods," *J. Pet. Tech.* (Jan. 1976) 86-94.

Holm, L.W. and Josendal, V.A.: "Discussion of Determination and Prediction of CO_2 Minimum Miscibility Pressure," *J. Pet. Tech.* (May 1980) 870-71.

Holm, L.W. and Josendal, V.A.: "Effect of Oil Composition on Miscible-Type Displacement by Carbon Dioxide," *Soc. Pet. Eng. J.* (Feb. 1982) 87-98.

Holm, L.W. and Josendal, V.A.: "Mechanisms of Oil Displacement by Carbon Dioxide," *J. Pet. Tech.* (Dec. 1974) 1427-36; *Trans.*, AIME, **257**.

Holm, L.W. and O'Brien, L.J.: "Carbon Dioxide Test at the Mead-Strawn Field," *J. Pet. Tech.* (April 1971) 431-42.

Huang, E.T.S. and Tracht, J.H.: "The Displacement of Residual Oil by Carbon Dioxide," paper SPE 4735 presented at the 1974 SPE Improved Oil Recovery Symposium, Tulsa, April 22-24.

Huie, N.C., Luks, K.D. and Kohn, J.P.: "Phase Equilibria Behavior of Systems Carbon Dioxide-n-Eicosane and Carbon Dioxide-n-Decane-n-Eicosane," *J. Chem. Eng. Data 18* (1973) 311-13.

Hurst, W.: "Determination of Performance Curves in Five-Spot Waterflood," *Pet. Eng.* (1953) **25**, B40-46.

Hutchinson, C.A. Jr.: "Reservoir Inhomogeneity Assessment and Control," *Pet. Eng.* (Sept. 1959) **31**, B19-26.

Hutchinson, C.A. Jr. and Braun, P.H.: "Phase Relations of Miscible Displacement in Oil Recovery," *AIChE J.* (1961) **7**, 64.

I

Improved Oil-Recovery Field Reports, SPE, Dallas (1975-1983).

J

Jacks, H.H., Smith, O.J.E., and Mattax, C.C.: "The Modeling of a Three-Dimensional Reservoir with a Two-Dimensional Simulator—The Use of Dynamic Pseudo Functions," *Soc. Pet. Eng. J.* (June 1973) 175-85.

Jacobs, W.L. and Bernard, G.G.: U.S. Patent No. 3,330,346 (1967).

Jacoby, R.H. and Rzasa, M.J.: "Equilibrium Vaporization Ratios for Nitrogen, Methane, Carbon Dioxide, Ethane and Hydrogen Sulphide in Absorber Oil-Natural Gas and Crude Oil-Natural Gas Systems," *Trans.*, AIME (1952) **195**, 99-110.

Jacquard, P.: "Calculs Numeriques de Desplacements de Fronts," *Proc.*, Seventh World Pet. Cong. (1963) Sec. II, PD6, paper 7.

Jardine, D. *et al.*: "Distribution and Continuity of Carbonate Reservoirs," *J. Pet. Tech.* (July 1977) 873-85.

Jenks, L.H., Campbell, J.B., and Binder, G.G. Jr.: "A Field Test of the Gas-Driven Liquid Propane Method of Oil Recovery," *Trans.*, AIME (1957) **210**, 34-39.

Johnson, J.P. and Pollin, J.S.: "Measurement and Correlation of CO_2 Miscibility Pressures," paper SPE 9790 presented at the 1981 SPE/DOE Enhanced Oil Recovery Sumposium, Tulsa, April 5—8; accepted for publication in *Soc. Pet. Eng. J.*.

Johnston, J.W.: "A Review of the Willard (San Andres) Unit CO_2 Injection Project," paper SPE 6388 presented at the 1977 SPE Permian Basin Oil and Gas Recovery Conference, Midland, TX, March 10-11.

Justen, J.J. *et al.*: "The Pembina Miscible Displacement Pilot and Analysis of Its Performance," *Trans.*, AIME (1960) **219**, 38-45.

K

Kane, A.V.: "Performance of a Large-Scale CO_2-WAG Project, SACROC Unit—Kelly Snyder Field," *J. Pet. Tech.* (Feb. 1979) 217-31.

Katz, D.L. *et al.*: *Handbook of Natural Gas Engineering*, McGraw-Hill Book Co., New York City (1959).

Kazemi, H.A., Vestal, C.R., and Shank, G.D.: "An Efficient Multicomponent Numerical Simulator," *Soc. Pet. Eng. J.* (Oct. 1978) 355-68.

Klinkenberg, L.J.: "Analog Between Diffusion and Electrical Conductivity in Porous Rocks," *Bull.*, GSA (1951) **62**, 559.

Kloepfer, C.V. and Griffith, J.D.: "Solvent Placement Improvement by Pre-Injection of Water, Lobstick Cardium Unit, Pembina Field," paper SPE 948 presented at the 1964 SPE Annual Meeting, Houston, Oct. 11-14.

Kniazeff, V.J. and Naville, S.A.: "Two-Phase Flow of Volatile Hydrocarbons," *Soc. Pet. Eng. J.* (March 1965) 37-44; *Trans.*, AIME, **234**.

Koch, H.A. Jr.: "High Pressure Gas Injection Is a Success," *World Oil* (Oct. 1956) **143**, 260.

Koch, H.A. Jr. and Hutchinson, C.A.: "Miscible Displacements of Reservoir Oil Using Flue Gas," *Trans.*, AIME (1958) **213**, 7-19.

Kolb, G.E.: "Several Parameters Affecting the Foam-Drive Process for the Removal of Water from Consolidated Porous Media," MS thesis, Pennsylvania State U., University Park (1964).

Koonce, K.T. and Blackwell, R.J.: "Idealized Behavior of Solvent Banks in Stratified Reservoirs," *Soc. Pet. Eng. J.* (Dec. 1965) 318–28; *Trans.*, AIME, **234**.

Koval, E.J.: "A Method for Predicting the Performance of Unstable Miscible Displacement in Heterogeneous Media," *Soc. Pet. Eng. J.* (June 1963) 145–54; *Trans.*, AIME, **228**; *Miscible Processes*, Reprint Series, SPE, Dallas (1965) **8**, 173–82.

Kuehm, H.G.: "Hawkins Inert Gas Plant: Design and Early Operation," paper SPE 6793 presented at the 1977 SPE Annual Technical Conference and Exhibition, Denver, Oct. 9–12.

Kulkarni, A.A. *et al.*: "Phase Equilibria Behavior of Carbon Dioxiden-Decane at Low Temperatures," *J. Chem. Eng. Data 19* (1974) 92–94.

Kyle, C.R. and Perrine, R.L.: "Experimental Studies of Miscible Displacement Instability," *Soc. Pet. Eng. J.* (Sept. 1965) 189–95; *Trans.*, AIME, **234**.

L

Lacey, J.W., Draper, A.L., and Binder, G.G. Jr.: "Miscible Fluid Displacement in Porous Media," *Trans.*, AIME (1958) **213**, 76–79.

Lacey, J.W., Faris, J.E., and Brinkman, F.H.: "Effect of Bank Size on Oil Recovery of the High Pressure Gas-Driven LPG-Bank Process," *J. Pet. Tech.* (Aug. 1961) 806–12; *Trans.*, AIME, **222**; *Miscible Processes*, Reprint Series, SPE, Dallas (1965) **8**, 215–21.

Lackland, S.D. and Hurford, G.T.: "Advanced Technology Improves Recovery at Fairway," *J. Pet. Tech.* (March 1973) 354–58.

Lane, L.C., Teubner, W.G., and Campbell, A.W.: "Gravity Segregation in a Propane Slug-Miscible Displacement Project—Baskington Field," *J. Pet. Tech.* (June 1965) 661–63.

Lantz, R.B.: "Rigorous Calculation of Miscible Displacement Using Immiscible Reservoir Simulators," *Soc. Pet. Eng. J.* (June 1970) 192–202.

Lapre, J.R.: "Reservoir Potential as a Function of the Geologic Setting of Carbonate Rocks," paper SPE 9246 presented at the 1980 SPE Annual Technical Conference and Exhibition, Dallas, Sept. 21–24.

Larson, R.G.: "Controlling Numerical Dispersion by Variably Timed Flux Updating in One Dimension," *Soc. Pet. Eng. J.* (June 1982) 399–408.

Larson, V.C., Peterson, R.B., and Lacey, J.W.: "Technology's Role in Alberta's Golden Spike Miscible Project," *Proc.*, Seventh World Pet. Cong. (1967) **3**, 533–44.

Latinen, G.A.: "Mechanism of Fluid Phase Mixing of Fixed and Fluidized Beds of Uniformly Sized Spherical Particles," PhD dissertation, Princeton U., Princeton, NJ (1951).

LeBlanc, R.J.: "Distribution and Continuity of Sandstone Reservoirs," *J. Pet. Tech.* (July 1977) 776–804.

Lee, B.D.: "Potentiometric Model Studies of Fluid Flow in Petroleum Reservoirs," *Trans.*, AIME (1948) **174**, 41–66.

Lee, J.L.: "Effectiveness of Carbon Dioxide Displacement Under Miscible and Immiscible Conditions," Research Report RR-40, Petroleum Recovery Inst., Calgary, Alta., Canada (March 1979).

Liphard, K.G. and Schneider, G.M.: "Phase Equilibria and Critical Phenomena in Fluid Mixtures of Carbon Dioxide plus 2, 6, 10, 15, 19, 23-hexamethyltetracosane up to 423°K and 100 MPa," *J. Chem. Thermodyn. 7* (1975) 805–14.

Lohrenz, J., Bray, B.G., and Clark, C.R.: "Calculating Viscosities of Reservoir Fluids From Their Compositions," *J. Pet. Tech.* (Oct. 1964) 1171–76.

M

Macon, R.B.: "Design and Operation of the Levelland Unit CO_2 Injection Facility," paper SPE 8410 presented at the 1979 SPE Annual Technical Conference and Exhibition, Las Vegas, Sept. 23–26.

Mahaffey, J.L., Rutherford, W.M., and Matthews, C.W.: "Sweep Efficiency by Miscible Displacement in a Five-Spot," *Soc. Pet. Eng. J.* (March 1966) 73–80; *Trans.*, AIME, **237**.

Marrs, D.G.: "Field Results of Miscible-Displacement Using Liquid Propane Driven by Gas, Parks Field Unit, Midland County, Texas," *J. Pet. Tech.* (April 1961) 327–32.

Marrs, D.G.: "Field Wide Miscible Displacement Project, Parks Field Unit, Midland Co., Texas," paper 1042G presented at the 1958 AIME Mid-Continent Section Petroleum Conference on Production and Reservoir Engineering, Tulsa, March 20–21.

Marrs, D.G.: "Miscible Drive Field Applications in the Parks Field," *J. Pet. Tech.* (May 1958) 20–21.

Marsden, S.S. Jr. *et al.*: "Use of Foam in Petroleum Operations," *Proc.*, Seventh World Pet. Cong., Mexico City (1967).

Marsden, S.S. and Khan, S.A.: "The Flow of Foam Through Short Porous Media and Apparent Viscosity Measurements," *Soc. Pet. Eng. J.* (March 1966) 17–25.

Martin, W.E.: "The Wizard Lake D-3A Pool Miscible Flood," paper SPE 10026 presented at the International Petroleum Exhibition and Technical Symposium, Beijing, China, March 19–22, 1982, accepted for publication.

Matheny, S.L.: "EOR Methods Help Ultimate Recovery," *Oil and Gas J.* (March 31, 1980) 79.

McLeod, J.G.F.: "Behavior of Enriched Gas Injected Alternately With Water in Horizontal Miscible Floods," *J. Cdn. Pet. Tech.* (April–June 1980) 51–57.

McCree, B.C.: "CO_2: How It Works, Where It Works," *Pet. Eng. Intl.* (Nov. 1977) 52–63.

McNeese, C.R.: "The High Pressure Gas Process and the Use of Flue Gas," paper A-62 presented at the Symposium on Production and Exploration Chemistry, American Chemical Soc., Los Angeles, March 31–April 5, 1963.

Meldau, R.F. and Simon, R.: "Phase Behavior in Miscible Systems of Solvent, Dry Gas, and Oil," paper SPE 226 presented at the 1961 SPE Annual California Regional Meeting, Bakersfield, Nov. 2–3.

Meldrum, A.H. and Nielsen, R.F.: "A Study of Three-Phase Equilibria for Carbon Dioxide-Hydrocarbon Mixtures," *Prod. Monthly* (Aug. 1955) 22–35.

Meltzer, B.D., Hurdle, J.M., and Cassingham, R.W.: "An Efficient Gas Displacement Project—Raleigh Field, Mississippi," *J. Pet. Tech.* (May 1965) 509–14.

Metcalfe, R.S. and Yarborough, L.: "Effect of Phase Equilibria on the CO_2 Displacement Mechanism," *Soc. Pet. Eng. J.* (Aug. 1979) 242–52.

Metcalfe, R.S.: "Effects of Impurities on Minimum Miscibility Pressures and Minimum Enrichment Levels for CO_2 and Rich-Gas Displacements," *Soc. Pet. Eng. J.* (April 1982) 219–25.

Metcalfe, R.S., Fussell, D.D., and Shelton, J.L.: "A Multicell Equilibrium Separation Model for the Study of Multiple Contact Miscibility in Rich-Gas Drives," *Soc. Pet. Eng. J.* (June 1973) 147–55.

Meyer, H.I. and Gardner, A.D.: "Mechanics of Two Immiscible Fluids in Porous Media," *J. of Applied Physics*, **25**, 11 (1954) 1400–05.

Miller, M.C. Jr.: "Gravity Effects in Miscible Displacement," paper SPE 1531 presented at the 1966 SPE Annual Fall Meeting, Dallas, Oct. 2–5.

Miscible Processes, Reprint Series, SPE, Dallas (1965) **8**.

Morel-Seytoux, H.J.: "Analytical-Numerical Method in Water-flooding Predictions," *Soc. Pet. Eng. J.* (Sept. 1965) 247–58.

Morel-Seytoux, H.J.: "Unit Mobility Ratio Displacement Calculations for Pattern Floods in Homogeneous Medium," *Soc. Pet. Eng. J.* (Sept. 1966) 217–27.

Morgan, J.T. and Thompson, J.E. Jr.: "Reservoir Study of the Camp Sand, Haynesville Field, Louisiana," paper SPE 7553 presented at the 1978 SPE Annual Technical Conference and Exhibition, Houston, Oct. 1–3.

Moscrip, R. III: "Butane Injection in San Pedro Block D," *Pet. Eng.* (Sept. 1958) B-10–24.

Mrosovsky, I., Wong, J.Y., and Lampe, H.W.: "Construction of a Large Field Simulator on a Vector Computer," *J. Pet. Tech.* (Dec. 1980) 2253–64.

Mungan, N.: "Carbon Dioxide Flooding—Fundamentals," *J. Cdn. Pet. Tech.* (Jan.–March 1981), 87–92.

Mungan, N.: "Certain Wettability Effects in Laboratory Water-floods," *J. Pet. Tech.* (Feb. 1966) 247–52; *Trans.*, AIME, **237**.

Mungan, N.: "Programmed Mobility Control in Polymer Floods," paper RR-1 presented at the Petroleum Recovery Institute, Calgary, Alta. (March 1968).

Muskat, M. and Wyckoff, R.D.: "A Theoretical Analysis of Waterflooding Networks," *Trans.*, AIME (1934) **107**, 62–76.

Muskat, M. and Wyckoff, R.D.: "An Approximate Theory of Water Coning in Oil Production," *Trans.*, AIME (1935) **114**, 144–63.

Muskat, M.: *Physical Principals of Oil Production*, McGraw-Hill Book Co. Inc., New York City (1949).

N

Newton, L.E. Jr. and McClay, R.A.: "Corrosion and Operation Problems, CO_2 Project, Sacroc Unit," paper SPE 6391 presented at the 1977 SPE Permian Basin Oil and Gas Recovery Conference, Midland, TX, March 10–11.

Nghiem, L.X., Fong, D.K., and Aziz, K.: "Compositional Modeling with an Equation of State," *Soc. Pet. Eng. J.* (Dec. 1981) 687–98.

Nobles, M.A. and Janzen, H.B.: "Application of a Resistance Network for Studying Mobility Ratio Effects," *Trans.*, AIME (1958) **213**, 356–58.

Nolen, J.S.: "Numerical Simulation of Compositional Phenomena in Petroleum Reservoirs," *Numerical Simulation*, Reprint Series, SPE, Dallas (1973) **11**, 269–84.

North American Production Capacity Data, Circular Z-57, National Fertilizer Development Center, TVW (Jan. 1975).

O

Offeringa, J. and van der Poel, C.: "Displacement of Oil from Porous Media by Miscible Liquids," *Trans.*, AIME (1954) **201**, 310–15; *Miscible Processes*, Reprint Series, SPE, Dallas (1965) **8**, 227–32.

Orlob, G.T. and Radhakrishna, G.N.: "The Effects of Entrapped Gases on the Hydraulic Characteristics of Porous Media," *Trans.*, A.G.U. (1958) **39**, 648.

Orr, F.M. Jr., Silva, M.K., and Lein, C.L.: "Equilibrium Phase Compositions of CO_2-Crude Oil Mixtures: Comparison of Continuous Multiple Contact and Slim-Tube Displacement Tests," paper SPE 10725 presented at the SPE/DOE Third Joint Symposium on Enhanced Oil Recovery, Tulsa, April 4–7, 1982, to be published in *Soc. Pet. Eng. J.*

Orr, F.M. Jr. and Silva, M.K.: "Equilibrium Phase Compositions of CO_2-Hydrocarbon Mixtures: Measurement by a Continuous Multiple Contact Experiment," paper SPE 10726 presented at the SPE/DOE Third Joint Symposium on Enhanced Oil Recovery, Tulsa, April 4–7, 1982, to be published in *Soc. Pet. Eng. J.*

Orr, F.M. Jr. and Taber, J.J.: *Displacement of Oil by Carbon Dioxide*, Annual Report for Oct. 1980–Sept. 1981 for the U.S. DOE Contract No. DE-AS19-80BC10331, Bartlesville, OK.

Orr, F.M. Jr., Yu, A.D., and Lein, C.L.: "Phase Behavior of CO_2 and Crude Oil in Low Temperature Reservoirs," *Soc. Pet. Eng. J.* (Aug. 1981) 480–92.

Outmans, H.D.: "Nonlinear Theory for Frontal Stability and Viscous Fingering in Porous Media," *Soc. Pet. Eng. J.* (June 1962) 165–76.

P

Palmer, F.S., Nute, A.J., and Peterson, R.L.: "Implementation of a Gravity Stable, Miscible CO_2 Flood in the 8,000-Ft Sand, Bay St. Elaine Field," paper SPE 10160 presented at the 1981 SPE Annual Technical Conference and Exhibition, San Antonio, Oct. 5–7.

Paulsell, B.L.: "Areal Sweep Performance of Five-Spot Pilot Floods," MS thesis, Pennsylvania State U., University Park (1958).

Peaceman, D.W. and Rachford, H.H. Jr.: "Numerical Calculation of Multidimensional Miscible Displacement," *Soc. Pet. Eng. J.* (Dec. 1962) 327–39; *Trans.*, AIME, **225**.

Pearson, D.G., Bray, J.A., and Strom, N.A.: "Simplified Approach for Assessing Gravity-Viscous Effects in Horizontal Steady-State Solvent Floods," paper No. 6823 presented at the 19th Annual Technical Meeting of the Petroleum Soc. of CIM, Calgary, Alta., May 7–10, 1968.

Peng, D.Y., and Robinson, D.B.: "A New Two-Constant Equation of State," *Ind. Eng. Chem. Fund.* (1976) **15**, 59–64.

Perkins, T.K. and Johnston, O.C.: "A Review of Diffusion and Dispersion in Porous Media," *Soc. Pet. Eng. J.* (March 1963) 70–84; *Trans.*, AIME, **228**; *Miscible Processes*, Reprint Series, SPE, Dallas (1965) **8**, 77–85.

Perkins, T.K. and Johnston, O.C.: "A Study of Immiscible Fingering in Linear Models," *Soc. Pet. Eng. J.* (March 1969) 39–46.

Perkins, T.K., Johnston, O.C., and Hoffman, R.N.: "Mechanics of Viscous Fingering in Miscible Systems," *Soc. Pet. Eng. J.* (Dec. 1965) 301–17; *Trans.*, AIME, **234**.

Perrine, R.L.: "A Unified Theory for Stable and Unstable Miscible Displacement," *Soc. Pet. Eng. J.* (Sept. 1963) 205–13; *Trans.*, AIME, **228**.

Perrine, R.L.: "Stability Theory and Its Use to Optimize Solvent Recovery of Oil," *Soc. Pet. Eng. J.* (March 1961) 9–16; *Trans.*, AIME, **222**.

Perrine, R.L.: "The Development of Stability Theory for Miscible Liquid-Liquid Displacement," *Soc. Pet. Eng. J.* (March 1961) 17–25.

Perry, G.E. and Kidwell, C.M.: "Weeks Island 'S' Sand Reservoir B Gravity Stable Miscible CO_2 Displacement," *Proc.*, **2**, Fifth Annual DOE Symposium on Enhanced Oil and Gas Recovery and Improved Drilling Recovery, Tulsa (1979).

Perry, G.E. *et al.*: "Weeks Island 'S' Sand Reservoir B Gravity Stable Miscible CO_2 Displacement, Iberia Parish, Louisiana," presented at the Fourth Annual DOE Symposium on Enhanced Oil and Gas Recovery, Tulsa, Aug. 29–31, 1978.

Perry, G.E.: "Weeks Island 'S' Sand Reservoir B Gravity Stable Miscible CO_2 Displacement, Iberia Parish, Louisiana," *Proc.*, ERDA Symposium on Enhanced Oil, Gas Recovery, and Improved Drilling Methods, Tulsa, Aug. 30–Sept. 1, 1977.

Perry, J.H.: *Chemical Engineers' Handbook*, third edition, McGraw-Hill Book Co. Inc., New York City (1950) 1705–11.

Peterson, A.V.: "Optimal Recovery Experiments with N_2 and CO_2," *Pet. Eng.* (Nov. 1978) 40–50.

Poettmann, F.H. and Katz, D.L.: "CO_2 in a Natural Gas Condensate System," *IEC* (1946) **38**, 530.

Poettmann, F.H.: "Vaporization Characteristics of Carbon Dioxide in a Natural Gas-Crude Oil System," *Trans.*, AIME (1951) **192**, 141–44.

Pontius, S.B. and Tham, M.J.: "North Cross (Devonian) Unit CO_2 Flood—Review of Flood Performance and Numerical Simulation Model," *J. Pet. Tech.* (Dec. 1978) 1706–16.

Pottier, J. *et al.*: "The High Pressure Injection of Miscible Gas at Hassi-Messaoud," *Proc.*, Third World Pet. Cong. (1967) **3**, 533–44.

Pozzi, A.L. and Blackwell, R.J.: "Design of Laboratory Models for Study of Miscible Displacement," *Soc. Pet. Eng. J.* (March 1963) 28–40; *Miscible Processes*, Reprint Series, SPE, Dallas (1965) **8**, 183–97.

Prats, M. *et al.*: "Prediction of Injection Rate and Production History for Multifluid Five-Spot Floods," *Trans.*, AIME (1959) **216**, 98–105.

Price H.S., Cavendish, J.C., and Varga, R.S.: "Numerical Methods of Higher-Order Accuracy for Diffusion-Convection Equations," *Soc. Pet. Eng. J.* (Sept. 1968) 293–303; *Trans.*, AIME, **243**.

Price, H.S. and Donohue, D.A.T.: "Isothermal Displacement Processes with Interphase Mass Transfer," *Soc. Pet. Eng. J.* (June 1967) 205–20.

R

Raimondi, P., Gardner, G.H.F., and Petrick, C.B.: "Effect of Pore Structure and Molecular Diffusion on the Mixing of Miscible Liquids Flowing in Porous Media," preprint 43 presented at the 1959 AIChE-SPE Joint Symposium, San Francisco, Dec. 6–9.

Raimondi, P. and Torcaso, M.A.: "Distribution of the Oil Phase Obtained Upon Imbibition of Water," *Soc. Pet. Eng. J.* (March 1964) 49–55; *Trans.*, AIME, **231**.

Raimondi, P., Torcaso, M.A., and Henderson, J.H.: "The Effect of Interstitial Water on the Mixing of Hydrocarbons During a Miscible Displacement Process," *Mineral Industries Experimental Station Circular* No. 61, Pennsylvania State U., University Park, Oct. 23–25, 1961.

Ramey, H.J. Jr. and Nabor, G.W.: "A Blotter-Type Electrolytic Model Determination of Areal Sweeps in Oil Recovery by In-Situ Combustion," *Trans.*, AIME (1954) **201**, 119–23.

Ramsey, E.H.: "Operating a Miscible Flood, Central Bistri Unit," *Producers Monthly* (Sept. 1961) 12–15.

Rathmell, J.J., Stalkup, F.I., and Hassinger, R.C.: "A Laboratory Investigation of Miscible Displacement by Carbon Dioxide," paper SPE 3483 presented at the 1971 SPE Annual Technical Conference and Exhibition, New Orleans, Oct. 3–6.

Raza, S.H. and Marsden, S.S.: "The Streaming Potential and the Rheology of Foam," *Soc. Pet. Eng. J.* (Dec. 1967) 359–68.

Reamer, H.H. and Sage, B.H.: "Phase Equilibria in Hydrocarbon Systems—Volumetric and Phase Behavior of the n-Decane-CO_2 System," *J. Chem. Eng. Data 8* (1963) 508–13.

Reamer, H.H. *et al.*: "Methane-Carbon Dioxide System in the Gaseous Region," *Ind. and Eng. Chem.* (1944) **36**.

Reid, T.B. and Robinson, H.J.: "Lick Creek Meakin Sand Unit Immiscible CO_2/Waterflood Project," *J. Pet. Tech.* (Sept. 1981) 1723–30.

Reitzel, G.A. and Callon, G.O.: "Pool Description and Performance Analysis Leads to Understanding Golden Spike's Miscible Flood," *J. Cdn. Pet. Tech.* (April–June 1977) 39–48.

Report and Decision on Review of Certain Solvent Flood Operations in the Pembina Cardium Pool, Alberta Oil and Gas Conservation Board, Calgary, 1967.

Rhodes, H.N.: "Secondary Recovery Booms in Western Nebraska," *Pet. Eng.* (May 1960) **32**, B-34.

Richardson, J.G., Harris, D.G., and Rossen, R.H.: "Synergy in Reservoir Studies," paper SPE 6700 presented at the 1977 SPE Annual Technical Conference and Exhibition, Denver, Oct. 9–12.

Richardson, J.G. *et al.*: "The Effect of Small, Discontinuous Shales on Oil Recovery," *J. Pet. Tech.* (Nov. 1978) 1531–37.

Robinson, D.B. and Mehta, B.R.: "Hydrates in the Propane-Carbon Dioxide-Water System," *J. Cdn. Pet. Tech.* (Jan.–March 1971) 33.

Roebuck, I.F. Jr. *et al.*: "The Compositional Reservoir Simulator: Case 1—The Linear Model," *Soc. Pet. Eng. J.* (March 1969) 115–30.

Ross, G.D. *et al.*: "The Dissolution Effects of CO_2 Brine Systems on the Permeability of U.K. and North Sea Calcareous Sandstones," paper SPE 10685 presented at the 1982 Joint SPE/DOE Symposium on Enhanced Oil Recovery, Tulsa, April 4–7.

Rowe, A.M. Jr.: "Internally Consistent Correlations for Predicting Phase Compositions for Use in Reservoir Composition Simulators," paper SPE 7475 presented at the 1978 SPE Annual Technical Conference and Exhibition, Houston, Oct. 2–4.

Rowe, A.M. Jr. and Silberberg, I.H.: "Prediction of Phase Behavior Generated by the Enriched-Gas Drive Process," *Soc. Pet. Eng. J.* (June 1965) 160–66.

Rowe, H.G., York, S.D., and Ader, J.C.: "Slaughter Estate Unit Tertiary Pilot Performance," *J. Pet. Tech.* (March 1982) 613–20.

Rump, M.W., Hare, M., and Porter, R.E.: *The Supply of Carbon Dioxide for Enhanced Oil Recovery*, report to the U.S. DOE by Pullman Kellogg, Houston, Contract No. EX-76-C-01-2515 (Dec. 1977).

Rushing, M.D. *et al.*: "Miscible Displacement with Nitrogen," *Pet. Eng.* (Nov. 1977) 26–30.

Rushing, M.D. *et al.*: "Nitrogen May Be Used for Miscible Displacement in Oil Reservoirs," *J. Pet. Tech.* (Dec. 1978) 1715–16.

Russell, T.F.: "Finite Elements with Characteristics for Two-Component Incompressible Miscible Displacement," paper SPE 10500 presented at the 1982 SPE Symposium on Reservoir Simulation, New Orleans, Jan. 31–Feb. 3.

S

Sage, B.H. and Lacey, W.N.: *Some Properties of the Lighter Hydrocarbons, Hydrogen Sulfide, and Carbon Dioxide,* Monograph on API Research Project 37, API (1955).

San Filippo, G.P. and Guckert, L.G.S.: "Development of a Pilot Carbon Dioxide Flood in the Rock Creek–Big Injun Field, Roane Co., West Virginia," *Proc.,* ERDA Symposium on Enhanced Oil, Gas Recovery, and Improved Drilling Methods, Tulsa, Aug. 30–Sept. 1, 1977.

Schnake, C.: "New Repressuring Method Promises to Triple Output," *World Oil,* (April 1958) **145,** 201.

Schneider, F.N. and Owens, W.W.: "Relative Permeability Studies of Gas-Water Flow Following Solvent Injection in Carbonate Rocks," *Trans.,* AIME (1976) **261,** 23-30.

Schneider, G.: "Phase Equilibria in Binary Fluid Systems of Hydrocarbons with Carbon Dioxide, Water, and Methane," AIChE Symposium Series **64** (1968) 9-15.

Schneider, G. *et al.*: "Phasengleichgewichte und kritische Erscheinungen in binaren Mischsystemen bis 1500 bar: CO_2 mit n-Octan, n-Undecan, n-Tridecan und n-Hexadecan," *Chem. Ing. Tech.* **39** (1967) 649-56.

Scott, D.S. and Dullien, F.A.L.: "Diffusion of Ideal Gases in Capillaries and Porous Solids," *AIChE J.* (1962) **8,** 113.

Secondary and Tertiary Oil Recovery Processes, Interstate Oil Compact Commission, Oklahoma City, Sept. 1974.

Secondary Recovery Railroad Commission Hearing Summaries as Extracted from the *Texas State House Reporter,* R.W. Bynum and Co. (Austin) (1960) 69.

Secondary Recovery Railroad Commission Hearing Summaries as Extracted from the *Texas State House Reporter,* R.W. Bynum and Co. (Austin) (1961).

Sessions, R.E.: "Small Propane Slug Proving Success in Slaughter Field Lease," *J. Pet. Tech.* (Jan. 1963) 31-36.

Sessions, R.E.: "How Atlantic Operates the Slaughter Flood," *Oil and Gas J.* (July 4, 1960) 91-98.

Settari, A., Price, H.S., and Dupont, T.: "Development and Testing of Variational Methods for Simulation of Miscible Displacement in Porous Media," *Soc. Pet. Eng. J.* (June 1977) 228-44.

Shah, D.O. and Schechter, R.S.: *Improved Oil Recovery by Surfactant and Polymer Flooding,* Academic Press, New York City (1977).

Sheely, C.Q.: "Description of Field Tests to Determine Residual Oil Saturation by Single-Well Tracer Method," *J. Pet. Tech.* (Feb. 1978) 194-202.

"Shell Starts Miscible CO_2 Pilot in Mississippi's Little Creek Field," *Oil and Gas J.* (Sept. 2, 1974) 52-53.

Shelton, J.L. and Schneider, F.N.: "The Effects of Water Injection on Miscible Flooding Methods Using Hydrocarbons and Carbon Dioxide," *Soc. Pet. Eng. J.* (June 1975) 217-26.

Shelton, J.L. and Yarborough, L.: "Multiple Phase Behavior in Porous Media During CO_2 or Rich Gas Flooding," *J. Pet. Tech.* (Sept. 1977) 1171-78.

Sherwood, T.K. and Pigford, R.L.: *Adsorption and Extraction,* McGraw-Hill Book Co. Inc., New York City (1952) 52-53.

Shum, Y.M.: "Use of the Finite-Element Method in the Solution of Diffusion-Convection Equations," *Soc. Pet. Eng. J.* (June 1971) 139-44.

Sigmund, P.M. *et al.*: "Laboratory CO_2 Floods and Their Computer Simulation," *Proc.,* 10th World Pet. Cong. (1979) **3,** 243-50.

Simlote, V.N. and Withjack, E.M.: "Estimation of Tertiary Recovery by CO_2 Injection—Springer A Sand, Northeast Purdy Unit," *J. Pet. Tech.* (May 1981) 808-18.

Simmons, J. *et al*: "Swept Areas After Breakthrough in Vertically Fractured Five-Spot Patterns," *Trans.,* AIME (1959) **216,** 73-77.

Simon, R. and Graue, D.J.: "Generalized Correlations for Predicting Solubility, Swelling, and Viscosity Behavior of CO_2-Crude Oil Systems," *J. Pet. Tech.* (Jan. 1965) 102-06.

Simon, R., Rosman, A., and Zana, E.: "Phase Behavior Properties of CO_2 Reservoir Oil Systems," *Soc. Pet. Eng. J.* (Feb. 1978) 20-26.

Simon, R. *et al*: "Compressibility Factors for CO_2-Methane Mixtures," *J. Pet. Tech.* (Jan. 1977) 81-85.

Singer, E. and Wilhelm, R.H.: "Heat Transfer in Packed Beds; Analytical Solution and Design Method; Fluid Flow, Solids Flow, and Chemical Reaction," *Chem. Eng. Prog.* (1950) **46,** 343.

Slobod, R.L.: "Use of Graded Viscosity Zone to Reduce Fingering in Miscible Phase Displacement," *Prod. Monthly* (Aug. 1960) 12.

Slobod, R.L. and Caudle, B.H.: "X-Ray Shadowgraph Studies of Areal Sweepout Efficiencies," *Trans.,* AIME (1952) **195,** 265-70.

Slobod, R.L. and Koch, H.A. Jr.: "High-Pressure Gas Injection—Mechanism of Recovery Increase," *Drill. and Prod. Prac.,* API (1953) 82.

Slobod, R.L. and Thomas, R.A.: "Effect of Transverse Diffusion on Fingering in Miscible-phase Displacement," *Soc. Pet. Eng. J.* (March 1963) 9-13.

Smith, A.S. and Brown, G.G.: "Viscosity of Ethane and Propane," *Ind. and Eng. Chem.* (1943) **35,** 705.

Smith, C.R.: *Mechanics of Secondary Oil Recovery,* Reinhold Pub. Corp., New York City (1966).

Smith, R.L.: "Sacroc Initiates Landmark CO_2 Injection Project," *Pet. Eng.* (Dec. 1971) 43-47.

Snedeker, R.A.: "Phase Equilibria in Systems with Supercritical Carbon Dioxide," PhD dissertation, Princeton U., Princeton, NJ (1955).

Sneider, R.M., Tinker, C.N., and Mackel, L.D.: "Deltaic Environment Reservoir Types and Their Characteristics," *J. Pet. Tech.* (Nov. 1978) 1538-45.

Sneider, R.M. *et al.*: "Predicting Reservoir Rock Geometry and Continuity in Pennsylvanian Reservoirs, Elk City Field, Oklahoma," *J. Pet. Tech.* (July 1977) 851-66.

Soave, G.: "Equilibrium Constants from a Modified Redlich-Kwong Equation of State," *Chem. Eng. Sci.* (1977) **27,** 1197-1203.

Sobocinski, D.P. and Cornelius, A.J.: "A Correlation for Predicting Water Coning Time," *Trans.,* AIME (1965) **234,** 594-600.

Spence, A.P. and Watkins, R.W.: "The Effect of Microscopic Core Heterogeneity on Miscible Flood Residual Oil Saturation," paper SPE 9229 presented at the 1980 SPE Annual Technical Conference and Exhibition, Dallas, Sept. 21-24, accepted for publication.

Spillette, A.G., Hillestad, J.G., and Stone, H.L.: "A High Stability Sequential Solution Approach to Reservoir Simulation," paper SPE 4542 presented at the SPE 48th Annual Meeting, Las Vegas, Sept. 30-Oct. 3, 1973.

Spivak, A.: "Gravity Segregation in Two-Phase Displacement Processes," *Soc. Pet. Eng. J.* (Dec. 1974) 619-27.

Spivak, A., Perryman, T.L., and Norris, R.A.: "A Compositional Simulation Study of the Sacroc Unit CO_2 Project," *Proc.*, Ninth World Pet. Cong., Tokyo (1975) **4**, 187.

Stalkup, F.I.: "Carbon Dioxide Miscible Flooding: Past, Present, and Outlook for the Future," *J. Pet. Tech.* (Aug. 1978) 1102-12.

Stalkup, F.I.: "Displacement of Oil by Solvent at High Water Saturation," *Soc. Pet. Eng. J.* (Dec. 1970) 337-48.

Stalkup, F.I.: "Status of Miscible Displacement," paper SPE 9992 presented at the 1982 SPE Intl. Pet. Exhibition and Symposium, Beijing, China, March 18-26, accepted for publication.

Stalkup, F.I.: "The Use of Phase Surfaces to Describe Condensing Gas Drive Experiments," *Soc. Pet. Eng. J.* (Sept. 1965) 184-88; *Trans.*, AIME, **234**.

Standing, M.B., and Katz, D.L.: "Density of Natural Gases," *Trans.*, AIME (1942) **146**, 140.

Statistics of Privately-Owned Electric Utilities in the United States, Federal Power Commission, Washington, D.C. (1972).

Steam Electric Plant Factors, National Coal Association.

Stewart, P.B. and Munjal, P.: "Solubility of Carbon Dioxide in Pure Water, Synthetic Sea Water, and Synthetic Sea Water Concentrates at $-5°$ to $25°C$ and 10- to 45-Atm Pressure," *J. Chem. and Eng. Data* (1970) **15**, No. 1, 67.

Stewart, W.C. and Nielsen, R.F.: "Phase Equilibria for Mixtures of Carbon Dioxide and Several Normal Saturated Hydrocarbons," *Prod. Monthly* (Jan. 1954) 27-32.

Stiles, W.E.: "Use of Permeability Distribution in Water Flood Calculations," *Trans.*, AIME (1949) **186**, 9-13.

Stright, D.H. Jr. *et al.*: "Carbon Dioxide Injection Into Undersaturated Viscous Reservoirs—Simulation and Field Application," *J. Pet. Tech.* (Oct. 1977), 1248-58.

Sturdivant, W.C.: "Pilot Propane Project Completed in West Texas Reef," *J. Pet. Tech.* (May 1959) 27-30.

Suder, F.E. and Calhoun, J.C. Jr.: "Waterflood Calculations," *Drill. and Prod. Prac.*, API (1949) 260.

T

Tables of the Error Function and Its Derivative, National Bureau of Standards, Applied Mathematics Series No. 41, Superintendent of Documents, U.S. Government Printing Office, Washington, DC (Oct. 1954).

Telkov, A.D.: "Solution of Static Problems in Coning with Account of Changes in Thickness of the Oil-Saturated Portion of a Formation," *Neft. Kohz.* (1971) **49**, No. 1, 34-37.

The Potential and Economics of Enhanced Oil Recovery, Lewin and Assocs. Inc., for U.S. ERDA, Washington, DC (April 1976).

Thomas, G.H., Countryman, G.R., and Fatt, I.: "Miscible Displacement in a Multiphase System," *Soc. Pet. Eng. J.* (Sept. 1963) 189-96; *Trans.*, AIME, **228**.

Thomas, G.W.: *Principles of Hydrocarbon Reservoir Simulation*, U. of Norway, Trondheim, Tapier (1977).

Thrash, J.C.: "Twofreds Field—A Tertiary Oil Recovery Project," paper SPE 8382 presented at the 1979 SPE Annual Technical Conference and Exhibition, Las Vegas, Sept. 23-26.

Tiffin, D.L.: "Effects of Mobile Water on Multiple Contact Miscible Gas Displacement," paper SPE/DOE 10687 presented at the 1982 SPE/DOE Symposium on Enhanced Oil Recovery, Tulsa, April 4-7, accepted for publication.

Tillman, R.W.: "Geology of El Dorado Field, A Deltaic Reservoir; An Application in Enhanced Oil Recovery," oral presentation at the 1977 SPE Annual Technical Conference and Exhibition, Denver, Oct. 8-12.

Tittle, R.M. and From, K.T.: "Success of Flue Gas Program at Neale Field," paper SPE 1907 presented at the 1967 SPE Annual Meeting, Houston, Oct. 1-4.

Todd, M.R.: "Modeling Requirements for Numerical Simulation of CO_2 Recovery Processes," paper SPE 7988 presented at the 1979 SPE California Regional Meeting, Ventura, April 18-20.

Todd, M.R., Cobb, W.M., and McCarter, E.D.: "CO_2 Flood Performance Evaluation for the Cornell Unit, Wasson San Andres Field," *J. Pet. Tech.* (Oct. 1982) 2271-82.

Todd, M.R. and Longstaff, W.J.: "The Development, Testing, and Application of a Numerical Simulator for Predicting Miscible Flood Performance," *J. Pet. Tech.* (July 1972) 874-82.

Todd, M.R., O'Dell, P.M., and Hirasaki, G.J.: "Methods for Increased Accuracy in Numerical Reservoir Simulators," *Soc. Pet. Eng. J.* (Dec. 1972) 515-30; *Trans.*, AIME, **253**.

Todd, M.R. *et al*: "Numerical Simulation of Competing Chemical Flood Designs," paper SPE 7077 presented at the 1978 SPE Symposium on Improved Methods for Oil Recovery, Tulsa, April 16-19.

Tomich, J.F. *et al.*: "Single-Well Tracer Method to Measure Residual Oil Saturation," *J. Pet. Tech.* (Feb. 1973) 211-18; *Trans.*, AIME, **255**.

Turek, E.A. *et al.*: "Phase Equilibria in Carbon Dioxide-Multicomponent Hydrocarbon Systems: Experimental Data and an Improved Prediction Technique," paper SPE 9231 presented at the 1980 SPE Annual Technical Conference and Exhibition, Dallas, Sept. 21-24; accepted for publication.

U

Unruh, C.H. and Katz, D.L.: "Gas Hydrates of Carbon Dioxide-Methane Mixtures," *Trans.*, AIME (1949) 83-86.

V

van der Poel, C.: "Effect of Lateral Diffusivity on Miscible Displacement in Horizontal Reservoirs," *Soc. Pet. Eng. J.* (Dec. 1962) 317-26; *Trans.*, AIME, **225**; *Miscible Processes*, Reprint Series, SPE, Dallas (1965) **8**.

Van Quy, N., Simandoux, P., and Corteville, J.: "A Numerical Study of Diphasic Multicomponent Flow," *Soc. Pet. Eng. J.* (April 1972) 171-84; *Trans.*, AIME, **253**.

W

Wadman, D.E., Mrosovsky, I., and Lamprecht, D.E.: "Reservoir Description Through Joint Geologic-Engineering Analysis," *J. Pet. Tech.* (July 1979) 933-40.

Walker, J.W. and Turner, J.L.: "Performance of Seeligson Zone 20B-07 Enriched-Gas Drive Project," *J. Pet. Tech.* (April 1968) 369-73.

Warner, H.R.: "An Evaluation of Miscible CO_2 Flooding in Waterflooded Sandstone Reservoirs," *J. Pet. Tech.* (Oct. 1977) 1339-47.

Warner, H.R. Jr. *et al.*: "University Block 31 Middle Devonian Reservoir Study: Part 1—History Match," *J. Pet. Tech.* (Aug. 1979) 962-70.

Warner, H.R. Jr., Hardy, J.H., and Davidson, C.D.: "University Block 31 Middle Devonian Reservoir Study: Part 2—Projections," *J. Pet. Tech.* (Aug. 1979) 971-78.

Warren, J.E. and Cosgrove, J.J.: "Prediction of Waterflood Performance in a Stratified System," *Soc. Pet. Eng. J.* (June 1964) 149–57; *Trans.*, AIME, **231**.

Warren, J.E. and Skiba, F.F.: "Macroscopic Dispersion," *Soc. Pet. Eng. J.* (Sept. 1964) 215–30; *Trans.*, AIME, **231**.

Watkins, R.L.: "The Development and Testing of a Sequential, Semi-Implicit Four Component Reservoir Simulator," paper SPE 10513 presented at the 1982 SPE Symposium on Reservoir Simulation, New Orleans, Jan. 31–Feb. 3.

Watkins, R.W.: "A Technique for the Laboratory Measurement of Carbon Dioxide Unit Displacement Efficiency in Reservoir Rock," paper SPE 7474 presented at the 1978 SPE Annual Technical Conference and Exhibition, Houston, Oct. 1–4.

Watts, R.J. *et al.*: "A Single CO_2 Injection Well Minitest in a Low-Permeability Eastern Carbonate Reservoir," *J. Pet. Tech.* (Aug. 1982) 1781–88.

Weber, K.J.: "Influence of Common Sedimentary Structures on Fluid Flow in Reservoir Models," *J. Pet. Tech.* (March 1982) 665–72.

Weimer, R.F.: "LPG-Gas Injection Recovery Process, Burkett Unit, Greenwood County, Kansas," *J. Pet. Tech.* (Oct. 1963) 1067–72.

Welker, J.R. and Dunlop, D.D.: "Physical Properties of Carbonated Oils," *J. Pet. Tech.* (Aug. 1963), 873–76.

West, J.M. and Kroh, C.G.: "Carbon Dioxide Compressed to Supercritical Values," *Pet. Eng.* (Dec. 1971), 63–66.

Whipple, A.P., Galloway, W.E., and Yancey, M.S.: "Seismic Stratigraphic Model of a Depositional Platform Margin Eastern Anadarko Basin, Oklahoma," oral presentation at the 1978 Annual Technical Conference and Exhibition, Houston, Oct. 1–4.

Wiebe, R. and Gaddy, V.L.: "Vapor Phase Composition of Carbon Dioxide-Water Mixtures at Various Temperatures and at Pressure to 700 Atmospheres," *J. Am. Chem. Soc.* (1941) **63**, 475–77.

Wilson, J.F.: "Miscible Displacement—Flow Behavior and Phase Relationships for a Partially Depleted Reservoir," *Trans.*, AIME (1960) **219**, 223

Wilson, K.: "Enhanced Recovery Inert Gas Processes Compared," *Oil and Gas J.* (July 31, 1978) 161–72.

Wizard Lake D-3A Pool—Engineering Study of Primary Performance and Miscible Flooding, Texaco Exploration Co. Application to Alberta Oil and Gas Conservation Board, Calgary (Dec. 1968).

Wright, H.T. Jr. and Die, R.R.: "Miscible Drive in the Seeligson Field," *J. Pet. Tech.* (May 1958) 13–16.

Y

Yang, H.W., Luks, K.D., and Kohn, J.P.: "Phase Equilibria Behavior of the System Carbon Dioxide-n-Butylbenzene-2-Methylnaphthalene," *J. Chem. Eng. Data* (1976) **21**, 330–35.

Yanosik, J.L. and McCracken, T.A.: "A Nine-Point Finite Difference Reservoir Simulator for Realistic Prediction of Unfavorable Mobility Ratio Displacements," *Soc. Pet. Eng. J.* (Aug. 1979) 253–62.

Yarborough, L.: "Application of a Generalized Equation of State to Petroleum Reservoir Fluids," paper presented at the I&EC Symposium on Equations of State in Engineering and Research, 176th ACS Natl. Meeting, Miami Beach, Sept. 10–15, 1978.

Yarborough, L.: "Vapor-Liquid Equilibrium Data for Multicomponent Mixtures Containing Hydrocarbon and Nonhydrocarbon Components," *J. Chem. Eng. Data* (1972) **17**, No. 2, 129.

Yellig, W.F.: "Carbon Dioxide Displacement of a West Texas Reservoir Oil," *Soc. Pet. Eng. J.* (Dec. 1982) 805–15.

Yellig, W.F. and Baker, L.E.: "Factors Affecting Miscible Flooding Dispersion Coefficients," *J. Cdn. Pet. Tech.* (Oct.–Dec. 1981) 69–75.

Yellig, W.F. and Metcalfe, R.S.: "Determination and Prediction of CO_2 Minimum Miscibility Pressures," *J. Pet. Tech.* (Jan. 1980) 160–68.

Young, L.C.: "A Finite-Element Method for Reservoir Simulation," *Soc. Pet. Eng. J.* (Feb. 1981) 115–28.

Young, L.C. and Stephenson, R.E.: "A Generalized Compositional Approach for Reservoir Simulation," paper SPE 10516 presented at the 1982 SPE Symposium on Reservoir Simulation, New Orleans, Jan. 31–Feb. 3; accepted for publication.

Youngren, G.K. and Charlson, G.S.: "History Match Analysis of the Little Creek CO_2 Pilot Test," *J. Pet. Tech.* (Nov. 1980) 2042–52.

Z

Zarah, B.Y., Luks, K.D., and Kohn, J.P.: "Phase Equilibria Behavior of Carbon Dioxide in Binary and Ternary Systems with Several Hydrocarbon Components," Symposium Series, AIChE, New York City (1974) **70**, No. 140, 91–101.

Zudkevitch, D., and Joffe, J.: "Correlation and Prediction of Vapor-Liquid Equilibria with the Redlich-Kwong Equation of State," *AIChE J.* (Jan. 1970) 112–19.

Author Index

Subject Index